WRITING THE HISTORY of MATHEMATICAL NOTATION: 1483–1700

Sister Mary Leontius Schulte

with an Introduction
by Albrecht Heeffer
and Bibliographic Notes
by Douglas R. Furman

Docent Press

DOCENT PRESS
Boston, Massachusetts, USA
www.docentpress.com

Docent Press publishes books in the history of mathematics and computing about interesting people and intriguing ideas. The histories are told at many levels of detail and depth that can be explored at leisure by the general reader.

The bibliographic sketch of Sister Mary Leontius Schulte reprinted from *Pioneering Women in American Mathematics: The Pre-1940 PhD's* by Judy Green and Jeanne LaDuke appears with the kind permission of the American Mathematical Society.

Cover design by Brenda Riddell, Graphic Details.

Produced with TEX. Textbody set in Baskerville.

Contents

List of Figures

Foreword

This book is one in the series of books published by Docent Press whose purpose is to highlight an exposition in the history of mathematics that may be of interest to today's readers but has faded into the fog of time. Rather than simply republishing the exposition, a Docent Press revival adds supporting material such as biography and bibliography of the author and a contemporary interpretation.

The exposition at hand is the 1935 doctoral dissertation of Sister Mary Leontius Schulte of the Sisters of Saint Francis, Rochester, Minnesota. Sister Leontius wrote her dissertation under the direction of Professor Louis C. Karpinski in the mathematics department of the University of Michigan. Docent Press extends many thanks to Sister Marilyn Geiger of the Sisters of Saint Francis for permission to republish Sister Leontius's work.

The supporting material in the book includes:

- a contemporary interpretation of the history of mathematical notation by Professor Albrecht Heeffer of Ghent University
- a biography of Sister Leontius by the archivist of the Sisters of Saint Francis, Sister Mary Lonan Reilly
- clips from and bibliographic notes regarding the original sources by Professor Douglas R. Furman of the State University of New York, Ulster

The text of the dissertation itself appears exactly as it was lithoprinted by Edwards Brothers, Inc. save a very small number of obvious typographical errors have been corrected. Editorial notes within the text such as references to figures are signified by square brackets. Footnotes authored by the editors are introduced by Ed. Footnotes without this introductory notation are those of Sister Leontius.

Interest in the history of mathematical notation, as with its allied studies, mathematical numeration and terminology, ebbs and flows as interest in firsts ebbs and flows. It is of passing note that recently, at least on history's timescale, two other books have been published on these topics: Stephen Christomalis' *Numerical Notation: A Comparative History* of 2010 [O29] and Joseph Mazur's *Enlightening Symbols: A Short History of Mathematical Notation and Its Hidden Powers* of 2014 [O57]. Mazur's book covers the history of notation as well as that of number figures and is a worthy descendent of Tropfke [O76]. Christomalis is laser-focused on systems of numeration and is worthy descendent of Pihan's *Exposé des Signes de Numération Usités Chez les Peuples Orientaux Anciens et Modernes* [O61]. Edward Scheinerman's *Mathematical Notation: A Guide for Engineers and Scientists* [O68] is not a history but collects in one place much of the mathematical notation seen in the field together with their LaTeX incantations.

Additional Reading on Mathematical Notation consists of a starter bibliography for readers interested in pursuing topics in mathematical notation, numeration, and terminology. The bibliographies of Christomalis and Mazur are far more extensive and are to be preferred.

Scott Guthery
Boston, MA
April, 2015

Note to Readers

A few words regarding this reprint are in order. The main enhancement that was made to Sister Leontius's work is this: whereas Sister Leontius, when depicting a particular mathematical notation, was limited to what could be achieved by means of a typewriter or occasionally a hand-drawn symbol, this current work predominantly utilized the scanned images of 15th, 16th and 17th century arithmetic texts that are available online. As almost 300 figures are included in Part III, it may be of interest to note a few guidelines that were used in selecting the 'clips' from the historic works that were included as figures in this work. This can be broken down into three categories. First, which edition of a work to use, secondly, when to insert a figure into the work and lastly, how much to clip from the historic work to comprise a figure to be inserted.

Regarding which edition of a work to use, in general, a hierarchy of preference was used in selecting which edition of a text, referenced by Sister Leontius, was used for a particular figure. Namely, in order of most preferred to least preferred:

- The exact edition that Sister Leontius referenced.
- An alternate edition, of the referenced work, that Sister Leontius utilized in her work and thus listed in her bibliography.
- The earliest edition of the work, that was accessible, online or otherwise, that contained an occurrence of the particular notation that was being discussed.

Of course, sometimes none of these criteria could be met. In that case, Sister Leontius's original format was reproduced, as close as possible, using modern word processing technology.

Now, regarding when to insert a figure into the work, it was decided to show as many figures as was practical, in the hope of not omitting some image that a reader would have liked to have seen. Of course this often had to be tempered by showing only a single image of a notation that Sister Leontius attributed to several authors. So three scenarios where an image would be searched for, to include as a figure are:

- Sister Leontius represented the notation by means of a typewriter (or hand-drawing).
- Sister Leontius mentions a specific edition and/or folio (or page) number.
- Sister Leontius mentions a work without specifying the edition (not necessarily in her bibliography).

Now regarding the process of determining what to include in the clips, from historical works, that would comprise a figure, this was not a particularly systematic process rather, each occurrence was considered individually. Sometimes a figure would just capture the image that Sister Leontius had depicted. More often though a figure would capture more than what Sister Leontius depicted or described. This allows the reader to appreciate more of the context that the notation occurred in. This could happen in several ways. Namely, possibly an entire problem or paragraph might be shown, or possibly an entire list of notations might be shown, not just the one in question. It is hoped that the reader will appreciate this 'wide-angle' approach.

I would like to mention there are three figures [2.31, 2.62 (used again in 4.6) and 3.6] which were not available online. But, I was able to view and photograph the associated texts, as they are part of the Plimpton Collection housed at the Rare Books and Manuscripts Library (RBML) at Columbia University. So it is with pleasure that I acknowledge the gracious assistance I received from two of the curators at the RBML, Jennifer B. Lee and Jane R. Siegel. I appreciated my visits to the RBML on many levels. Not only did I gain access to the same editions that Sister Leontius used, but I was very likely handling the exact copy that she had used for her work!

There are three books that may be helpful, and add to one's enjoyment, while reading this work. In particular, D. E. Smith's *History of Mathematics: Volume II Special Topics of Elementary Mathematics.* Originally published in 1925 it was republished by Dover Publications in 1958. Smith discusses many of the archaic algorithms for the fundamental arithmetic operations. This may help the reader to follow some of the computations which are exhibited in this current work. Additionally, Sister Leontius makes several references to two works: Cajori's *A History of Mathematical Notations,* originally published in 1928 (vol. 1) and 1929 (vol. 2) and republished by Dover Publications in 1993 (two volumes bound as one) and D. E. Smith's *Rara Arithmetica* originally published in 1908 with a fourth edition published in 1970 by Chelsea Publishing (which includes De Morgan's *Arithmetical Books* originally published in 1847).

In closing, I want to express my gratitude to my wife, Susan, whose love, encouragement and support has been steadfast over the years.

Douglas R. Furman
Rock Hill, NY
April, 2015

Acknowledgments

Sister Leontius included the following Acknowledgments at the beginning of her dissertation.

> The writer wishes to express in a very special manner her gratitude to Dr. Louis C. Karpinski, Professor of Mathematics at the University of Michigan, for his guiding interest, kindly encouragement,and generous assistance through-out the development of this study. To other members of the Faculty of Mathematics the writer offers her thanks for their careful and patient reading and criticism of the manuscript.
>
> Acknowledgment is made of the generous aid given by the librarians of the Rare Book Division of the Library of the University of Michigan. Gratitude is due also to Dr. David Eugene Saith and Dr. George A. Plimpton of New York for placing at the writer's disposal the invaluable material found In their private libraries.
>
> Ann Arbor, Michigan
Feast of the Sacred Heart
June 8, 1934.

PART I

Introduction

Chapter 1

On the Historiography of Mathematical Notations

1.1 Introduction

Anyone involved with or interested in mathematics must have had several occasions in which he or she wondered how and why specific symbols were introduced. Our plus and minus signs, how long have they been in use? Who invented brackets? Were exponents known outside of Europe? Why do we use x, y and z for variables and a, b, c for constants? And if one takes the effort to look up the sources for such innovations in mathematical notations, one will inevitably arrive at Florian Cajori. For a book written in 1928–9, this classic work remains remarkably popular, still being shelved and in print in the Dover edition, and now even available for digital reading. Lesser known, but equally important for the historian of mathematics is Tropfke's earlier work on elementary mathematics. It goes beyond the history of notations and also discusses the history of elementary concepts and mathematical operations across cultures. While both these classic works are often used and quoted, a critical assessment of their value today and in the past, was lacking.

Now, the edition of the previously unpublished dissertation by Sister Leontius (1935), filling several gaps in the surveys by Tropfke and Cajori, presents the opportunity to assess the importance of these works, to place them in their proper historical context and to evaluate their relevance today. These are the aims for the first part of this book.

We will first discuss Johannes Tropfke and show how innovative his thematic approach was to the history of mathematics in a time when the historiography was dominated by chronological, encyclopedic approaches concentrating on persons rather than mathematical concepts, as the four-volume work by Moritz Cantor [12]. Then we will place the classic work by Florian Cajori in the context of the history of science as an emerging scientific discipline in the USA. We will demonstrate that Cajori, as well as David Eugene Smith were indebted to Tropfke and that mathematics education was a concern and an inspiration shared by the three of them. With the additions by Sister Leontius, we will then critically assess the whole project from a contemporary perspective. Our main objection is their internalist approach to historiography which accepts implicitly or more explicitly that the development towards symbolic mathematics was inevitable and depended on some internal mechanisms and intrinsic processes. However, they were not alone in this view, and we will show that this was the prevailing view in the historiography of mathematical notations and symbolism, up to the 1970s. We will conclude with a survey of some contemporary approaches to mathematical symbolism from different disciplines, such as semiotics, cognitive science, experimental psychology and historical epistemology.

1.2 Tropfke's Thematic History of Mathematics

1.2.1 Short biography of Johannes Tropfke (1866–1939)

Tropfke has always been closely connected with Berlin. He was born as a Berliner, proud to be a Berliner and lived all his live in Berlin. He was born and he died in the house that his grandfather Theodor Tropfke had built in 1827 in the Marienstrasse 14, near the Jewish quarter, just North of the River Spree. A plate at the house today pays tribute to Tropfke for his contributions to mathematics (see Figure 1.1). The same street also housed the composers Glinka and Sibelius in the nineteenth century and Angela Merkel during the 1980s. In 1907 he became a Berlin municipal executive (stadtverordneter), a position he held until the end of WW1.

FIGURE 1.1: A plaque at Tropfke's family house in Berlin pays tribute for his contributions to mathematics. (Photograph by the author)

The young Tropfke had a carefree youth and he showed an aptitude for mathematics, physics, botany and zoology and was well versed in Latin and Greek.[1] He was educated in a system where the most able students were encouraged and spurred on by rewards.

[1]The best biographical data we have about Tropfke is from his own hand in a *Festschrift* published for the occasion of 25 years Kirschner college [48]. J. E. Hofmann, the historian of mathematics who wrote Tropfke's obituary for the anti-Semitic nazi journal *Deutsche Mathemtik* [33], freely borrowed from this source.

At the end of the tenth grade he was given a copy of Hankel's *History of mathematics in antiquity and the Middle Ages* [23]. This work would have a profound influence on the sixteen year old, including his multi-cultural approach to the history of mathematics (as discussed below).[2]

He studied mathematics under Lazarus Fuchs at the Martin-Luther-Universität in Halle-Wittenberg. His PhD dissertation deals with elliptic functions and integrals (*Zur darstellung des elliptischen integrales erster gattung*, published in 1889). After his PhD, Tropfke began teaching and was involved in education for the rest of his life. He started with his research in the history of mathematics while teaching which proved to be a difficult task. The main problem he faced was access to original sources in the history of mathematics. Conscientious as he showed to be, Tropfke was not content with secondary sources as Cantor's work, and was well aware of the bias in the treatment of non-Western mathematics by Cantor and others. However, not being affiliated with a university he was posed with great difficulties in access to Berlin libraries. This problem was less of an issue for his second edition through his cooperation with Julius Ruska and Heinrich Wieleitner.

FIGURE 1.2: A painted portrait of Johannes Tropfke (from [48]).

In 1912 he founded the Kirschner college in Berlin which he directed until 1932. In 1930 he received the Ackermann-Teubner prize for the second edition of his work. In

[2][48], p. 6: "Bei der Versetzung nach Obersekunda erhielt ich eine Schulprämie: Hrm. Hankel, Zur Geschichte der Mathematik in Altertum und Mittelalter, das erst gute fünf Jahre vorher erschienen war und der Wissenschaft neue Wege wies. In ihm liegt die feste Grundlage für meine ganze spätere wissenschaftliche Tätigkeit, die mich heute noch in Arbeit und Schaffen hält".

1938 he was elected a full member of the International Society for the History of Science. On 6 August 1939 he was awarded the silver Leibniz medal from the Prussian Academy of Science. Apart from his books, he also published some articles in German journals as *Jahresbericht der Deutschen Mathematiker-Vereinigung* and *Elemente der Mathematik* as well as an article on Archimedes in *Osiris*.

1.2.2 The editions

The first edition of his *Geschichte der elementar-mathematik in systematischer darstellung* was published with Von Veit in Leipzig and consisted of two volumes. The first appeared in 1902 and was titled *Das Rechnen und Algebra*, the second in 1903 with the title *Geometrie. Logarithmen. Ebene trigonometrie. Sphärik und sphärische trigonometrie. Reihen. Zinseszinsrechnung. Kombinatorik und wahrscheinlichkeitsrechnung. Kettenbrüche. Stereometrie. Analytische geometrie. Kegelschnitte. Maxima und minima.* The motivation of the author for his work was the lack of a systematic treatment of specific subjects in the history of mathematics suitable for teaching.

The first edition was followed by a second one from 1920 onwards, published by Walter De Gruyter in Berlin and Leipzig, expanding to seven separate volumes. Also its title was expanded with a special mention of technical terms: *Geschichte der Elementar-Mathematik: in systematischer Darstellung mit besonderer Berücksichtigung der Fachwörter*. The first volume was published in 1920, vol. 1 *Rechnen*. It was followed in 1921 by vol. 2. *Allgemeine Arithmetik*. In 1922 vol. 3: *Proportionen, Gleichungen*. In 1923, vol. 4: *Ebene Geometrie*, and vol. 5: 1. *Ebene Trigonometrie*; 2. *Sphärik und sphärische Trigonometrie*. In 1924 appeared vol. 6, *Analysis, analytische Geometrie* and vol. 7: *Stereometrie, Verzeichnisse*. In the foreword of the 1920 edition Tropfke expresses his gratitude to Gustaf Eneström from Sweden, editor of the journal *Bibliotheca mathematica*, which contained many short notes on historical works by his international peers including Florian Cajori, Louis Karpinski and David Eugene Smith from the USA, Thomas Heath from the UK, H.G. Zeuthen from Denmark, Henri Bosmans from Belgium, Paul Tannery from France and Gino Loria from Italy. Another improvement of the second edition was the inclusion of several Arabic and Indian sources, facilitated by the assistance of Julius Ruska, an orientalist from Heidelberg who published his own *Zur ältesten arabischen Algebra und Rechenkunst* in 1917.

In 1929 it was decided that a third edition was in order to update the seven volumes once again. However, Tropfke never saw his project finished. Only the first four volumes were published anew: vol 1. *Das Rechnen* in 1930, vol. 2 *Allgemeine Arithmetik* in 1933, vol 3. *Proportionen, Gleichungen* in 1937. Vol. 4. *Ebene Geometrie* was published after his death in 1940. The manuscripts of the updates for the last three volumes were lost when Berlin became occupied in 1945. For the second volume, Tropfke was assisted 'by a young collaborator from Munich', Kurt Vogel, who would pursue his project later on.

Tropfke never achieved a position as *Privatdozent* or university professor, and continued to teach mathematics at the Kirschner college. This was an era in which it was not uncommon for distinguished mathematicians to teach mathematics at the gymnasium level, such as Karl Weierstrass (1815 – 1897) and Hermann Grassmann (1809 – 1877). He was one of the first to recognize the value of the history of mathematics for the teaching of mathematics. In a publication preceding the first edition of his book, he provides an outline of a program corresponding with his later work and explicitly states that this has a use in the curriculum [83].

If we compare the subsequent editions we indeed see a continuous and significant expansion of the material. The 1902 edition on arithmetic consisted of 120 pages, the part on algebra 209 pages. The volume on arithmetic was split up in 1920 in a part on calculation of 177 pages and theoretical arithmetic of 221 pages. The number of footnotes increased from 1200 in 1902 to no less than 2600 in the first two volumes of 1920-1 edition. In the third edition the first volume grew to 222 pages and the second to 266.

During the 1970's it was decided that the legacy of Tropfke deserved another update. The task was undertaken by Kurt Vogel at the Institut für Geschichte der Naturwissenschaften at the University of Munich, assisted by Karin Reich and Helmut Gericke. The plan was to publish the new edition in three volumes 1) *Arithmetic and algebra*, 2) *Geometry* and 3) *Analysis*. Only the first volume appeared in 1980. Kurt Vogel was then 92 years old. The organisation, structure and approach of the last edition significantly differs from the Tropfke editions.

Tropfke was well regarded by his peers. While Archibald's review of the 1920-1 edition was somewhat lukewarm [2], Karpinski was already more favourable and wished his work would be translated, because no similar one was available in English [43]. David Eugene Smith's three reviews on the third edition showed high respect "the work will continue to stand as the most helpful one of its kind, for beginners in the study of the history of mathematics, that we have. The particular value lies not in its style or in the grasp of the large problems so much as in its mass of detail and its very extensive bibliography" [77], [78], [79].

1.2.3 Tropfke's approach

As with most historians of mathematics at the beginning of the twentieth century, Tropfke's example was Cantor's 'master work' [12]. Despite his great admiration for Cantor, Tropfke expressed his dissatisfaction with his book since its use in mathematics education required a lot of effort in collecting material from the chronological arrangement of this book. Hence the subtitle of his own work, articulating his new approach "in a systematic treatment":[3]

> The predominantly historical arrangement of Cantor's work presents the reader, who wants to teach about a specific topic, with great difficulties, since a systematic treatment of the material requires a compilation from the contents of various chapters. Monographs are not yet available in large numbers, and also these still treat such a broad subject area that the answer to our question becomes only slightly easier. A work that distinguishes itself should provide quick access to such and so subject, serving as a kind of

[3]From the foreword of the first edition, p. i: "Die vorwiegend historische Anordnung in Cantor's Werk bereitet dem Leser, der sich über ein Thema unterrichten will, große Schwierigkeiten, da die stufenweise Entwicklung des gesuchten Stoffes aus den verschiedensten Kapiteln mit Hilfe des Inhaltsverzeichnisses zusammengetragen werden muß. Monographien sind noch nicht in größerer Zahl vorhanden; zum Teil stellen auch diese noch ein so umfassendes Sachgebiet dar, daß das Beantworten der gestellten Frage nur wenig erleichtert wird. Für ein Werk, das im stände sein soll, schnell Auskunft über diesen oder jenen Punkt zu geben, das also als eine Art Nachschlagewerk dienen kann, ist deshalb die systematische Anordnung durch aus vorzuziehen".

FIGURE 1.3: The last volume of the second edition of Tropfke's monumental work (a copy owned by and with the autography of Otto Neugebauer).

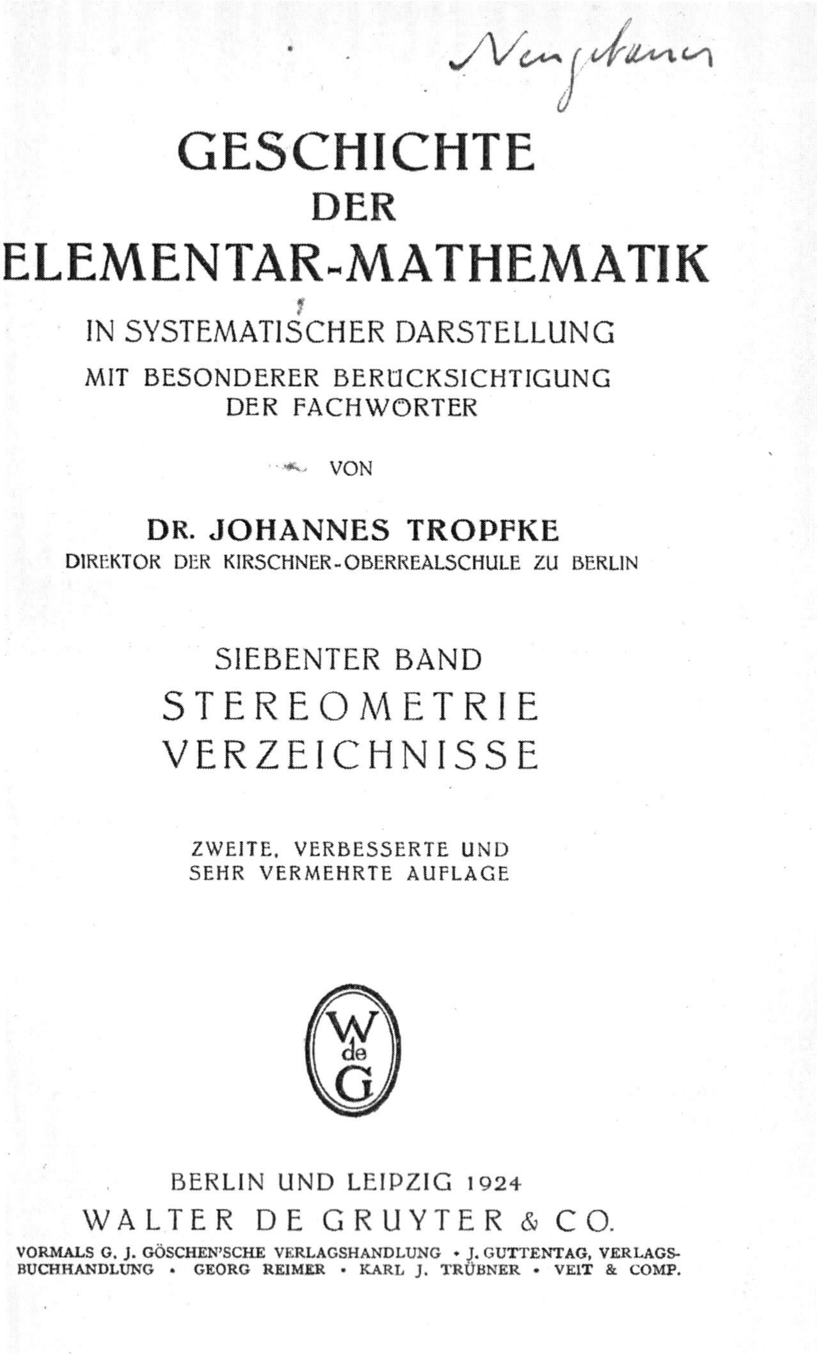

Neugebauer

GESCHICHTE
DER
ELEMENTAR-MATHEMATIK

IN SYSTEMATISCHER DARSTELLUNG
MIT BESONDERER BERÜCKSICHTIGUNG
DER FACHWÖRTER

VON

DR. JOHANNES TROPFKE
DIREKTOR DER KIRSCHNER-OBERREALSCHULE ZU BERLIN

SIEBENTER BAND
STEREOMETRIE
VERZEICHNISSE

ZWEITE, VERBESSERTE UND
SEHR VERMEHRTE AUFLAGE

BERLIN UND LEIPZIG 1924
WALTER DE GRUYTER & CO.
VORMALS G. J. GÖSCHEN'SCHE VERLAGSHANDLUNG • J. GUTTENTAG, VERLAGSBUCHHANDLUNG • GEORG REIMER • KARL J. TRÜBNER • VEIT & COMP.

reference work, which requires therefore a systematic arrangement as the preferable format.

One should not underestimate the degree of innovation and the importance of Tropfke's approach. It rendered a completely new way of looking at the history of mathematics. None of the works previously published on the history of mathematics followed his systematic treatment of topics on mathematics. Only after the publication of the second edition in 1920 did similar works start to appear. The first one was David Eugene Smith with his two-volume *History of Mathematics* [75], [76]. While Smith's first volume still adopts the chronological approach to history, he does not hide the fact that his second volume was inspired by Tropfke ([76], iii):

> The general plan adopted in the preparation of this work is that of presenting the subject from two distinct standpoints, the first, as in Volume I, leading to a survey of the growth of mathematics by chronological periods, with due consideration to racial achievements; and the second, as in Volume II, leading to a discussion of the evolution of certain important topics. To attempt to fuse these two features and thus to carry them along together has often been attempted. characterizes, for example, the monumental treatise of Montucla and, to a large extent, that of Cantor. For the teacher, however, this plan is not satisfactory, and the excellent work of Tropfke is an example of the tendency to break away from the mere chronological recital of facts. Long experience in teaching the subject in colleges has convinced the author that a general historical presentation is desirable for the purpose of relating the development of mathematics to the development of the race, of revealing the science as a great stream rather than a static mass, and of emphasizing the human element, but that this ought to lead to a topical presentation by which the student may understand something of the life history of the special subject which he may be studying, whether it be the elementary theory of numbers, the methods of calculation, the solution of equations, the functions of trigonometry, the common symbolism in use, the various types of elementary geometry, the early steps in the calculus, or one of the various other important topics of elementary mathematics.

As Tropfke, Smith was equally motivated by his experience as a teacher to adopt the topical treatment for his second volume. In his bibliography Smith lists Tropfke's work as "the best history of elementary mathematics".

The second to adopt Tropfke's thematic treatment was Cajori. In his preface to the first volume, Cajori writes that he got the idea for the book from the mathematician Eliakim Hastings Moore (1862 – 1932). While Cajori does not mention Tropfke as an inspiration, his abundant references to Tropfke clearly testify that not only he was aware of this work, but that it must have been a model for his own. Not only the thematic approach inspired Cajori, but Tropfke was also a pioneer in his historical study of notations. Already the first edition included a comprehensive treatment of symbols such as the signs for the unknown and powers of the unknown. A table in the second volume provided an overview of the notations in 9 historical works (p. 191). This was extended to 12 works with alternative notations in the second edition (vol. II, p. 112) and the third edition (vol. II, p. 141). Cajori copied this table in his own treatment on

FIGURE 1.4: **Tropfke's overview of notations for the powers of the unknown, from manuscripts to printed books (from the 3rd edition, vol. II, p. 141.)**

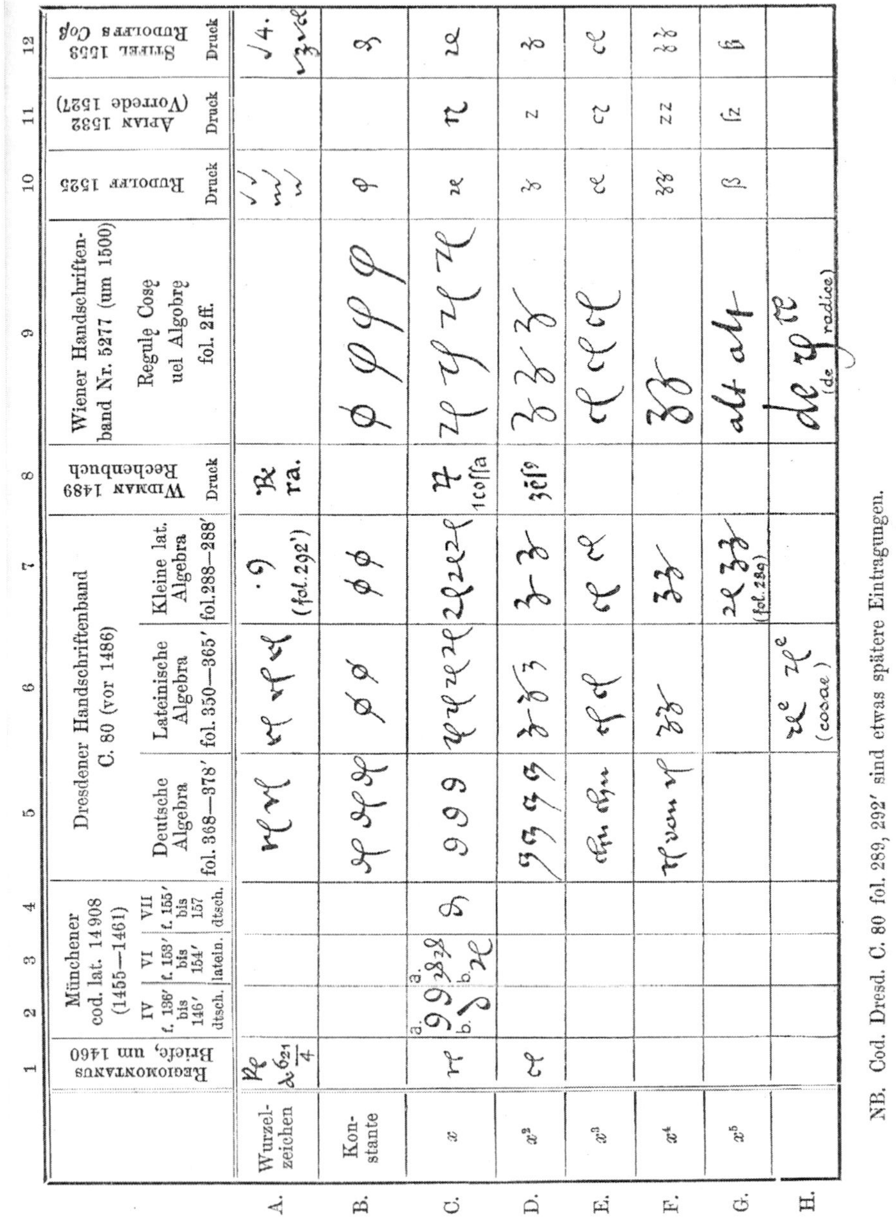

Signs and powers ([11], p. 339) (see figure 1.4). Comparing the two it is immediately obvious that Tropfke's treatment of this particular subject is more extensive and includes manuscript notations as well as typeset notations in printed works. Cajori vastly expanded the treatment of mathematical notations but would focus mainly on printed works. Schulte's additions to Tropfke and Cajori also concern printed works only.

1.2.4 Multi-cultural sources

With the second edition, Tropfke included discussions on the mathematical achievements of cultures different from the Ancient Greek–Roman–European tradition. These included Hindu mathematics, Arabic mathematics and also Old-Babylonian mathematics (newly studied by the group of Otto Neugebauer at Göttingen). While this was not new within a general history of mathematics, as Cantor's *Geschichte* testifies, Tropfke took a different approach by integrating these foreign sources within his topical treatment of subjects such as numeration, basic operations and algebraic notations. Such an integration lead to a much more balanced approach than many of the commentaries on mathematical achievements of Indians and Arabs by European historians from the end of the nineteenth century.

However, whenever explanations were required, scholars became divided into two opposing camps, which we could call the believers and the non-believers. Non-believers did not grant Indian mathematicians the status of original thought. Indian knowledge must have stemmed from the Greeks, the cradle of Western mathematics, or even mathematics as such.[4] The major non-believer was Moritz Cantor who takes every opportunity to point out the Greek influences on Hindu algebra. Some examples: the Indians learned algebra through traces of algebra within Greek geometry;[5] or Brahmagupta's solution to quadratic equations has Greek origins.[6]

The dichotomy between believers and non-believers dates back to the first translations of Sanskrit sources of the classic period of Indian mathematics at the beginning of the nineteenth century. Best known is the English orientalist Henry Thomas Colebrooke, who previously published his *Sanskrit Grammar* (1805) and undertook the task of translating three classics of Indian mathematics, the *Brāhmasphuṭasiddhānta* of Brahmagupta (628) and the *Līlāvatī* and the *Bījagaṇita* of Bhāskara II (1150) [15]. At once European historians had something to reflect upon. In a period when mathematics was hardly practiced in Europe and in the Islam regions, there appeared to have existed this Indian tradition in which algebraic problems were solved with multiple unknowns, in which zero and negative quantities were accepted and in which sophisticated methods were used to solve indeterminate problems. In general, nineteenth-century historians showed an admiration for the Hindu tradition.

[4] The reception of Indian mathematics by Western historians is more extensively discussed in [29] including Cantor's attribution of Indian linear problems to the *Epanthema* in ancient Greek sources. The example clearly shows how historical investigation can be misled through prejudice.

[5] "Spuren griechischer Algebra müssen mit griechischer Geometrie nach Indien gedrungen sein und werden sich dort nachweisen lassen", ([12], II, 562).

[6] "So glauben wir auch deutlich die griechische Auflösung der quadratischen Gleichung, wie Heron, wie Diophant sie übte, in der mit ihr nicht bloss zufällig übereinstimmenden Regel des Brahmagupta zu erkennen", ([12], II, 584).

Lesser known is an earlier translation of the *Bījagaṇita* by Edward Strachey (1774 – 1832), based on a Persian translation from 1634. We quote from the foreword of his book published in 1813 ([80], 1):

> It is known that there are Sanscrit books on astronomy and mathematics. Wether the science they contains is of Hindoo origin and of high antiquity, or is modern and borrowed from foreign sources, is a question which has been disputed. Some of the advocates for the Hindoos have asserted their pretensions with a degree of zeal which may be termed extravagant; and others among their opponents have with equal vehemence pronounced them to be imposters, plagiaries, rogues, bind slaves, ignorant, etc. etc.
>
> My object in the following paper is to support the opinion that the Hindoos had an original fund of science not borrowed from foreign sources. I mean to infer also, because of the connexion of the sciences and their ordinary course of advancement, that the Hindoos had other knowledge besides what is established by direct proof to be theirs, and that much of what they had, must have existed in early times.

The believers were not convinced by accidental resemblances between Greek and Hindu solution methods and did not see why Indian mathematics could not have been an independent development. In particular, Hankel ([23], 204), in the book which had such a profound influence on the young Tropfke, touches the sore spot when he writes:[7]

> That by humanist education deeply inculcated prejudice that all higher intellectual culture in the Orient, in particular all science, is risen from Greek soil and that the only mentally truly productive people have been the Greek, makes it difficult for us to turn around the direction of influence for one instant.

Soon after Kern published the Sanskrit edition of the *Āryabhaṭīya* [46], the French orientalist Léon Rodet was the first to provide a translation in a Western language (1877, published in [65],). Rodet wrote several articles and monographs on Indian mathematics and its relation with earlier and later developments in the Arab and Western world, published in the French *Journal Asiatiques*. He is the scholar who displays the most balanced and subtle views on the relations between traditions. In particular, his appraisal of Hindu and Arabic algebra as two independent traditions is still of value today (see [25] for an assessment). He certainly was a believer. Concerning Āryabhaṭa's inadequate approximation of the volume of a sphere (prop. 7), he writes somewhat cynically that if Āryabhaṭa got his knowledge from the Greeks, then apparently he chose to ignore Archimedes.[8]

George Thibaut, who translated several Sanskrit works on astronomy, such as Varāhamihira's *Pañcasiddhāntikā* [81], also wrote an article on Indian mathematics and

[7] "Das uns durch die humanistische Erziehung tief eingeprägte Vorurtheil, dass alle höhere geistige Cultur im Orient, insbesondere alle Wissenschaft aus griechischem Boden entsprungen und das einzige geistig wahrhaft productive Volk das griechische gewezen sei, kann uns zwar einen Augenblick geneigt machen, das Verhältniss umzukehren".

[8] "Mais elle a, pour l'histoire des mathématiques, d'autant plus de valeur, parce qu'elle nous démonstre que si Āryabhaṭa avait reçu quelque enseignement des Grecs, il ignorait au moins les travaux d'Archimède", [65], 409).

astronomy in the *Encylopedia of Indo-Aryan Research*. Concerning influences from Greek mathematics, he takes a middle position. In discussing Hindu algebra he writes that "in all these correspondences does Indian algebra surpass Diophantus".[9] As on the origins of Indian mathematics, he points out that Indian algebra, especially indeterminate analysis, is closely intertwined with its astronomy. As he argued on the Greek roots of Indian "scientific" astronomy, his evaluation is that Indian mathematics is influenced by the Greeks through astronomy. However, he adds that several arithmetical and algebraic methods are truly Indian ([82], 76-8).

Despite the existence of several works and studies which should provide sufficient counterbalance for Cantor's position as a non-believer—including Tropfke—his views remained influential well into the twentieth century. We may say that the "humanist prejudice" is still alive today. The myth that Greek mathematics is our (Western) mathematics has become intertwined with our cultural identity so strongly that it becomes difficult to understand intellectual achievements within mathematics foreign to the Greek tradition.

1.2.5 The topics

We provide an overview from the thematic divisions of the first two books of the third edition, relevant to the additions by Schulte. The first volume on calculation covered the following subjects:

- Numbers in general
    1. Terms for numbers (hundred, thousand, . . .)
    2. Signs for numbers (cuneiform, hieroglyphs, Greek, Indian, Hindu-Arabic)
- Units
    1. Units of time
    2. Divisions of angles
    3. Decimal units
- Whole numbers
    1. Caculation with whole numbers
        (a) Mental calculation
        (b) Literal calculation
            i. The general art of calculation
            ii. Operations (addition, subtraction, multiplication and division)
            iii. Approximate calculation
        (c) Calculating with terms (*bennanten Zahlen*)
    2. Properties of whole numbers (prime numbers, divisibility, perfect numbers, . . .)
    3. Tables (Old-Babylonian, Egyptian, squares, cubes, . . .)

[9]"In allen diesen Beziehungen erhebt sich die indische algebra erheblich über das von Diophant Geleistete" ([81], 73).

- Fractions
    1. General fractions (unit fractions, Greek, Arabic, Roman, ...)
    2. Decimal fractions
- Applied arithmetic
    1. The rule of three (Egyptian, Indian, Arabic, Fibonacci, Welsche praktik)
    2. Calculation of interest (Greek, Roman, Fibonacci, Stevin, ...)
    3. Periodic calculation (Pacioli, Widman, Tartaglia)
    4. Profit and loss calculation
    5. Calculation of discount(Pacioli, Leibniz)
    6. Tare weight calculation (Köbel, Grammateus, Widman, Ries)
    7. Calculation of mixtures (Greek, Arabic, Fibonacci, Widman)
    8. Companies (Indian, Arabic, Fibonacci, Pacioli)
    9. Exchange (Pacioli, Grammateus, Tartaglia)

And the second volume, which in the third edition was split into a second and third volume, contained the following subjects:

- Algebraic expressions
    1. Overview
    2. History of modern symbols and notations
    3. The introduction of general signs for quantities
- The term 'algebra'
- The development of the number concept
    1. The number one
    2. The number zero
    3. The infinite
    4. Fractions
    5. Irrational numbers
    6. Negative numbers
    7. Complex numbers
- Algebraic operations
    1. Addition, subtraction, multiplication and division
    2. Powers of unknowns (names of powers, calculating with powers)
    3. Roots of unknowns (concepts, terms, calculating with roots)

- Proportions
    1. Theory of proportions
    2. Expressions and terms
- Equations
    1. Overview. Concepts. Known and unknown quantities
    2. Linear equations in one unknown
    3. Linear equations in several unknowns
    4. Quadratic equations
        (a) Regular quadratic equations
        (b) Quadratic equations in several unknowns
    5. Cubic equations
    6. Biquadratic equations
    7. Equations in a degree higher than four
    8. Indeterminate equations
- Logarithms

Schulte's additions to Tropfke concern mostly the subjects of the first volume on calculation (numeration, the four operations of arithmetic, and fractions) and three subjects from the second volume (on proportions, powers and roots), while she avoids operations that "belong in the realm of algebra".

1.3 Cajori's History of Mathematical Notations

1.3.1 Biography of Florian Cajori (1859–1930)

Florian Cajori was born in Switzerland as the son of an engineer and emigrated to the USA when he was sixteen years old. He was naturalized as an American citizen and attended the State Normal School in Wisconsin. He studied science at the University of Wisconsin where he received a B.S. in 1883 and a M.S. degree in 1886. In 1894 he got a PhD from Tulane university where he was teaching mathematics for several years. After his PhD he taught physics and later mathematics at the Colorado college at Colorado Springs. He was appointed the chair of history of science at the University of California in 1918 until his retirement in 1929. With this assignment he became, together with George Sarton, one of the pioneers of the history of the exact science in the USA. He wrote several influential works, such as *The History of Mathematics* in 1894, *A History of Elementary Mathematics with Hints on Methods of Teaching* in 1896, and a *History of Physics* in 1899. He is best known for his work in two volumes *A History of Mathematical Notations*, published in 1928-1929 [10], [11]. This book became a classic reference work for a long time and only gained popularity when it was republished unaltered by Dover Press in 1993.

FIGURE 1.5:
Portrait of
Florian Cajori
(1859–1930).

Cajori published more than a hundred articles on mathematical notations in *Isis*, *The Mathematical Gazette*, *The American Mathematical Monthly*, *Science*, *The Monist*, the *American Journal of Mathematics*, *The Mathematics Teacher*, *The Scientific Monthly*, and the *Transactions of the American Mathematical Society*. Archibald compiled a list of 286 publications by Cajori in his obituary published in *Isis* [3]. Cajori always held an interest in the value of the history of mathematics for mathematics education.

1.3.2 Cajori's approach

While Cajori's first published note on mathematical notations goes back to 1886 [6], it is only from 1903 that he starts publishing on the subject with some regularity. His most productive period is from 1920 to 1925 in which he published 80 articles, most of them on mathematical notations. In his foreword to the first volume he reveals that the manuscript was completed in August 1925, but that it was decided to split the book in two volumes ([10], v). The first volume is concerned with notation in elementary mathematics, the second with higher mathematics until the nineteenth century. The published book includes many material from his articles. This is apparent from subjects such as his entry on *Signs of equality* (§260 – §269), which is an edited version of his article in *Isis* on the same subject [8], or on authors such as Johann Heinrich Rahn (§194), which is an abbreviated version of his article on Rahn in the *American Mathematical Monthly* [9].

It seems that Cajori struggled somewhat with the organisation of the book. Like Tropfke, he aimed for a topical organisation, but in his publications he mostly worked on individual authors and works. This ambiguity is reflected in the structure of the book. The first volume deals with 1) numerical notation, 2) elementary arithmetic and algebra, and a small section on 3) geometry. The major part of the first volume (256 p.), on arithmetic and algebra, is split in two sections. The first deals with individual authors and the second is a topical survey. However, the second part still contains several notes which are on specific authors, such as on Descartes or Herigone. Comparing Cajori with Tropfke, it becomes clear that Tropfke adopted a more systematic approach, researching the sources from a topical perspective. Cajori instead, had to fit his research notes in a format which was not the one he started with.

Furthermore, as noted by Sarton [68], the chronologies in Cajori's survey do not always fit. Bhāskara II (12th century) is treated before Al-Khwārizmī (9th century), Planudes (14th century) before Fibonacci (13th century) and Estienne de la Roche (1520) before Pietro Borgi (1484).

The second volume concerns higher mathematics, being arithmetic and algebra again (122 p.) analysis (188 p.), a short part on geometry (14 p.) and a section on teaching history of mathematics. However, the intended distinction between elementary and advanced mathematics becomes meaningless when we find a large section on the evolution of the dollar mark in volume 2 (pp. 15-29).

1.4 A Critical Assesment of Schulte's 'Overlooked Works'

While the additions of Sister Leontius and the works of Tropfke and Cajori are precious to historians of mathematics, we must place her whole enterprise in a critical light. From a current point of view, there are several problems with her approach: 1) the sources she added to the compilation of mathematical notations were not always the most relevant ones, 2) the notations she added were limited to the basic operations of arithmetic, 3) she ignored the third edition of Tropfke's work of which the two first volumes appeared two years before she completed the dissertation, 4) together with Tropfke and Cajori, she assumed that nothing important happened before 1483, and 5) also together with her two predecessors, she did not provide any explanation how mathematical notations evolve.

1.4.1 The relevance of the 'overlooked' authors

Sister Leontius's selection of the hundred and one works overlooked by Cajori and / or Tropfke was motivated more by direct availability rather than by their high relevance. She had easy access to Karpinski's impressive personal collection of rare books and the even larger collections of David Eugene Smith and George A. Plimpton. Some works on her list, such as the one by Nicolaus Felner von Winnsheim (D30),[10] Conrad Feme (D31) or Girjka Gorla Z Gorlssteyna (D41) are extremely rare, but their contents do not contribute very much to our understanding of the evolution of mathematical notations.

1.4.2 Only basic notations

While the historical development of mathematical notations become most interesting when dealing with algebra and calculus, Sister Leontius only included finger reckoning, basic operations, proportions, roots and aggregation signs in her topical overview. Concerning algebra, only the powers of the unknown were included. This misses out on much of the development of the late sixteenth and the whole of the seventeenth century.

1.4.3 Why ignoring Tropfke's new edition?

As a third critical note, we are surprised to see that she only used the second edition of Tropfke's work. By the time Sister Leontius finished her dissertation in 1935, Tropfke's third edition was already several years on the book shelves, the first volume since 1930 and the second in 1933. Smith's reviews of the third edition were published before she finished her text [77], [78], so if she was not aware of the publication, at least Karpinski must have been and she chose to ignore it. Several of her additions were already anticipated by Tropfke himself. In the third edition he does occasionally refer to the 'overlooked authors' of Schulte, such as Cataneo (vol. 2, p. 176) and Ghaligai (vol 2; p. 32) though we must add that Tropfke used secondary sources, as by Smith and Cajori to include these comments. The 1980 edition by Vogel, long after Schulte's dissertation, included more of these missing works, such as the full coverage of Cataneo (D19), Ciruelo (D23), Coutereels (D26), Curtius (D27), Ghaligai (D38), Girard (D39), Noviomagus (D65), Suevus (D88) and Tunstall (D91).

1.4.4 The dark age syndrome

A fourth word of criticism, which is also applicable to Tropfke and Cajori, is what I would call their 'dark age syndrome'. Despite the wealth of historical material in

[10] The bibliography of Schulte's dissertation begins on page 213.

these two classic works and the additions by Schulte, their overview of medieval sources contain very large gaps. Between two well-known books on medieval mathematics, Fibonacci's *Liber Abbaci* (1202, 1212) and Pacioli's *Summa* (1494), Tropfke only discusses operations from *De numeris datis* by Jordanus (c. 1250). Cajori only quotes from the *Algorismus proportionum* by Oresme (c. 1370). Both refer to the correspondence between Regiomontanus and Bianchini in 1462-3. Schulte herself draws the line at 1483, with Prosdocimo de Beldomandi's *Algorismus de Integris*, as the oldest printed work in her overview, and does not use any earlier sources. The rich tradition of mathematical practice of abbaco culture, remains untouched.

The abbaco tradition ('abbaco' or 'abbacus' spelled with a double 'bb' to distinguish it from the material calculation aid, the abacus) is situated in the late medieval cities of Northern Italy, the south of France and the Catalan region in the North of Spain and was flourishing between 1300 and 1500. Abbaco mathematical practice constitutes a mathematical culture separate from university education, supported by abbaco masters, lay surveyors and merchants. Basic knowledge of arithmetic and practical geometry was disseminated through abbaco schools and more advanced subjects such as algebra through master-apprentice relationships. Mathematical treatises produced within this culture, were written in vernacular and had a secretive character. They were eagerly collected by bankers such as the Strozzi family, which secured the survival of more than two hundred manuscripts to this day.

We must admit that, despite the pioneering work by Pietro Cossali [16] and Guillaume Libri [54], the abbaco tradition was not very well known at the beginning of the 20th century. However, the best placed scholar knowledgeable about this tradition at that time was Schulte's supervisor Karpinski himself. Not only did he publish on Medieval algorisms and Latin translations of Arabic algebra [42], it was Karpinski who already in 1910 reported on an important manuscript from the abbaco tradition, a copy of Maestro Benedetto's *Trattato di practicha d'arismetrica* in the Plimpton collection (Plimpton 189) [41]. In 1929 he published on another important manuscript by Jacopo da Firenze [44], which Jens Høyrup recently identified as the earliest vernacular treatise on algebra [36]. So, Karpinski was well aware of the existence of the abbaco tradition—citing even Cossali—but considered it but a continuation of the work of Fibonacci. He states that the treatise by Jacopo da Firenze (as also cited by Høyrup ([36], 4):

> like the similar arithmetic of Calandri, marks little advance on the arithmetic and algebra of Leonard of Pisa. The work indicates the type of problems which continued current in Italy during the thirteenth to the fifteenth and even sixteenth centuries, stimulating abler students than this Jacob to researches which bore fruit in the sixteenth century in the achievements of Scipione del Ferro, Ferrari, Tartaglia, Cardan and Bombelli.

Certainly from the point of the history of mathematical notations, ignoring the contribution of this tradition was a missed opportunity. Comparing the notations of the early printed works on mathematics with the manuscripts from the abbaco and the cossic tradition would have been an interesting project to be proposed to Sister Leontius. Both Tropfke and Cajori were conscious of the particular issues in the transition of mathematical symbolism from manuscript to print, but apart of noting possible limitations in typesetting, neither of them undertook a systematic study on this subject. Even today, the subject has only been partially treated [55], [30].

What *has* been established during the past few decades is that the abbaco tradition is not just an addendum to Fibonacci's great book, but that it constituted an important mathematical culture on its own. It has been argued by Høyrup that Fibonacci was more likely part of this tradition that he was its initiator [35]. It has also become clear that mathematical notations adopted by the abbaco tradition much depended on material practices from the Maghreb region in the Arabic world, in particular calculations on the *takht* (dust-board) or *lawḥa* (clay board). The practice has been studied by Mahdi Abdeljaouad [1] and its influence on the abbaco tradition has been demonstrated by Høyrup [39] and Oaks [57]. One such example of a symbol inherited from the Maghreb practices and still used today is the fraction bar.

1.4.5 Descriptive, not explanatory

A final word of critique concerns the descriptive-only character of the endeavors of Tropfke, Cajori and Schulte. Their topical arrangement of mathematical notations in the course of history provides us with an excellent reference to sources, but little or no explanation is given on how and why changes in notations occur. Their work raises many questions: what is the function of mathematical notations? Why do they change over time? Why is one system of notations preferred over another? Does the introduction of symbols increase our understanding of mathematics? Is a symbolic representation of a problem solution better than a solution written down in words? How do we interpret mathematical symbols? These and other explanatory questions are avoided by the three authors. However, Schulte in her introduction, reveals her ambition to venture beyond the descriptive nature of the works by Tropfke and Cajori (p. 7, emphasis mine):

> The new material should complete Cajori's *A History of Mathematic al Notations*,1 v. I, 1928, and certain sections of Tropfke's *Geschichte der Elementar Mathematik* 2, v. I, II, III, 2nd. ed., 1921, not only by noting the mathematical notations of writers who have been given little or no mention in these works, but *also by pointing out phases of the treatment of the arithmetic material which have been passed over in silence by these authors.*

So, her ambition with this project was not just to provide a chronological and topical overview as Tropfke and Cajori did, but also aspired "pointing out phases" in the development of mathematical notations. In retrospect we must conclude that her contribution amounted only to filling the gaps in the surveys of Tropfke and Cajori.

Possibly, we are expecting too much from the pioneering works on mathematical notations during the first half of the twentieth century. Later authors did attempt at an explanation on how symbolism emerged in mathematics. In the next section we will give an overview of the most common view on the development of symbolism in algebra, as it was in the emergence of symbolic algebra where the most important development took place. This approach we will call the internalistic account. In a last section we will then discuss some more recent, non-internalistic positions on the function of symbolism from different disciplines.

1.5 Internalist Accounts of the History of Mathematical Symbolism

The studies by Tropfke, Cajori and Schulte on the history of mathematical notations provide an internalistic account. They start from contemporary notions and symbols, as for example the equation sign, and then look for its occurrences in historical sources on mathematics. Special attentions is given to first occurrences and precedence. To follow the example of the equation sign, Cajori writes ([8] 116):

> To some readers this choice of the equality symbol may seem so evidently a foreordination, that they cannot readily understand how any other symbol could have been proposed as a serious rival. They experience a very perceptible shock when informed that Recorde's symbol [where it first appeared] had several competitors and that it nearly perished in its fierce struggle for existence.

Cajori then continues to discuss 'competing signs' going back to the *Arithmetica* of Diophantus. The implicit supposition in such an approach is that the meaning behind the sign, of the concept of equality, remains unchanged and that only the symbols are replaced by choice.[11] Furthermore, the meaning of the concept behind the sign appears to be our current conception. The historical approach in which all evolutions are seen to be functional towards the development of our current understand is called Whigg history. This approach to historiography accepts implicitly or more explicitly that the development towards symbolic algebra was inevitable and depended on some internal mechanisms and intrinsic processes. The history of algebra is particularly important for the development of mathematical notations, as it is in the context of algebra that most of our mathematical symbolism developed. The equations sign discussed by Cajori appeared for the first time in Robert Recorde's book on algebra and not in a work on basic arithmetic. Let us therefore look somewhat closer to different views on the emergence of symbolic algebra.

Moritz Cantor whose *Vorlesungen* [12] had a profound influence on early twentieth-century historians of mathematics, attributes, for the early period, high importance to the Latin works of Fibonacci and Jordanus. He believed that the vernacular tradition of practical arithmetic and algebra did not produce any men capable of understanding the works of these two giants. Cantor assumes this to be true for most of the 14th and 15th century. When dealing with the sixteenth century, extraordinary importance is given to the *Arithmetica* of Diophantus [24], [71], [62]. The idea that symbolic algebra originated with Diophantus was fabricated by humanist mathematicians after Regiomontanus's Padua lecture of 1464 [63]. As a consequence of their reform of mathematics, humanist writers distanced themselves from "barbaric" influences and created the myth that all mathematics, including algebra descended from the ancient Greeks [34]. Later writers such as Petrus Ramus (D74), Jacques Peletier [60], François Viète [85] and Christopher Clavius [14] participated in a systematic program to set up sixteenth-century mathematics on Greek foundations. The late discovered *Arithmetica* of Diophantus was taken as an opportunity by Viète to restore algebra "which was spoiled and defiled by the barbarians" to a fictitious pure form. To that purpose he devised a new vocabulary of Greek

[11] For an example how conceptions of basic operations may differ between cultures, see section 1.6.5 below.

terms to cover up the Arab roots of algebra "lest it should retain its filth and continue to stink in the old way" ([47], 318). The reality was that, with some exceptions, ancient Greek mathematics was more foreign to European mathematics than Indian and Arabic influences which were well digested within the vernacular tradition [26].

The most quoted theory on the development of mathematical symbolism is Nesselmann's 19th century study on "Greek algebra" [59], which argues that algebra went through a historical development in three stages: rhetorical, syncopated and symbolic algebra. This tripartite distinction has become such a commonplace depiction of the development of algebraic symbolism that modern-day authors even fail to mention their source (e.g., [5] p. 201; Flegg and Hay; [74]). The repeated use of Nesselmann's distinction in three *Entwickelungstufen* on the stairs to perfection is odd because it should be considered a highly normative view which cannot be sustained within our current assessment of the history of algebra. Its use in present-day text books can only be explained by an embarrassing absence of any alternative models. There are several problems with Nesselmann's approach as I have argued before [28]: the chronology just does not fit with historical evidence, the syncopated algebra of Diophantus appears to be an invention of medieval scribes, and there is ample counter evidence which challenges the chronology in three phases.

Contemporary scholars, such as Rafaella Franci and Laura Toti Rigatelli [19] [20] or Van der Waerden [87] narrate the story of the history of algebra from their internal dynamics. Early European algebra, as inherited from the Arabs, recognized six types of equations. Instead of dealing with the general form of a quadratic equation, the first Latin translations distinguishes three cases depending on the sign of the coefficients and three cases with one or two terms missing. Each case had its own solution method. Double solutions were not recognized except for the case of two positive roots. Early abbacus masters extended the list of six to include higher degree cases most of which could be reduced to the original six. During the fourteenth century Maestro Dardi of Pisa expands this to no less than 198 cases [84], [21]! Already from the fourteenth century abbacus masters started experimenting with irreducible cases of higher degree. False rules were given for special cases of the cubic equation [37]. However, Maestro Dardi gives some examples of cubic equations with a correct solution derived from numerical examples. These histories of algebra then focus on the sixteenth-century breakthrough of Scipione del Ferro in solving the depressed cubic and the feud between Tartaglia and Cardano for publishing a general solution for the cubic equation. It then moves to symbolism introduced by Viète and the general approach to problems. With a mention of Girard, these developments culminate into the quest for the fundamental theorem of algebra moving well into the eighteenth century. All this is presented as a continuous flow of necessary logical development. Each step is the necessary next move in the logical puzzle of the history of symbolic algebra.

Van der Waerden goes to great lengths of demonstrating this continuity back to the earliest Greek mathematics. In an earlier publication, Van der Waerden ([87], 116) uses the so-called *Bloom of Thymaridas* to connect algebra with the Pythagoreans. He goes as far as to claim that "we see from this that the Pythagoreans, like the Babylonians, occupied themselves with the solution of systems of equations with more than one unknown". We have elsewhere demonstrated that these claims cannot be sustained and that these should be understood as eloquently stated by Hankel, a by"humanist education deeply inculcated prejudice that all higher intellectual culture, in particular all science, is risen from Greek soil" [28].

Jacob Klein, a student of Heidegger and interpreter of Plato, wrote a long treatise in 1936 on the number concept starting with Plato and the development of algebra from Diophantus to Viète. It became very influential for the history of mathematics after its translation into English in 1968. Klein goes even further than Van der Waerden or Franci and Rigatelli in their internalistic account of history. For Klein it is not the evolution of solution methods for solving equations which follows some logical path but the ontological transformation of the underlying concepts within an ideal Platonic realm. He restricts all other possible understandings of the emergence of symbolic algebra by formulating his research question as follows: "What transformation did a concept like that of *arithmos* have to undergo in order that a 'symbolic' calculating technique might grow out of the Diophantine tradition?" ([47], 147). According to Klein it is ultimately Viète who "by means of the introduction of a general mathematical symbolism actually realizes the fundamental transformation of conceptual foundations" ([47], 149). Klein places the historical move towards the use of symbols with Viète and thus ignores important contributions by the abbaco masters, by Michael Stifel ([72], [73]), Cardano (D16, [13]) and the French algebraist such as Peletier [60], Buteo [4] and Gosselin [22]. The new environment of symbolic representation provides the opportunity to "the ancient concept of *arithmos*" to "transfer into a new conceptual dimension" ([47], 185). As soon as this happens, symbolic algebra is born: "A soon as 'general number' is conceived and represented in the medium of species as an 'object' in itself, that is, symbolically, the modern concept of 'number' is born". ([47], 175). It is hard to understand why a philosophy like this, rooted in German idealism, where concepts realize themselves with the purpose to advance mathematics, is so appealing to modern historians looking for an explanation for the emergence of symbolic algebra.

The three different approaches to the history of algebra are exemplified by three historians of mathematics. Cantor believes in a continuous development from ancient Greek notions of number and proof to modern algebra, only obscured during the medieval period in which the old masters were not fully understood. Van der Waerden and Franci see a historic realization of the logical development from quadratic equations to cubic and higher degree ones towards a theory of the structure of equations. Klein discerns a realization of symbolic algebra in a necessary ontological transformation of the underlying number concept. All three share the idea that there is some internal necessity and logic in the historical move towards symbolic algebra. But all pass by at the fundamental historical changes that took place in the context in which medieval algebra matured: the mercantile centers of northern Italy and the French Provence region. Contextual factors related to the the socio-economical context of that time, scribal practices and the didactic models of the abbaco schools influenced the practices of mathematics including the development towards a symbolic mathematics. We will now discuss some new approaches to the emergence of symbolic algebra.

1.6 Contemporary Approaches to the History of Mathematical Notations

For a broader answer to the questions on the role of symbolic notations that we stated before, an interdisciplinary perspective is in order. We will now give an overview of the major insights from four fields that contribute to our current understanding of mathematical symbolism: 1) semiotics, 2) extended cognition theory, 3) cognitive science and 4) historical epistemology.

1.6.1 Semiotic approaches

There exists a general theory about signs and their interpretation, which is called semiotics. While semiotic approaches to mathematical symbolism may seem a natural consequence of this discipline, they have been developed in rather recent times and mostly in connection to mathematics education and cultures of mathematics.

Charles Sanders Peirce proposed a triadic model of semiotics consisting of the signifying element, the object and the interpretant. Other models have been used in which the sign is formed by the union of the signifying element and the signified (the object it represents) as by Ferdinand de Saussure in *On Signifying*. The connection between them is considered arbitrary and conventional.

Brian Rotman, has been the main proponent of a semiotic approach to mathematics. He recently developed a semiotic model [67], in which a *Subject* imagines actions and an *Agent* enacts instructions by writing down ideal signs. Semiotic approaches are very general in nature and provide little explanation on how specific signs get their meaning in a mathematical context. However, Rotman [66] devoted a semiotic study on the symbol zero which he considers as a giant conceptual step as the meta-linguistic completion of the place value system, and compares it with other semiotic breakthrough as the symbolic function of paper money and the vanishing point in Renaissance perspective.

Semiotic approaches have recently been popular in mathematics education as they follow naturally from the theory of Vygotsky [86] which emphasizes socio-cultural and social constructivist theorisations of mathematics and mathematics education. Some examples of this approach are Paul Ernest [18] and Louis Radford [61]. Ernest summerizes the semiotic research program as follows ([18], 68):

> A semiotic perspective of mathematical activity provides a way of conceptualising the teaching and learning of mathematics driven by a primary focus on signs and sign use. In providing this viewpoint it offers an alternative to psychological perspectives that focus exclusively on mental structures and functions. It also rejects any straightforwardly performance or assessment focussed perspectives concerned only with student behaviours. Instead it offers a novel synthesis that encompasses but also transcends these two types of perspective, driven by a primary focus on signs and sign use in mathematics. Beyond the traditional psychological concentration on mental structures and functions 'inside' an individual it considers the personal appropriation of signs by persons within their social contexts of learning and signing. Beyond behavioural performance this viewpoint also concerns patterns of sign

use and production, including individual creativity in sign use, and the underlying social rules, meanings and contexts of sign use as internalised and deployed by individuals.

1.6.2 The extended cognition approach

The extended cognition approach starts from the idea that humans delegate complex cognitive tasks to an external environment. It has been recently proposed that humans are probably not biologically equipped for formal mathematics, and that symbolism can be considered as an external environment which allows us to delegate complex cognitive tasks (e.g. [17]). The idea that symbolic representations should be interpreted as a form of extended or embodied cognition is most strongly advocated by a seminal work of Lakoff and and Núñez, *Where Mathematics Comes From*. Their central point concerning symbols is formulated as follows ([49], 49):

> The meaning of mathematical symbols is not in the symbols alone and how they can be manipulated by rule. Nor is the meaning of symbols in the interpretation of the symbols in terms of set-theoretical models that are them elves uninterpreted. Ultimately, mathematical meaning is like everyday meaning. It is part of embodied cognition.

In times when symbolism was not sufficiently developed, the external environment facilitating calculations consisted of material artifacts, such as the abacus or the *takht* (dust-board) and the *lawḥa* (clay board), we previously mentioned. Also the use of hands facilitated calculations and memory. We find some far-going applications of this in Renaissance mathematics. I will provide an illustration of extended cognition by means of an explanation in one of the most popular arithmetic books of sixteenth century Germany, by Adam Riese [64]. It went through more than hundred editions. In the book, Riese describes procedures for solving typical problems by some dedicated rule. Problems and rules had often a specific name. On such type of problem was called the *Regula Virginum*, referring to young girls that accompanied adults to a tavern. The bill was divided in proportion to the type of company: male adults, woman adults and young girls, as for example $7 : 5 : 3$. In a more formal description:

> N men, woman and children drink in a tavern. They share the bill, men paying a, woman b and children c. The total is d. How many are there or each?

Riese describes a procedure for the *Regula Virginum* with two types of persons. The total bill amounts to 54 and there are 12 persons. The adults have to pay 5 each and the children 3 each. However, the numbers of adults and children is unknown. In algebraical terms (which Riese does not use) the problem amounts to the equations:

$$x + y = 12$$
$$3x + 5y = 54$$

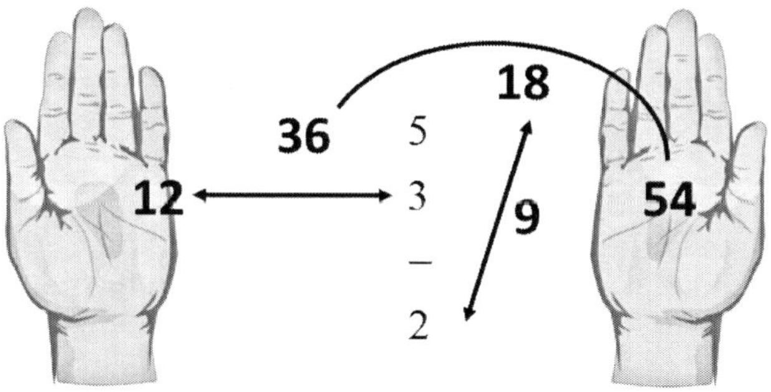

$$x + y = 12$$
$$3x + 5y = 54$$

FIGURE 1.6: An embodied procedure for solving a typical Renaissance problem, following an explanation by Riese [64]).

Adam Riese describes the procedure as follows:[12]

> Lay down before you at the left hand the number of persons [12]. At the right hand how much each has drunk [54]. And in the middle, how much each person each had to pay for himself [5&3]. Then make the money of the largest every where equal, and then multiply the smallest payment with the number of persons [3×12], and take this from what they have been drinking [$54 - 36 = 18$]. What then remains is the quantity which had to be divided.

By referring to the hands and laying down the values, this clearly accounts as an embodied procedure: laying down the known quantities of the problem (visualization) and assigning specific values and operation to each hand (memorization). The specific physical operations that have to be performed following the procedure correspond with the following mathematical operations:

$$x = 12 - y$$
$$3(12 - y) + 5y = 54$$

which resolves in the following solution:

$$y = \frac{54 - 36}{5 - 3}$$

In the procedure proposed by Riese, the body parts stand for *any possible* problem value (N and d). In the terminology of Lakoff and Núñez, the body parts have a metonymic function. The hands can be used to represent any possible problem value. By following the embodied procedure a solution to the problem can be attained. This principle they have termed "The fundamental metonymy of algebra". So, while the embodied procedure proposed by Riese for didactic reasons, does not use any symbolic representations it shares the same metonymic function as the equations we have shown.

[12] [64], f.135: "Schreib vor dich gegen der lincken handt die anzahl de Personen. Gegen der rechten handt / wie viel sie vertruncken / und in die mitte / wie vil einjegliche Person / jeglichs geschlechts in sonderheit gibt. Darnach mach das gelt dem menigsten uberall gleich / als dann multiplicir das kleinest an der bezahlung mit den Personen / und nimb von dem das sie vertruncten haben/ Das da bleibt ist die zahl / welche getheilt sol werden". The specific values of the problem are shown between square brackets.

Such examples provide strong evidence that we indeed have to look at mathematical symbols as an extension of our cognition.

1.6.3 Cognitive science

In his first volume, Cajori included a curious quote by the English mathematician William Oughtred (1574 – 1660) on his use of symbols ([11], 199) also cited by Sarton in his review of Cajori's book [68] ([58], p 20):

> This manner of setting downe theoremes, whether they be Proportions, or Equations, by Symboles or notes of words, is most excellent, artificiall, and doctrinall. Wherefore I earnestly exhort every one, that desireth though but to looke into these noble Sciences Mathematicall, to accustome themselves unto it: and indeede it is easie, being most agreable to reason, yea even to sence. And out of this working may many singular consectaries be drawne: which without this would, it may be, for ever lye hid.

Oughtred was an able mathematician who made extensive use of symbols in his books. Cajori spends no less than 12 pages on the notations he used and earlier published a book on him [7]. Interesting from a current point of view is that Oughtred refers to the perceptual aspect of symbols "agreeable to reason, even to sense". Current research from experimental psychology indeed suggests that symbolism in mathematics heavily depends on visual perception or even perception of movement [50].

Recent studies in cognitive science imply that modern symbolism, as it is taught and practiced today, relies on the visual processing of elements in a way which is similar to the interpretation of diagrams. Spatial organization, directionality, grouping and mental operations like "picking up" and moving elements across symbolic expressions, appear to be crucial for our understanding of mathematical symbolism. We will limit our discussion to some results from David Landy and Robert L. Goldstone, that support this new interpretation of symbolic understanding.

Ever since Simon and Newell coined the term 'physical symbol systems hypothesis' (PSSH) [56], the idea that a symbolic system has the necessary and sufficient means for general intelligent action, has been strongly debated. This philosophical debate is of no concern to us now, but the underlying mechanism of symbol grounding is. How do symbols—supposedly abstract and arbitrary—get their meaning? Or, placed within the context of this study, what underlying processes provide meaning to symbolic expressions in mathematics? In 2007, Landy and Goldstone [51] formulated a weaker but fruitful revision of the PSSH, that "symbols are not arbitrary, unconstrained tokens but rather are represented and processed using space and perceptually organized groups. This conception of physical symbols makes them far more constrained .." than generally considered from a formalist view point. Over a period of five years they have conducted several studies that gives strong support to such thesis. Two such experiments are interesting for the way we interpret symbols.

The first concerns the role of proximity relations in the understanding of a symbolic expression. Landy and Goldstone tested the effects of four types of proximity relations on the interpretation of symbolic expressions: physical space, common region,

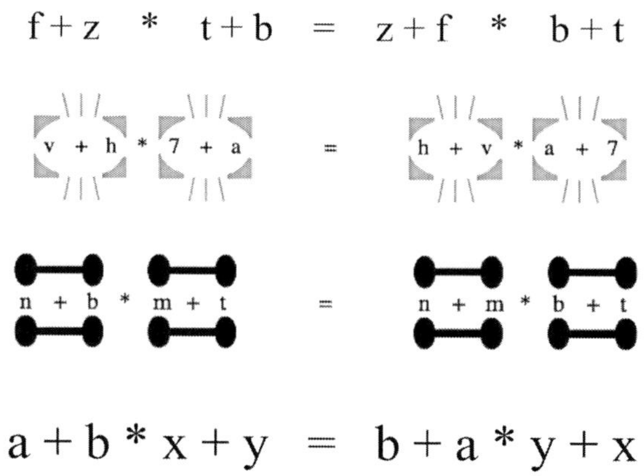

FIGURE 1.7: Four types of proximity relations that influence the interpretation of symbolic expressions: physical space, common region, connectedness, and alphabetic proximity, from [50]).

connectedness, and alphabetic proximity (see Figure 1.7).[13] These proximity relations can be consistent, inconsistent, or neutral with respect to the operator structure. In the examples shown in Figure 1.7, they are inconsistent. The experiments show an important difference in accurate interpretation between consistent and inconsistent perceptual grouping. If symbolic expression were purely sentential, there would be no difference in interpretation with subtle proximity relations as physical spacing and, more surprisingly, alphabetic proximities. The results provide strong support for the thesis that visual and perceptual factors play an important role in our interpretation of symbolic expressions.

Obviously, the inconsistencies shown in the examples can be avoided by using brackets as in $f + (z \times t) + b = z + (f \times b) + t$, but bracket symbols themselves use the mechanism of proximities which is here under investigation. Furthermore, symbols for aggregation of terms were introduced only in the late seventeenth century. In absence of an unambiguous notation system, abbaco masters employed schemes like the crosswise multiplication of surd binomials, avoiding inconsistent interpretations.

A second curious effect is studied by Landy and Goldstone, demonstrating that subjects were systematically affected in their interpretation of simple linear equations by the simultaneous perception of actual motion. Simple linear equations can be mentally solved by 1) moving individual terms across the equation sign and changing their sign, and/or 2) removing the coefficient of the unknown by moving it below the fraction bar on the other side of the equation sign. When random dots are shown moving in the background of such an equation (as in Figure 1.8) the direction of the movement can be consistent or inconsistent with the shifting of terms to be performed. The results of the experiments show that the effect is significant and systematic, growing larger with higher mathematical skills, and being strongest for problems involving simpler operations such as addition and multiplications, rather than subtractions and divisions.

That the perception of actual motion of random dots influences the interpretation of symbolic equations again supports the idea that modern symbolism still relies on tangible practices which preceded the introduction of symbolism.

[13]Proximity relations are studied in [52], [51], with a fuller treatment in [53] and [50].

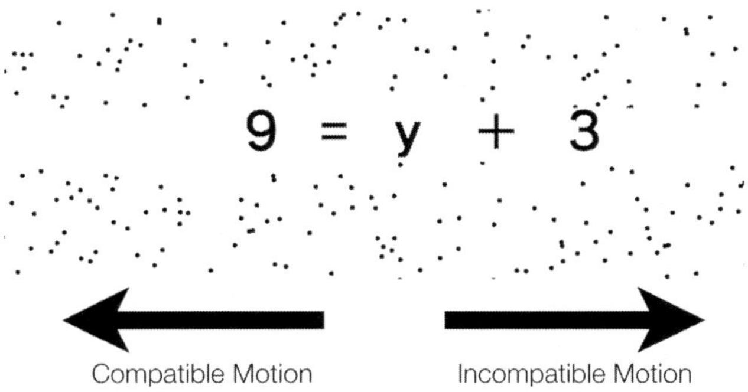

FIGURE 1.8: An experiment demonstrating that random dots moving on the background can influence our interpretation of symbolic expressions, (from [50]).

1.6.4 Historical epistemology

Finally, the discipline of historical epistemology in which I am myself involved. Historical epistemology tries to determine the meaning of historical concepts and the criteria of validity of historical methods for the actors involved. Applied to the history of mathematics, it is senseless to speak about 'correct' or 'incorrect' methods or conceptualisations from our current understanding of mathematics, as historical epistemology aims to establish the meaning of concepts within a specific historical and social context. Several studies have been devoted to the history of mathematical notations from such starting point. I will discuss three that demonstrate the importance of this discipline to the historiography of mathematical notations.

The meaning of historical concepts

In a recent article, Jens Høyrup gives us a methodological reflection on the different shades of meanings of mathematical concepts in their historical context and how these conceptual structures seem to transfer between different mathematical cultures [38]. Høyrup uses lexical analysis and accounts for socio-economical structures within a mathematical culture to determine differences in meaning. The examples he discusses concerns the operations of elementary arithmetic, such as addition, subtraction, multiplication and division in Old-Babylonian mathematical tablets. One such example is subtraction for which he distinguishes no less than eight different meanings (see Table 1.1). For those who find this extensive list of meanings somewhat far-fetched, let us remind that one of the concrete meanings of subtraction, *nasāḫum* or "tearing off", brought Høyrup to the idea that Old-Babylonian algebra was based on concrete and tangible operations known from surveyors practices. This new interpretation, different from Neugebauer and others, has led to a completely new view on mathematics from this era.

Now, consider the approach of Cajori to finding the earliest symbol for "minus" [10], 7):

> 10. Besides the principles of addition and multiplication, Babylonian tablets reveal also the use of the principle of subtraction, which is familiar to us in the Roman notation XIX (20 − 1) for the number 19. Hilprecht has collected ideograms from the Babylonian tablets which he has studied, which represent the number 19. We reproduce his symbols in Figure 2. In each of these

Term	Translation	Meaning
nasāḫum/z i	to tear out	an identity-conserving concrete removal, inverse operation of "appending"
ḫarasum	to tear off	another concrete removal, preferred in a few texts (while certain others have a tendency to "cut off" from lines and "tear out" from areas)
tabālum	to cut off	another concrete removal, used occasionally about what can "justly" be removed
šutbum	to withdraw	sometimes used within arguments of "false position" about the removal of the "due" fraction
A *eli* B *d itter*/A ugu B d d i r i g	A over B, d it goes beyond	comparison of two different concrete magnitudes, necessarily of the same kind
d i r i g	the going-beyond	excess by previous operation
A *ana* B d *imti*/l a l	A to B, d it becomes smaller	"comparison the other way round", used when the text format or other considerations require it
bêrum	to single out	rarely occurring inverse operation of heaping, separating the sum into constituents

Table 1.1: **Different meanings of the subtraction operation in Old-Babylonian mathematics distinguished by Høyrup, [38].**

twelve ideograms (Fig. 2), the two symbols to the left signify together 20. Of the symbols immediately to the right of the 20, one vertical wedge stands for "one" and the remaining symbols, for instance Y ≻, for LAL or "minus"; the entire ideogram represents in each of the twelve cases the number 20 − 1 for 19.

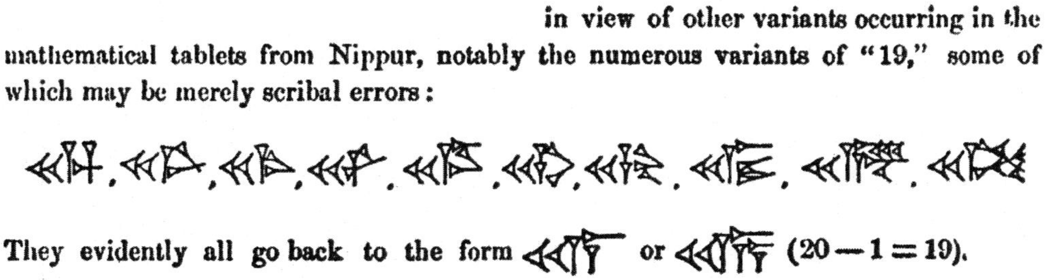
in view of other variants occurring in the mathematical tablets from Nippur, notably the numerous variants of "19," some of which may be merely scribal errors:

They evidently all go back to the form or (20—1 = 19).

FIGURE 1.9: The Old-Babylonian "symbol for minus" according to Cajori, (from [10], 6).[14]

Cajori looked at one rather peculiar meaning of subtraction, l a l, as he there found an idiogram that in his view best approached our current use of the subtraction sign. In contrary, the detailed analysis of the concept of subtraction in Old-Babylonian tablets by Høyrup shows a variety of meanings, used consistently in different contexts. Representing the subtraction operation by a single sign obscures our understanding of these important differences.

The symbolic revolution

A recent epistemological study by Michel Serfati [69] is one of the few aimed at determining the factors that constitute symbolism in mathematics. Serfati distinguishes six patterns of symbolism necessary and sufficiant for the development of operational symbolism at the beginning of the seventeenth century:[15]

1. **The representation of the "unknown" or "required"**. Dealing with the kinds of notations of the unknown and its powers as treated by Tropfke, Cajori and Schulte. Though the notations may seem simple in nature, Serfati shows that they can involve surprising features.

2. **The dialectic of indeterminancy** (the representation of the "given"). Given is for example the use of a in $y = ax$. François Viète is often considered as the father of modern algebra. This somewhat does injustice to the achievements of Cardano, Stifel and Peletier who brought major advances which were put to practice by Viète. But his own invention, was to distinguish between "the arbitrary and the fixed, or the one and the multiple or even, maybe more significantly, the unspecified and the singular", as formulated by Serfati. In his *Isagoge*, Viète (1591) introduced vowels for the unknowns and consonant letters for the given. This allowed to express general equations with arbitrary coefficients and moved algebra from problem solving to the study of symbolic equations.

[14] Ed. The caption of this figure in Cajori is "Fig. 2.—Showing application of the principle of subtraction."
[15] Based on his own summary in [70]

3. **Structuration by assemblers** (the representation of elementary operative instructions: addition, division, extraction of roots, etc.), with the signs +, $\div$, $\cdot$, −. Thus the basic notation: $2 \cdot x + 1$.

4. **The ambiguity of order** (the representation of the succession and *entanglement* of instructions). This pattern is coextensive to both dividing and gathering signs in the symbolic text, *via* brackets or *vinculum* for instance, that is via *punctuation* signs, also denoted *aggregation* signs. The interpretation of $f + z \times t + b = z + f \times b + t$ as $f + (z \times t) + b = z + (f \times b) + t$ discussed in the previous section, falls under this pattern

   Serfati emphasizes the conceptual importance of a tree-structure underlying any symbolic text (it structures the mathematical thought). The two possible readings of the example above, one can represented by two different tree-structures (as shown in Figure 1.10).

FIGURE 1.10: **Parsing trees for two interpretations of the ambiguous $2 \cdot x + 1$.**

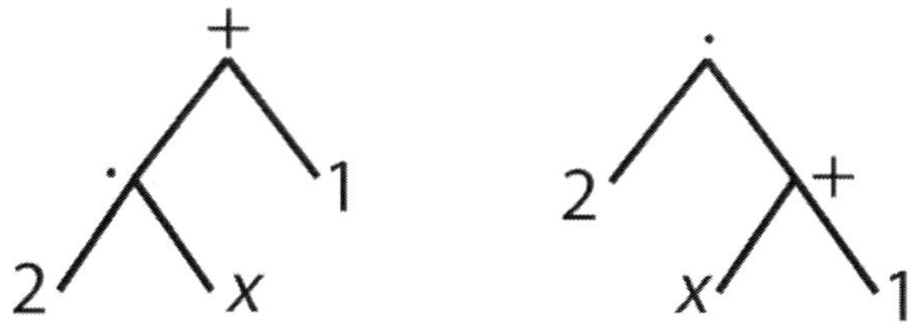

He further notes that the parsing tree can be read in two ways (from the root or from the leaves), corresponding with two (theoretical) epistemological positions of the mathematical subject, the one by the author and the other by the reader.

A more complex example is provides is taken from Descartes in the *Géométrie* in his solution to the Pappus problem leading to an equation of the second degree:

$$m - \frac{n}{\xi} + \sqrt{m \cdot m + o \cdot x - \frac{p}{m} x \cdot x}$$

The parsing tree of this equation is shown in Figure 1.11:

FIGURE 1.11: **Parsing tree for a quadratic equation of the second degree.**

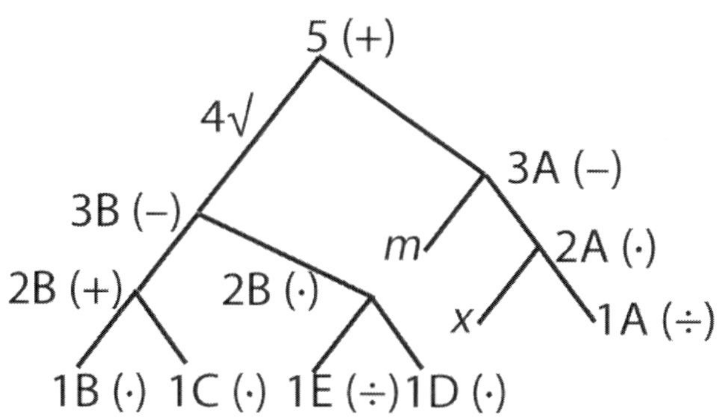

Signs of aggregation, as different kinds of brackets, progressively opened up the possibility of the successive execution of a significant number of various instructions. Such complex combinations of notations were beyond the scope of abbaco and early cossic algebra.

In relation to this fourth pattern, Serfati notes that this opens up the possibility of two different interpretations of symbolism: as a *procedure* or as an *object*.

5. **Representation of relationship** (for example by means of the signs = (Recorde), $>$ (Harriot), and $\cong$ (Leibniz)). As already discussed, the equality sign was established by Robert Recorde (Recorde 1557), and used by Descartes' (*Géométrie*, 1637). It is surprising that relationships were symbolized rather late in the process.

6. **Representation of compound concepts**, based on the example of the representation of the sequence of powers (squares, cubes, sursolids or quadratics, etc.).

   This is an important breakthrough which originates with Descartes. The cossic system of notations clearly had its limitations. Stifel could express 𝔷 as the square of the unknown. However, he could not express the square of 1 𝔷 $+1$ ꝛe $+2$ in a similar structure, as we do today in the Cartesian system, by simply replacing A with A^2. The missing mechanism, according to Serfati, is substitution. The cossic expressions ꝛe and 1 𝔷 $+1$ ꝛe $+2$ can not be freely substituted for one another in the symbolic expression of a square. Even the more simple expression such 3 ꝛe $+5$ could not be raised symbolically to a square. With the Cartesian exponents however, any possible symbolic expressions could be raised to a higher power, without having to perform the calculation, as Stifel had to do. The importance of the introduction of such seemingly 'natural' expressions cannot be underestimated. However, Cajori almost a century ago, did appreciate the importance of this innovation as he stated "there is perhaps no symbolism in ordinary algebra which has been as well chosen and is as elastic as the Cartesian exponents" ([10], 360)

The patterns of symbolic representation, as described by Sefati, provide a systematic framework for analyzing and understanding the functions of mathematical notations. In fact, the overview of the mathematical notations by Tropfke, Cajori and Schulte could be reorganized in catageories following these patterns.

The diagrammatic character of mathematical symbolism

Finally, in my own epistemological research into the mathematics of abbaco masters, I have formulated the hypothesis that the earliest form of mathematical symbolism had a justificational function, in the same way as certain schemes were used for demonstrating the validity of certain operations in arithmetic [31].

Let us start with an anonymous and unpublished family of 15th-century manuscripts which can justly be described as the first European text in which the author intentionally uses some form of symbolism: rhetorical (*per scritura*) and 'symbolic' (*figuratamente*) demonstrations.[16] While the 'symbols' used in the symbolic part are not new—some of them also appear in previous manuscripts—the explicit distinction between the two

[16] This family of manuscripts and their dependence is described in [27].

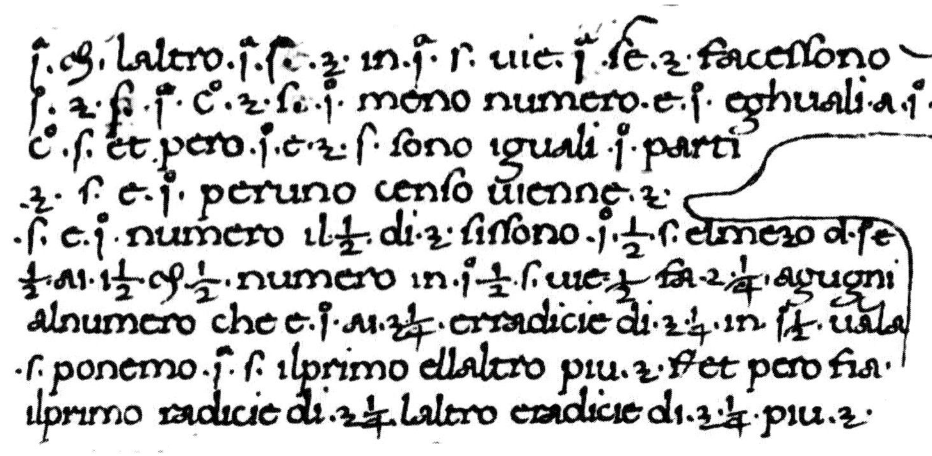

FIGURE 1.12: The 'symbolic' solution to a problem (used with permission from BL, Add. 10363, f. 60r).

methods is rather unique. I do not know of any other abbaco text in which this is made so explicit.

The second problem of chapter 33 asks to find two numbers given certain conditions. In modern symbolism the problem amounts to the equation $x(x+3) = 1$ (Florence, BNF, Magl. Cl. XI. 119, f. 58v). After the problem enunciation the author asks to solve this by algebra ("Farenlo per la cosa") and also to show it 'symbolically' (*figuratemente*). First, the symbolic version is given (as shown in Figure 1.12).

Several remarks about this fragment of the manuscript are appropriate here. It describes the solution to the problem in a terse format which corresponds with the solution repeated in the text following this fragment. Firstly, for the modern reader, this may seem very different from algebraic symbolism. However, we should consider this fragment within its specific context. The main text of this treatise (or at least this chapter on algebra) does not use any abbreviations or ligatures for the unknown (cosa), the square of the unknown (censo), the addition and subtraction operations, and the roots. All these mathematical terms for which 'symbols' are used with increasing frequency during the 15th century, are written in full words in the discursive part. The author therefore wants to make a clear distinction between a rhetorical solution—still the standard practice in abbaco algebra—and a 'symbolic' version of the solution which occasionally and unsystematically crops up in such texts. It is a formal way to make clear the distinction between two ways of presenting the solution to an algebraic problem. So, even if the "symbolic" solutions still look a lot like rhetorical algebra to us, for the author at least there is a fundamental difference between the two.

Secondly, the "symbolic" solution is 'boxed', a scribal practice to indicate that the fragment is outside of the text. Figure 1.12 shows only a line at the right-hand side of the box, but some instances in other copies show the boxes more prominently. This represents the author's clear separation between the discursive explanation and the solution *per figura*.

A third observation, which can be drawn from comparing the different manuscript copies, is that the 'symbolic' solution differs greatly between the copies. The BL copy shown in Figure 1.12 is rather conservative. We know that it is at least two generations

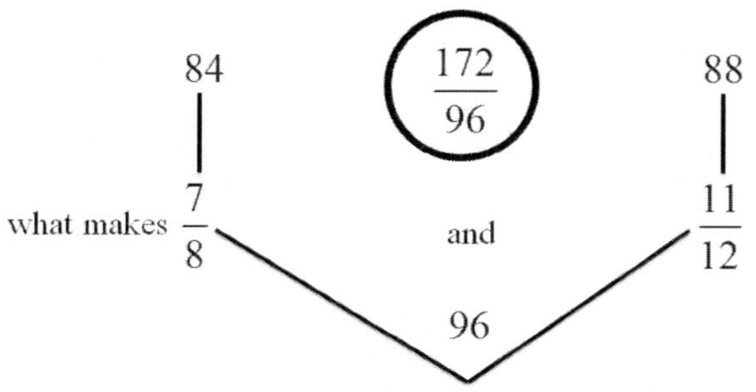

FIGURE 1.13: A justification scheme for the operation of joining two broken numbers (following Vat. Lat. 4826 f. 14v).

away from the lost archetype. The operations 'più' and 'meno' are not abbreviated as in other copies and most likely expanded from a previous copy by the scribe. Professional scribes did not always understand the contents and meaning of abbaco manuscripts and especially the 'symbolic' parts are not faithful reproductions. One scribe (Florence, Biblioteca Mediceo-Laurenziana, Ash. 343) even omits these parts. This scribe did not understand the sense of writing down the same solution twice. He did not understand the principle distinction our author wanted to make.

This brings us to our main point: what was the purpose of solving a problem in two ways, *per scritura* and *figuratamente*? I would like to draw a parallel with the non-discursive character of justification schemes used in abbaco arithmetic. The author considers the 'symbolic' solution as an additional non-discursive justification of the algebraic solution in the text. This belief is supported by a remark between the 'symbolic' and rhetorical solution: "I showed this symbolically as you can understand from the above, not to make things harder but rather for you to understand it better. I intend to give it to you by means of writing as you will see soon".[17] This quotation is very important as it shows that the author understands the two solutions as two different approaches to present an algebraic solution. Moreover he makes explicit his motivation to do so: "to understand it better". The "symbolic" solution adds something to the discursive one. If the author would consider the "symbolic" solution on the same level as the discursive, it would make no sense to go through the trouble of solving all the problems in this chapter twice. The "symbolic" solution adds to the understanding of the discursive solution in the text. It thus functions as a justification for the discursive part. The author appeals to some intuitive kind of understanding in the same way as the justification schemes are being used for the operations on fractions (as shown in Figure 1.13).

The abbaco tradition was concerned with teaching methods for calculating with Hindu-Arabic numerals. Not able to rely on authoritative texts, abbaco masters devised schemes for the justification of the correctness of procedures and algorithms. Such schemes appear as non-discursive elements in the texts and provide justification of discursive descriptions of the operations. The appearance of early symbolic solutions in a box outside the text can also be considered as non-discursive elements, providing a

[17]Quoted from BNF, Magl. Cl. XI. 119, f. 58r: "Ora io telo mostrata figurativamente come puoi comprendere di sopra bene ch'è lla ti sia malagievole ma per che tulla intenda meglio. Io intende di dartela a intendere per scrittura come apresso vedrai".

justification for rhetorically described algebraic procedures. These findings correspond closely with the experiments in cognitive science discussed earlier. In our interpretation, early symbolism function as representations of concrete operations as are the justification schemes. Even the modern mind, trained in the interpretation and manipulation of symbolic structures, relies on perceptive processes in which spatial grouping, directionality and mental movement play an important role. We thus find a strong parallel between the phylogenesis and psychogenesis of symbolism, between the historical emergence of symbolism in mathematical practice and the development process of twelve to fourteen year old students acquiring the skills of symbolic reasoning. Such strong correspondence has important consequences for the epistemology of mathematical practice as well as for mathematics education.

1.6.5 Conclusion

This book dealt with the historiography of mathematical notations. Inspired by Hankel's book [23], awarded to him when he was a dedicated 16-year old student, Tropfke started the project of rewriting the history of mathematics from a topical viewpoint. In doing so over a period of forty years, Tropfke drastically changed the way we look at the history of mathematics. Instead of following the biographical and chronological approach to history by Cantor and others, Tropfke approached the history of mathematics from basic concepts, operations and notations. He was motivated in doing so because of his experience as a teacher. He believed that the history of mathematics is important to mathematics education, but existing historiography was not suited to the task. Tropfke was also ground-breaking in the way he integrated non-Western mathematics in his topical study.

This new way of looking at history of mathematics and the recognition of its relevance for mathematics education had an important influence on the early historiography of mathematics in the USA. We have tried to demonstrate that Florian Cajori, Louis Karpinski and David Eugene Smith are all indebted to Tropfke's legacy, while this is not always expressed explicitly in their works. Cajori vastly expanded on the historical study of mathematical notations while Smith concentrated more on specific historical methods. All shared the interest in the use of history of mathematics in mathematics education. As a student of Karpinski, also sister Leontius fits in this picture. With her "Additions in Arithmetic, 1483–1700" she continued the project of Tropfke and Cajori. Cajori's two-volume work is such an extensive study that it immediately became a classic since its publication in 1929. Hardly anyone has tried to replace Cajori's study with an updated one. The all-encompasing multi-cultural study by Georges Ifrah comes to mind, but is limited to numbers and numbers systems only [40].

We have been critical in our assessment of Tropfke, Cajori and Schulte for a contemporary understanding on the emergence and importance of mathematical symbolism. However, this does not do injustice to the importance of these works in their historical context. The study and the historiography of mathematical notations is as pertinent as it was when Tropfke started his project. We argued that a contemporary study of mathematical notations needs an interdisciplinary approach, integrating what we have learned from semiotics, cognitive studies and historical epistemology in a setting where we study mathematical practice in different cultures and time periods.

References

[1] Abdeljaouad, Mahdi (2002) "Le manuscrit mathématique de Jerba: Une pratique des symboles algébriques maghrébins en pleine maturité". *Quaderni de Ricerca in Didattica del Gruppo di Ricerca sull'Insegnamento/Apprendimento delle Matematiche*, 11, 110-173

[2] Archibald, Raymond Clare (1922) Review of *Geschichte der Elementar-Mathematik in Systematischer Darstellung mit Besonderer Berucksichtigung der Fachworter*, by J. Tropfke, *The American Mathematical Monthly*, Vol. 29, No. 1 (Jan., 1922), pp. 16-17.

[3] Archibald, Raymond Clare (1922) "Florian Cajori 1859-1930", *Isis*, Vol. 17, No. 2, pp. 384-407.

[4] Buteo, Johannes (1559) *Ioan. Buteonis Logistica, quæ & Arithmetica vulgò dicitur, in libros quinque digesta ... Eiusdem, ad locum Vitruuij corruptum restitutio, qui est de proportione lapidum mittendorum ad balistæ foramen, libro decimo*, Lugduni: Apud G. Rouillium.

[5] Boyer, Carl B. (1989) *A History of Mathematics* (revis. U.C. Mertzbach), New York.

[6] Cajori, Florian (1886) "History of Arabic notation", *Journal of Education*, vol. 8, pp. 201-203; pp. 225-228.

[7] Cajori, Florian (1916) *William Oughtred: a Great Seventeenth-century Teacher of Mathematics*, Chicago: The Open Court Publishing Company.

[8] Cajori, Florian (1923) "Mathematical Signs of Equality", *Isis*, Vol. 5, No. 1, pp. 116-125.

[9] Cajori, Florian (1924) "Rahn's Algebraic Symbols", *The American Mathematical Monthly*, Vol. 31, No. 2, pp. 65-71.

[10] Cajori, Florian (1928) *A history of Mathematical Notations, Vol I. Notations in Elementary Mathematics*, Chicago: Open Court.

[11] Cajori, Florian (1929) *A history of Mathematical Notations, Vol II. Notations Mainly in Higher Mathematics*, Open Court, Chicago.

[12] Cantor, Moritz (1880 - 1908) *Vorlesungen über Geschichte der Mathematik* (4 Vols., 1880-1908), Vol. I (1880); Von den ältesten Zeiten bis zum Jahre 1200 n. Chr., Vol. II (1892); Von 1200-1668, (2nd ed. Teubner: Leipzig, 1894, 1900).

[13] Cardano, Girolamo (1545) *Hieronymi Cardani Artis magnæ, sive, De regulis algebraicis lib. unus: qui & totius operis de arithmetica, quod opus perfectum inscripsit, est in ordine decimus*, Norimbergæ: Per Ioh. Petreium excusum.

[14] Clavius, Christopher (1608) *Algebra Christophori Clavii Bambergensis e Societate Iesv*, Avrelianæ Allobrogvm, Excudebat Stephanus Gamonetus, Anno M. DCIX.

[15] Colebrooke, Henry Thomas (1817) *Algebra with Arithmetic and Mensuration from the Sanskrit of Brahmagupta and Bhāscara*. John Murray, London (Sandig Reprint Verlag, Vaduz, Lichtenstein 2001).

[16] Cossali, Pietro (1797-1799) *Origine, trasporto in Italia, primi progressi in essa dell'algebra. Storia critica di nuove disquisizioni analitiche e metafisiche arricchita*, (2 vols), Parma.

[17] De Cruz, Helen and Johan De Smedt (2012) "Mathematical symbols as epistemic actions", *Synthese*, (Online first) DOI: 10.1007/s11229-010-9837-9.

[18] Ernest, Paul (2006) "A semiotic perspective of mathematical activity: the case of number", *Educational Studies in Mathematics*, 61, pp. 67-101.

[19] Franci, Rafaella and Rigatelli, Laura T. (1979) *Storia della teoria delle equazioni algebriche*. Milan: Mursia.

[20] Franci, Rafaella and Rigatelli, Laura T. (1985) "Towards a history of algebra from Leonardo of Pisa to Luca Pacioli", *Janus*, 72 (1-3), pp 17-82.

[21] Franci, Rafaella (2001) *Maestro Dardi (sec. XIV). Aliabraa argibra. Dal manoscritto I.VII.17 della Biblioteca Comunale di Siena*, volume 26 of Quaderni del Centro Studi della Matematica Medioevale. Università di Siena, Siena.

[22] Gosselin, Guillaume (1577) *Gvlielmi Gosselini Cadomensis Bellocassii De arte magna, seu de occulta parte numerorum, quae & algebra, & almucabala vulgo dicitur, libri qvatvor. In quibus explicantur aequationes Diophanti, regulae quantitatis simplicis, [et] quantitatis surdae*, Parisiis: apud Aegidium Beys, M.D. LXXVII.

[23] Hankel, Herman (1874) *Zur Geschichte der Mathematik in Alterthum und Mittelalter*, Leipzig: Teubner.

[24] Heath, Thomas Little (ed.) (1885) *Diophantus of Alexandria: A Study in the History of Greek Algebra*, Cambridge, Ma., [second edition, 1910, With a supplement containing an account of Fermat's Theorems and Problems connected with Diophantine analysis and some solutions of Diophantine problems by Euler (New York, 1964)].

[25] Heeffer, Albrecht (2007) "A Conceptual Analysis of Early Arabic Algebra", in T. Street, S. Rahman and H. Tahiri (eds.) *The unity of science in the arabic tradition: science, logic, epistemology and their interactions*, Heidelberg: Springer, pp. 89-128.

[26] Heeffer, Albrecht (2007) "The tacit appropriation of Hindu algebra in renaissance practical arithmetic", *Gaṇita Bhārāti*, 29 (1-2), pp. 1-60.

[27] Heeffer, Albrecht (2008) "Text production reproduction and appropriation within the abbaco tradition: a case study", *Sources and Commentaries in Exact Sciences*, 9, pp. 211-256.

[28] Heeffer, Albrecht (2009) "On the Nature and Origin of Algebraic Symbolism" in B. Van Kerkhove (ed.) *New Perspectives on Mathematical Practices. Essays in Philosophy and History of Mathematics*, Singapore: World Scientific Publishing, 2009, pp. 1-27.

[29] Heeffer, Albrecht (2010) "The Reception of Ancient Indian Mathematics by Western Historians" in B.S. Yadav and Man Mohan (eds.) *Ancient Indian Leaps into Mathematics*, Birkhäuser Verlag, 135-152.

[30] Heeffer, Albrecht (2012)"Surmounting obstacles: circulation and adoption of algebraic symbolism", *Philosophica*, 87, (4) pp. 5-25.

[31] Heeffer, Albrecht (2014) "Epistemic Justification and Operational Symbolism", *Foundations of Science* (Online First, DOI 10.1007/s10699-012-9311-x).

[32] Heeffer, Albrecht and Maarten Van Dyck (eds.) (2010) *Philosophical Aspects of Symbolic Reasoning in Early-modern Mathematics*, Studies in Logic 26, London: College Publications.

[33] Hoffman, Joseph Ehrenfried (1939) "Johannes Tropfke: (14. X. 1866 bis 10. XI. 1939)", *Deutsche Mathematik*, 6, pp. 113-118.

[34] Høyrup, Jens (1996) "The formation of a myth: Greek mathematics—our mathematics", in Goldstein, C., Gray, J., and Ritter, J. (eds.) *L'Europe mathéematique. Mathematical Europe*, pp. 103-119, Paris: Éditions de la Maison des sciences de l'homme.

[35] Høyrup, Jens (2004) "Leonardo Fibonacci and abbaco culture a proposal to invert the roles" *Revue d'Histoire des Mathématiques*, 11, 23-56.

[36] Høyrup, Jens (2007) *Jacopo da Firenze's Tractatus Algorismi and Early Italian Abbacus Culture*, Science Networks. Historical Studies, 34, Basel: Birkhäuser.

[37] Høyrup, Jens (2009) "What did the abbacus teachers aim at when they (sometimes) ended up doing mathematics? An investigation of the incentives and norms of a distinct mathematical practice", in Bart Van Kerkhove (ed.) New perspectives on mathematical practices. Essays in philosophy and history of mathematics, Singapore: World Scientific Publishers, pp. 47-75.

[38] Høyrup, Jens (2009) "How to transfer the conceptual structure of Old Babylonian Mathematics: solutions and inherent problems. With an Italian Parallel", Beitrag zur Tagung Zur Übersetzbarkeit von Wissenschaftssprachen des Altertums Johannes Gutenberg-Universität Mainz, 27-29 Juli 2009.

[39] Høyrup, Jens (2010) "Hesitating progress: the slow development toward algebraic symbolization in abbacus and related manuscripts, c. 1300 to c. 1550" in A. Heeffer and M. Van Dyck (eds.) 2010, pp. 3-56.

[40] Ifrah, Georges (1981) *Histoire Universelle des Chiffres* (Paris). Translated by David Bellos, E.F. Harding, Sophie Wood and Ian Monk. *Universal History of Numbers: From Prehistory to the Invention of the Computer*, Harville Press, London, 1998.

[41] Karpinski, Louis Charles (1910) "An Italian algebra of the fifteenth century", *Bibliotheca Mathematica*, 3rd series, 11 (1910-11), 209-219.

[42] Karpinski, Louis Charles (1915) *Robert of Chester's Latin Translation of the Algebra of Al-Khwarizmi*, Macmillan Company, New York (Reprinted by University of Michigan Press, Ann Arbor, 1933).

[43] Karpinski, Louis Charles (1922) Review of *Geschichte der Mathematik in Systematischer Darstellung mit besonderer Berucksichtigung der Fachworter*, by J. Tropfke, *The American Mathematical Monthly*, Vol. 29, No. 9 (Oct., 1922), pp. 349-351.

[44] Karpinski, Louis Charles (1929) "The Italian Arithmetic and Algebra of Master Jacob of Florence, 1307". *Archeion*, 11, 170-177.

[45] Kaye, George Rusby (1927-1933) *The BakhshālīManuscript. A Study in Mediaeval Mathematics*, (3 parts in 2 Vols.). Vol. 1: Calcutta: Archaeological Survey of India (1927); Vol. 2, Delhi (1933) (Reprint, New Delhi: Cosmo Publications, 1981).

[46] Kern, Hendrik (1875) *The Āryabhaṭīya, with the commentary Bhaṭadīpikā of Paramādiçvara*, Leiden: Brill.

[47] Klein, Jacob (1968) *Greek mathematical thought and the origin of algebra*. MIT Press, Cambridge, MA. Translated by E. Brann.

[48] Kirschner Schule (ed.) (1937) *Festschrift zur 25 Jahrfeier der Kirschner schule, Berlin*, Berlin: Kirschner Schule.

[49] Lakoff, George and Rafael E. Núñez (2000) *Where Mathematics Comes From: How the Embodied Mind Brings Mathematics into Being*, New York: Basic Books.

[50] Landy, David (2010) "Toward a physics of equations", in A. K. Goel, M. Jamnik & N. H, Narayanan (eds.) *Diagrammatic Representation and Inference: 6th International Conference, Diagrams 2010*, Portland, OR, August 2010 Proceedings, New York: Springer, pp. 160-166.

[51] Landy, David and R. L. Goldstone (2007) "Formal notations are diagrams: Evidence from a production task", *Memory and Cognition*, 35(8), 2033-2040.

[52] Landy, David and R. L. Goldstone (2007) "Grounding symbol structures in space: Formal notations as diagrams". In D. S. McNamara & J. G. Trafton (eds.) *Proceedings of the 29th Annual Conference of the Cognitive Science Society*, Austin, TX: Cognitive Science Society, pp. 425-410.

[53] Landy, David and R. L. Goldstone (2007) "How abstract is symbolic thought?" *Journal of Experimental Psychology: Learning, Memory, and Cognition*, 33(4), pp. 720-733, doi:10.1037/0278-7393.33.4.720

[54] Libri, Guillaume (1838) *Histoire des sciences mathématiques en Italie, depuis la Renaissance des lettres jusqu'à la fin du dix-septième siècle* (4 vols.), Paris: Jules Renouard.

[55] Loget, François (2012) "Printers and Algebraists in Mid-16th Century France", *Philosophica*, 87, pp. 85-116.

[56] Newell, Allen and Herbert A. Simon (1976) "Computer Science as Empirical Inquiry: Symbols and Search", *Communications of the ACM*, 19 (3), pp. 113-126.

[57] Oaks, Jeffrey (2012) "Algebraic notation in medieval Arabic", *Philosophica*, 87, pp. 27-83.

[58] Oughtred, William (1632) *The circles of proportion and the horizontal instrument: The former shewing the maner how to work proportions both simple and compound: and the ready and easy resolving of quæstions both in arithmetic, geometrie, & astronomie: and is newly increased with an additament for navigation. All which rules may also be wrought with the penne by arithmetic, and the canon of triangles. The later teaching how to work most quæstions, which may be performed by the globe: and to delineat dialls upon any kind of plaine. Invented, and written in latine by W.O. Translated into English, and set out for the public benefit, by William Forster*, London: Printed by Augustine Mathewes, and are to bee sold by Nic. Bourne at the Royall Exchange.

[59] Nesselmann, Georg Heinrich Ferdinand (1842) *Versuch einer kritischen Geschichte der Algebra*. Vol 1. *Die Algebra der Griechen*, Berlin. [Reprinted Frankfurt, 1969]

[60] Peletier, Jacques (1554) *L'Algebre de Jacques Peletier départie en deux Livres*, A Lion: Par Ian de Tournes, MDLIIII.

[61] Radford, Louis (1998) *On Signs and Representations: a Cultural Account*, School of the Sciences of Education, University Laurentienne, Ontario: Sudbury.

[62] Rashed, Roshdi(ed.) (1984) *Diophante. Les Arithmétiques, tome III (livre IV) et tome IV (livres V, VI, VI)*. Texte établit et traduit par Roshdi Rashed, Paris: Les Belles Lettres.

[63] Regiomontanus, Johannes (1537). *Alfraganus, Continentur in hoc libro. Rvdimenta astronomica alfragani. Item Albategni vs alstronomvs peritissimvs de motv stellarvm, ex obseruationibus tum proprijs, tum Ptolomaei, omnia cu demonstratiōibus Geometricis & Additionibus Ioannis de Regiomonte. Item Oratio introductoria in omnes scientias Mathematicas Ioannis de Regiomonte, Patauij habita, cum Alfraganum publice praelegeret. Eivsdem utilissima introductio in elementa Euclidis*. Nürnberg. (Facsimile published in Schmeidler (1972), pp. 43-53).

[64] Riese, Adam (1522) *Rechenung auff der linihen und federn in zal, mass und gewicht auff alleley handierung*,

[65] Rodet, Léon (1879) "Leçons de calcul d'Āryabhaṭa", *Journal Asiatique*, 13 (7), 393-434.

[66] Rotman, Brian (1987) *Signifying Nothing: The Semiotics of Zero*, London: Routledge.

[67] Rotman, Brian (2000) *Mathematics as Sign: Writing, Imagining, Counting*, Stanford: Stanford University Press.

[68] Sarton, George (1929) "A History of Mathematical Notations by Florian Cajori, Review by George Sarton" *Isis*, 12 (2) (May, 1929), pp. 332-336.

[69] Serfati, Michel (2005) *La révolution symbolique. La constitution de l'écriture symbolique mathématique*, Paris: Éditions Petra.

[70] Serfati, Michel (2005) "Symbolic revolution, scientific revolution: mathematical and philosophical aspects" in A. Heeffer and M. Van Dyck (2010) *Philosophical Aspects of Symbolic Reasoning in Early-Modern Mathematics*, Studies in Logic 26, London: College Publications, pp. 103-122.

[71] Sesiano, Jacques (1982) Books IV to VII of Diophantus' Arithmetica in the Arabic translation attributed to Qustā ibn Lūqā, Heidelberg: Springer Verlag.

[72] Stifel, Michael (1544) *Arithmetica Integra*, Norimbergae: Petreius.

[73] Stifel, Michael (1554) *Die Coss Christoffs Rudolffs: Mit schönen Exempeln der Coss*, Königsperg in Preussen.

[74] Struik, Dirk (1967) *A Concise History of Mathematics*, New York: Dover. [revised ed. 1987]

[75] Smith, David Eugene (1923) *History of Mathematics*, vol. I, Boston: Ginn and company (Dover Reprint, 1958).

[76] Smith, David Eugene (1925) *History of Mathematics. Special Topics of Elemenatry Mathematics*, vol. II, Boston: Ginn and company (Dover Reprint, 1958).

[77] Smith, David Eugene (1931) Review of *Geschichte der Elementar-Mathematik*. Erster Band, Rechnen. by Johannes Tropfke, *The American Mathematical Monthly*, Vol. 38, No. 6 (Jun. - Jul., 1931), pp. 331-334.

[78] Smith, David Eugene (1934) Review of *Geschichte der Elementar-Mathematik in systematischer Darstellung mit besonderer Berucksichtung der Fachworter*. by Johannes Tropfke, *The American Mathematical Monthly*, Vol. 41, No. 1 (Jan., 1934), pp. 38-39.

[79] Smith, David Eugene (1937) Review of Geschichte der Elementar-Mathematik. by Johannes Tropfke, *The American Mathematical Monthly*, Vol. 45, No. 3 (Mar., 1938), pp. 177-179.

[80] Strachey, Edward (1813) *Bija Ganita*, London: Glendinning.

[81] Thibaut, George; Dvivedin, Mahamahopadhyaya Sudhākara (1889)*The Panchasiddhantika. The Astronomical Work of Varāhamihira*, Benares.

[82] Thibaut, George, "Astronomie: Astrologie und Mathematik", in G. Bühler (ed., later F. Kielhorn), *Grundriss der Indo-Arischen Philologie und Alterumskunde*, Vol. 3 (9), Strassburg: Tübner.

[83] Tropfke, Johannes (1899) "Erstmaliges Auftreten der einzelnen Bestandteile unserer Schulmathematik", Wissenschaftliche Beilage zum Jahresbericht des Friedrichs-Realgymnasiums zu Berlin.

[84] Van Egmond, Warren (1983) "The algebra of Master Dardi of Pisa", *Historia Mathematica*, 10, pp. 399-421.

[85] Viète, François (1591) *Francisci Vietae In artem analyticem isagoge : Seorsim excussa ab Opere restituae mathematicae analyseos, seu algebra nova*, Turonis: apud Iametium Mettayer typographum regium.

[86] Vygotsky, L. (1978) *Mind in Society*, Cambridge: Harvard University Press, Massachusetts.

[87] van der Waerden, Bachtel Leendert (1985) *A history of algebra from al-Khwārizmī to Emmy Noether*, Heidelberg: Springer.

PART II

Biography of Catherine Mary Schulte

Catherine Mary Schulte (1901–2000)

Born in 1901 of German parentage in a small Wisconsin town on the shore of Lake Michigan, Catherine Schulte was the youngest of a family of eight children.[18] Growing up on a farm which she loved for its friendliness and beautiful environment, Catherine's formal education was varied and not too promising so far as its ability to promote learning leading to higher studies in the future.

She explained that she attended several schools during her childhood: the "little school" for a year and a half followed by four years in the "big school" conducted by religious women from Manitowoc, Wisconsin. Then she transferred to the public school to gain additional credentials required for high school entrance. This setting was more challenging to her. Miss Alice Brady, the teacher, took a special interest in her progress and even became a life long personal friend.

High school necessitated a train ride on Monday morning, then a room with some "wonderful ladies" (friends of the family) until Friday afternoon when she rode the train back to her small town of Cleveland and to the family farm. The high school was "progressive" for its time and included involvement in clubs based on civic issues, history and the study of Latin. Music lessons also occupied some of her time. Another important aspect of this experience was that she made many friends of her own age and activities, unavailable while living at home since the older children had found other interests or had moved beyond the family circle by the time Catherine was in high school.

Catherine seemed not to have vocational goals during these years, perhaps due to the era in which she lived. Opportunities for young women were limited with the exception of marriage, teaching in elementary schools, and a limited number of similar domestic involvements.

Near the close of her senior year in high school the principal met with each student to evaluate individual scholastic achievements, to select candidates for valedictorian and salutatorian, and to consult each about the future. At that time, though vocational plans were vague, Catherine was considering a course in home economics, probably a year or two that would enable her to earn her living in the future.

Fate intervened; during late summer her high school principal appeared at the farm with his niece and two Franciscan Sisters from the College of Saint Teresa (CST)—a Catholic women's institution—in Winona, Minnesota. Catherine and Viola Stengel, the niece, had received the coveted honors at their high school, and the College was offering scholarships to them to attend there when classes would begin in a few weeks. Some rapid calculations and much conversation convinced the two young women and their elders that the additional college finances were indeed possible.

Within weeks they had taken the train to Winona and were registering for classes. Catherine continued to favor home economics, but early in the second year she found it no longer challenging and changed to chemistry. In some ways that field may have been too challenging. In their senior year several of the women failed including Catherine who had completed minors in mathematics, English, and French and had met all other requirements. Thus she graduated with her class in May, 1923.

[18] This narrative is written by Sister Mary Lonan Reilly, Congregational Archivist, and is taken from information in personal files of Sister Leontius located in the Congregational Archives. Included is a transcript of an oral interview dated April 10, 1991.

That fall she began teaching in the schools of the Franciscan Sisters which eventually led to her decision to join the Congregation. Less than three years after that point Sister Mary Leontius as she was then known, had finished the necessary training and once again was happily teaching in a high school classroom convinced that this was to be her role for many years to come. Professionally she was highly successful and an asset to the Sisters there who through their joint efforts achieved accreditation of that school during its opening year. Once again her life changed because others had noticed the potential she possessed not only for her own personal development but also for the future of the Congregation. She was contacted and told that she was to prepare to go to Ann Arbor, Michigan to pursue a higher degree in mathematics.

Since she had only a minor in the field, there were make-up courses to complete in addition to the prescribed advanced work. One of the professors became her mentor, noted her interest in the history of mathematics, and suggested that she might like to continue in that branch of the major. Sister was pleased with the suggestion; after two academic years and five summer sessions of study, she had completed all required courses and had passed the comprehensive examinations.

Freed to follow her mentor's advice in the choice of a dissertation topic, Sister Leontius turned in earnest to the necessary research and then writing of "Additions in Arithmetic, 1483–1700." This became the focus of her life.

Not only had her major professor Dr. Louis C. Karpinski acquired a very large personal library with many holdings related to the topic but he also knew two former colleagues who had interest in the field. Though both lived and worked in New York, they were willing to share with her pertinent resources from their private libraries.

Finally, she had a good grasp of the languages needed having grown up in a family with a German background, earned a college minor in French, and could read some Italian — all of which were needed in investigating and interpreting these resources.

And so the dissertation was completed; with degree in hand upon graduation in 1935 Sister Leontius returned to Winona and the College of Saint Teresa prepared to teach in the Department of Mathematics and convinced that now this was to be her life's work. Her methods and personality were both well received by her students and she enjoyed the challenges of each semester. She was even able to admit that although she had not earlier foreseen that her field of studies might be mathematics, she was very happy not to be teaching either home economics or chemistry.

During the first decade her courses included those common to the field: algebra, trigonometry, analytic geometry and some upper division courses such as differential equations and number theory. Then the college began to educate young women not only for graduate work but also for positions as teachers at the secondary level. Sister Leontius was asked to prepare to teach the methods courses at all levels from beginning elementary school through college level.

That was followed in the 1960s by the trend toward "Modern Mathematics" a revised view of the field for which few teachers were prepared. The Chair of the Department asked Sister Leontius to be a pioneer in the use of televised instruction for the many teachers who needed assistance in this new area as well as for the parents of the students who were unfamiliar with the terms and the concepts.

Sister Leontius had been prepared in part by her work with student teachers and by receiving National Science Foundation Grants for attendance at programs at three summer conferences. During two summer sessions (1963 & 1964), she and others in the department assisted in eight week programs presented at the College of Saint Teresa and sponsored by the National Science Foundation. She had also directed a pilot program at a parish school near the college. The televised series, however, proved to be challenging despite this excellent background.

She needed to write the scripts, go through practice sessions, and then present the program on live TV. With the assistance of a television technician, faculty members from the drama department, and others, Sister Leontius prepared and presented the educational television series entitled "Parents Ask About Arithmetic" in half hour segments over a period of nineteen weeks in a broadcasting station located forty-five miles from her home base. She also decided to include visuals and newer materials being used such as an abacounter, magneboards, and the like.

The program was highly successful and led to a series of fifteen 28-minute films "Teaching Mathematics K–6" to be viewed and studied by in-service as well as prospective teachers. All were supplemented with film notes. Again, Sister Leontius was the script writer and narrator with a faculty member from the drama department handling the visuals and coordination of the programs. Of course, a teacher's manual would be helpful, so Sister Leontius wrote and published one to accompany the films.

In 1969, Sister Leontius attended sessions at Colombia University learning what materials should be included in a Math Lab; she came back to Winona and prepared a similar lab in Roger Bacon Hall. Soon faculty from other colleges as well as teachers from the local area were welcomed to examine the results of her work, then to return to their home institution and establish their own labs to be used by students and teachers of their institutions.

Although the activities described above were labor intensive, and required contacts with other institutions as well as with her students and with teachers in area schools, Sister Leontius had always been interested in local, area, and statewide activities in her discipline during her long career at the College, among them are the following:

- Sigma XI, Michigan Chapter 1934 ff
- National Council of Teachers of Mathematics, 1934 ff
- Minnesota Council of Teachers of Mathematics, 1934 ff
  - Program Director for many years
  - Service Honor, Honorary Membership, and donation in her name to the MCTM building fund then in progress, 1977
- Chair of the CST Mathematics Department, 1948–1954
- S.E. Section of Minnesota Council, 1934–1970
- American Mathematical Association, MN section
  - Executive Board, 1954–1955
- Instructor in College Upward Bound Program, 1967
- Cuisenaire-Gattegno Consultant, 1968

- Outstanding Educator of America, 1975

- Served as workshop teacher for in-service sessions in Wisconsin as well as at several sites throughout Minnesota

- Conducted seminars and/or summer programs at the College for area teachers

Sister Leontius also participated in events and activities beyond her academic field and received several honors recognizing that involvement. During her first years in Winona, she helped to organize a chapter of the Legion of Mary at CST and as moderator of the group was key to its on-going success during her years there. Members visited the aged and others in need or alone, ill, or confined to their homes within walking distance of the College. She was honored on the organization's twentieth anniversary in 1968 by those who were currently members of the group.

Other honors included: Citation in 1973 from the CST Alumnae Association which read in part "... a woman concerned with others, willing to serve others, and completely dedicated to her college, her community, her church, and all of God's people." It went on to note her qualities of "the patient concerned teacher who always had time to give the extra help, the encouragement needed" and her "devotion to the aged, the sick, and the lonely" through her work with the Legion.

Then in 1975, her name appeared on the listing of Outstanding Educators of America. In 1977 the service honor from the MCTM was acknowledged by the Executive Secretary of the national board of that group in these words:

> ... that the professional efforts of Sister Leontius have been untiring and that many mathematic teachers are the leaders of today because of her influence in the classroom or in the Council sponsored activities. [19]

Following her retirement from teaching in 1978 Sister Leontius volunteered in the Department. Later she assisted in the College Relations Office until 1989 when the College closed and the Sisters moved to other locations. Sister Leontius chose to reside at the Congregation Center, Assisi Heights in Rochester, Minnesota where she died on March 20, 2000. She had spent more that 73 years as a Franciscan, the majority of them at the College and in the teaching of mathematics. Humble with a great sense of humor she loved life, her discipline, her students, and those with whom she lived and worked.

[19] Letter in her file from Executive Secretary, MCTM to Executive Secretary of NCTM dated February 9, 1977.

The following biographic sketch of Sister Mary Leontius Schulte is reprinted from *Pioneering Women in American Mathematics: The Pre-1940 PhD's* by Judy Green and Jeanne LaDuke, published American Mathematical Society.

> Sister Mary Leontius Schulte was born Catherine Mary Schulte in Cleveland, Wisconsin, the youngest of seven surviving children of Mary (Wagner) and Joseph Schulte, both of whom had some grade school education and were farmers. Catherine Schulte attended primary school in Cleveland and graduated from high school in Manitowoc in 1919. That fall she entered the College of Saint Teresa in Winona, Minnesota, with the intention of studying home economics. While there she played hockey, tennis, and basketball, and became interested in mathematics. In 1923 she received her bachelor's degree, with a major in chemistry and minors in mathematics, French, and English. After her graduation Schulte taught mathematics at high schools in Minnesota; she was at Holy Trinity High School in Rollingstone 1923–24, Saint John High School in Rochester 1924–26, and Saint John High School in Caledonia 1927–28. She entered the Sisters of Saint Frances in Rochester on January 19, 1926, and made her profession of vows in 1927.
>
> Sister Mary Leontius Schulte began her graduate work at the University of Michigan in the summer session of 1928 and became an instructor at the College of Saint Teresa that fall. She took graduate courses in mathematics and physics during the four summers of 1928–31 and received her master's degree in September 1931. Sister M. Leontius continued teaching except for the two academic years 1932–34, when she was in residence full time at the University of Michigan to finish the course work for her PhD. During these two years she lived at Mercy Hospital in Ann Arbor. She was enrolled during the summer of 1934, when her dissertation was accepted, and the degree was conferred in March 1935. Sister M. Thomas à Kempis Kloyda, her slightly older colleague at the college, had a similar educational and professional history, which included working with the same dissertation advisor.
>
> Sister M. Leontius was instructor and associate professor at the College of Saint Teresa until 1948 when she was promoted to full professor. She retired as professor emeritus in 1975. During her years on the faculty she was particularly involved with issues concerning mathematics education and was a participant at NSF institutes and conferences at Ball State Teachers College 1959, Montclair State College 1961, Oklahoma State University 1961, Washington State University 1962, and Teachers College, Columbia University, 1969. In the summer of 1966 she attended a Cuisenaire workshop in Denver. Sister Leontius did a television series of nineteen telecasts, *Parents Ask About Arithmetic,* that were shown in Rochester in 1961–62, and fifteen kinescope recordings, *Teaching Mathematics K-6,* shown in St. Paul the next year. Among the mathematics and educational organizations in which Sister Leontius participated were the Minnesota Council of Teachers of Mathematics, where she was a district board member 1970–75, and the MAA, where she was on the executive committee of the Minnesota Section 1953–54 and 1955–56 and was chairman of the section 1954–55. As a faculty member she was a sponsor of

the Legion of Mary, which ministered to the sick: she served as a pastoral minister and visited shut-ins.

Sister Leontius received an alumnae citation from the College of Saint Teresa in 1973. She remained there for several years after her retirement and did volunteer work in the college offices. After the college closed in 1987, Sister Leontius lived at the Motherhouse, Assisi Heights, Rochester, Minnesota. She died in 2000 at age ninety-eight in Rochester.

PART III

Additions in Arithmetic, 1483–1700, to the Sources of Cajori's "History of Mathematical Notations" and Tropfke's "Geschichte Der Elementar-Mathematik"

Chapter 1

Introduction to Overlooked Notations

1.1 Purpose of this Study

It will be the purpose of the present study, after pointing out that the standard works on arithmetical notations are demonstrably incomplete, to supplement these works with the results of further research, particularly among the arithmetics of the sixteenth and seventeenth centuries. The new material should complete Cajori's *A History of Mathematic al Notations,*[1] v. I, 1928, and certain sections of Tropfke's *Geschichte der Elementar Mathematik*[2], v. I, II, III, 2nd. ed., 1921, not only by noting the mathematical notations of writers who have been given little or no mention in these works, but also by pointing out phases of the treatment of the arithmetic material which have been passed over in silence by these authors. In particular, the attempt has been made to increase the source material of a period when notations were undergoing rapid change, the sixteenth and early seventeenth centuries. These studies have been undertaken with a view to increasing the usefulness of these works to teachers of elementary mathematics, to historians of this field, to historians in general, and to students of mathematics.

There is no questioning the fact that Cajori's *Notations,* Tropfke's *Geschichte,* and D. E. Smith's *Rara Arithmetic*[3] (Boston, 1908) are the fundamental works on the history of arithmetic after the invention of printing. The first two are widely cited and used; yet they present only a selected list of authors, without reference either to the number of editions or to other indications of the influence of the given works. Even when authors not cited by Cajori or by Tropfke use notations that are the same as, or similar to, the notations of the authors who are cited by them, it is highly desirable to have this fact noted as a matter of record. In the library at the University of Michigan there is a collection which permits this extension. Some one hundred works have been investigated. Of these some merit more attention than has been given them by the above historians, others have not been treated at all. Further than this, there are numerous different editions

[1] Hereafter referred to as Cajori, *Notations*

[2] Hereafter referred to as Tropfke, *Geschichte.* Several references are to v. I, II, 3d. ed., 1930, 1933

[3] Hereafter referred to as D. E. Smith, *Rara.*

of the works cited by both Cajori and Tropfke which have been examined and which have been found to contain variations worthy of note. With the resources of the University of Michigan supplemented by a study in the libraries of New York City, the Library of George A. Plimpton, the Library of Columbia University, the Library of David Eugene Smith (Teachers College, Columbia University) and the New York Library, the writer has been able to add materially to the sources utilized by the standard works on mathematical notations.

For the completion of material given by Cajori, it seems desirable,to use the method employed by him, having a preliminary discussion of the important individual writers followed by a detailed study of notations involving not only these individual writers but numerous others. It has been deemed best to separate the topical surveys of Cajori and Tropfke, supplementing the former's material on notations and supplementing the latter's material on methods. It is believed that in this parallel study of the two works by Cajori and Tropfke, the limitations and also the strong points of both works will be emphasized.

The particular group of writers selected for more detailed study (No. 1, 19, 29, 32, 33, 37, 47, 52, 54, 56, 63, 67, 74, 82, 85, 89, 91, 100) has been chosen chiefly because of numerous indications of their relatively greater importance. This is evidenced by the number of editions of their works, by translations, and by the use of their material by other writers. Ten of this group of writers produced works of at least five editions, and yet seven are not cited by Cajori and eight not by Tropfke. In fact, the textbooks in arithmetic of seven writers of this group are 'centenarians'[4], but only two of their authors, Gemma Frisius and Oronce Finé, are cited by both Cajori and Tropfke. The authors of the remaining five 'centenarians', Hemeling in Germany, Feliciano, Lapazzaia, and Figatelli in Italy, and Savonne in France, produced works which were favorably received by their contemporaries, if one makes his judgment on the basis of their circulation over this long period of time. The influence of these 'centenarians' can hardly be disputed.

Of the numerous further sources not cited by either Cajori or Tropfke, some works were widely used which may have had limited editions, as was the work of Leonard of Pisa to whom numerous references are made by later writers. We have in mind also such authors as Cataneo, Peverone, Bassi, and Ghaligai.

It is difficult to ascertain what principles Cajori used in his selection of authors. A number of the writers drawn upon in the following pages have been chosen because of the continued use of their works over a long period of time, as noted above. The works of such authors as Tagliente, Hodder, and Reisch, though not 'centenarians', passed through many editions; and yet these are cited by neither Cajori nor Tropfke. Some authors such as Feliciano, Hemeling, Köbel, Albert, Finé, and Gemma Frisius deserve more than their one reference or at most three references in Cajori. Again, for certain authors (Clavius, Tartaglia, Cardan, Bombelli, Vieta, Descartes, Girard, Harriot, Rudolff, Rahn, and bcheubel) who are included in Cajori, particularly important further notational references are given. The including of still others may have its justification in the fact that their works continued to be published a century or more after the appearance of the first edition; for example Bassi and Hemeling. There are also those works which

[4]Textbooks with a long life. See account given by L. C. Karpinski, "Arithmetic Centenarians,", *Scripta Mathematica,* Nov. 1935, pp. 54–40. These textbooks continued to be published over a period of one hundred years.

are worthy of addition because of their wide use in translations in other countries; the work of Ortega is a case in point.

For the most part, references are made to the peculiarities in symbolism not noted by Cajori and to the treatment of arithmetic material not noted by Tropfke in works rich in notation and relatively influential. For instance, in the prefaces of many works may be found references to the influence of earlier or contemporary writers. This is the case in the arithmetic of the Italian, Lorenzo Forest (No. 35), in which among the writers included are such names as Tartaglia, Ghaligai, Sfortunati, Cataneo, Lapazzaia, Orteba, Clavius, and Peverone, all listed in the present bibliography, but of whom only two are mentioned in Tropfke. About the same time, Hulsius, in his *Tractatus Primus Instrumentorum* (Frankfurt, 1604, U. of M.), makes reference to these writers: Peurbach, Gemma Frisius, Clavius, Ghaligei, Cataneo, and Waser, of whom the last three are not found in Tropfke. In an earlier work, Antich Rocha[5] in his *Arithmetica* (Barcelona, 1565) has likewise consulted among others, the authors Scheubel, Finé, Ramus , and Gemma Frisius. Again, the work of Gruffenried (No. 42.), although it itself is not so well known, embodies the labors of eminent predecessors, such as Cardan, Clavius, Rudolff, Gemma Frisius, Glareanus, Alsted, Hamus, and Jacob.

A modern writer on notations, J.W.L. Glaisher[6], mentions these early Italian authors: Borgi, Calandri, Pellas, Pacioli, Tagliente, Feliciano, Sfortunati, Cataneo, Ghaligai, and Tartaglia. Of these Calandri, Tagliente, and Cataneo are not noted in Cajori, while Pellos and Feliciano are given only one reference. It is important, then, to have a more complete list of sources, one which will include these writers who are of almost equal prominence with Borghi and Pacioli.

Another contemporary writer, J. Pelseneer[7,8] (Brussels), states in *Isis* that it would be interesting to illustrate by new examples the conclusions of Cajori in his *Mathematical Notations*. His brief note brings to light a point which is in danger of being obscured by the mass of data in Cajori, namely that there was a great body of mathematical symbols in evidence at the end of the sixteenth century and in the seventeenth century which had a wide circulation—especially in the works of the secondary authors.

Finally, it should be noted that Cajori dismisses the important topic of finger and counter reckoning in a very few lines, while this material was widely used and had considerable influence upon arithmetical terminology. The desirability of this study finds its justification in the fundamental importance of Cajori's treatise as a work of reference. The editor of *Isis* himself, George Sarton[9], reviews it thus:

> Cajori's History of *Mathematical Notations* will remain a standard work for many years to come.

---

[5]D. E. Smith, *Rara,* p. 316.

[6]J.W.L. Glaisher, "On the Early History of Signs + and − and on the Early German Arithmeticians", *Messenger of Mathematics,* v. 51 (1921), pp. 77, 78.

[7]J. Pelsener, "A propos de notations mathematiques", *Isis,* v. 17, 1932, p. 259.

[8]Ed. Jean Pelsener (1903–1985) received a PhD from the Free University in Brussels and was visiting fellow at Harvard and Berkeley University. He became professor of History of Science at the Free University in Brussels and acted as the Belgian commissioner of the International Committee for the History of Science.

[9]G. A. Sarton, Review in *Isis,* v. 13, 1929, pp. 129, 130.

Speaking of Cajori and his work, R. C. Archibald[10] writes:

> His monumental work on Mathematical Notations, original in conception, must be of enduring value.

In a similar way, Tropfke's *Geschichte* is also so important as a work of reference that a discussion of the sources is highly desirable.[11]

1.2 Summary of Additions to Our Knowledge of Notations Not Mentioned by Cajori or by Tropke

In view of the fact that one-half of the authors listed in the present bibliography are not included by Cajori or by Tropfke considered together, and since approximately two-thirds and almost three-fourths are omitted by Cajori and Tropfke respectively, it is desirable to complete their lists of sources. It is of paramount importance to note that a very large number of the writers of the sixteenth century given by D. E. Smith in his *Rara Arithmetica* is missing entirely in Cajori. Incidentally, too, I have made references to some works not cited in D. E. Smith's very complete bibliography, but available in the University of Michigan Library and the G. A. Plimpton Library in New York (No. 26, 30, 31.).

A true estimate of D. E. Smith's bibliography has been given by L. C. Karpinski[12] when he writes:

> Bibliographically the *Rara Arithmetica* will always be an authority in so far as arithmetical books of the period are concerned.

In this work D. E. Smith has amassed a total of over four hundred and fifty authors who wrote arithmetics in the fifteenth and sixteenth centuries. Including editions, there are more than one thousand different items. Of these fifteenth and sixteenth century authors, Mr. Plimpton's library includes a majority, and certainly the more important ones. Cajori and Tropfke have used less than one-half of the writers in Dr. Smith's bibliography. The present study involves the study of an appreciable further group of the sixteenth-century authors, making it far more inclusive of the sixteenth-century works available in American libraries.

---

[10] R. C. Archibald, "Florian Cajori 1859-1930", *Isis,* v. 17, 19

[11] See the review of Tropfke's work, *Bibliotheca Mathematica,* 3d. ed., v. 4, 1903, pp. 213–218. Also *Archeion,* v. 13, 1931, pp. 102–3; v. 15, 1933, p. 456.

[12] Karpinski, "A Unique Collection of Arithmetics", *Popular Science Monthly,* Sept. 1910, pp. 226–233.

The following authors are not mentioned by Cajori or Tropfke:[13]

4, 5, 6, 7, 8, 9, 12, 15, 19, 21, 22, 23, 26, 27, 30, 31, 32, 35, 36, 41, 42, 43, 44, 46, 50, 51, 53, 55, 56, 57, 58, 59, 60, 61, 62, 66, 68, 69, 72, 77, 80, 82, 84, 87, 88, 89, 92, 94, 97, 98, 99, 100, 101.

The following authors are not mentioned by Cajori:

4, 5, 6, 7, 8, 9, 10, 12, 13, 14, 15, 19, 20, 21, 22, 23, 26, 27, 30, 31, 32, 35, 36, 40, 41, 42, 43, 44, 46, 50, 51, 52, 53, 54, 55, 56, 57, 58, 59, 60, 61, 62, 66, 68, 69 , 71, 72, 77, 80, 82, 84, 87, 88, 89, 92, 94, 97, 98, 99, 100. 101.

The following authors are not mentioned by Tropfke:

1, 3, 4, 5, 6, 7, 8, 9, 12, 15, 19, 21, 22, 23, 26, 27, 29, 30, 31, 32, 35, 36, 38, 41, 42, 43, 44, 46, 47, 48, 50, 51, 53, 55, 56, 57, 58, 59, 60, 61, 62, 63, 65, 66, 68, 69, 72, 76, 77, 80, 82, 84, 85, 86, 87, 88, 89, 91, 92, 93, 94, 97, 98, 99, 100, 101.

[13] Numbers refer to the entries in the bibliography.

Chapter 2

Completion of Cajori's Notations: Notations Used by Individual Writers

2.1 German: Johannes Husswirt (No. 52.) (1501)

Husswirt's *Enchiridion,* written in Latin, bears in the colophon the spelling *huswirt,* with the date MCCCCCI [2.1].[1]

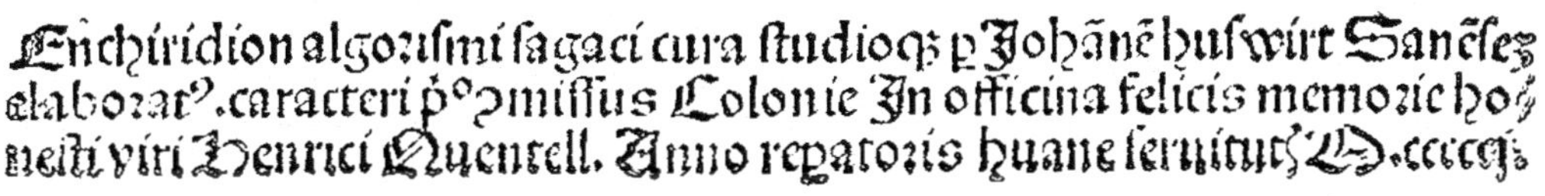
Enchiridion algorismi sagaci cura studioq; p Johãnẽ huswirt Sanēses
elaborat⁹. caracteri p⁹ mississus Colonie In officina felicis memorie ho
nesti viri Henrici Quentell. Anno reparatoris humane seruitutis M.ccccci.

FIGURE 2.1: Husswirt's colophon. No. 52, 1501, f. 20, r.

This book is not mentioned by Cajori, although it would appear to be noteworthy because it had five subsequent editions, and was published with historical notes by Professor Wildermuth, at Tübingen, in 1865. De Morgan[2] calls it 'a short treatise, apparently one of the earliest printed in Germany on the Arabic system.' It is an evidence of the fact that chapters on doubling and halving persisted well into the sixteenth century.

[1] ED. 'Johane huswirt' appears in the top line of the colophon. The date is at the very end of the colophon and reads, 'M.ccccci.'.

[2] Augustus De Morgan, *Arithmetical Books from the Invention of Printing to the Present Time,* (London, 1847), p. 5.

The number 1 is like a printed j [2.2],

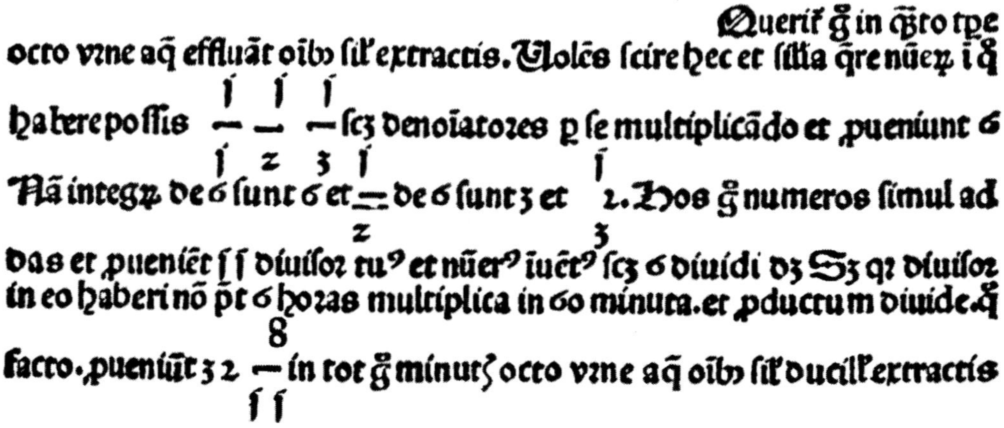

FIGURE 2.2: Husswirt's number 1. No. 52, 1504, f. 20, r.

and as D. E. Smith[3] remarks, Husswirt gives four names to the tenth character, 0 [2.3], which still has no generally accepted name in the English language:

FIGURE 2.3: Husswirt's constructs for zero.[4] No. 52, 1504, f. 3, r.

These four names appear in Sacrobosco's *Algorismus*.[5] Husswirt recommends the regular form X in checking results by casting out the 9's and 7's, although he uses it but twice (ff. 4, r., 10, r.) [2.4, 2.5].

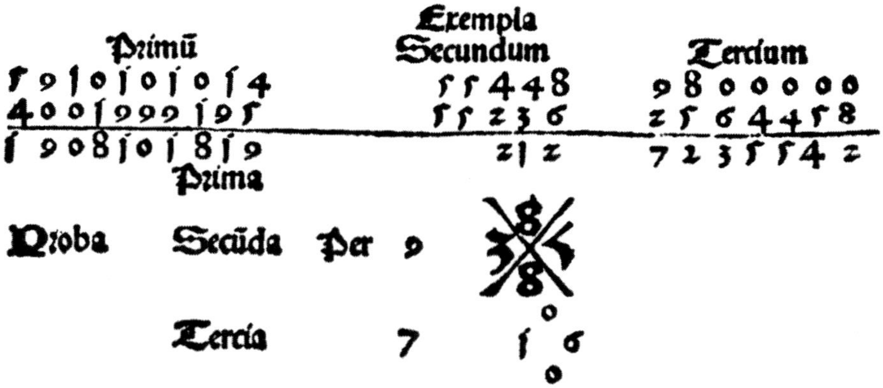

FIGURE 2.4: Husswirt's casting out 9s and 7s. No. 52, 1504, f. 4, r.

[3] D. E. Smith, *Rara,* p. 76.

[4] Ed. The third word (*theca*), fourth word (*circulus*), fifth word (*cifra*), and seventh and eighth words taken together (*figura nihili*) are constructs used for zero.

[5] Ed. Sacrobosco (c. 1195 – c. 1256), a.k.a. John of Hollywood, was a British monk and a mathematics professor at the University of Paris.

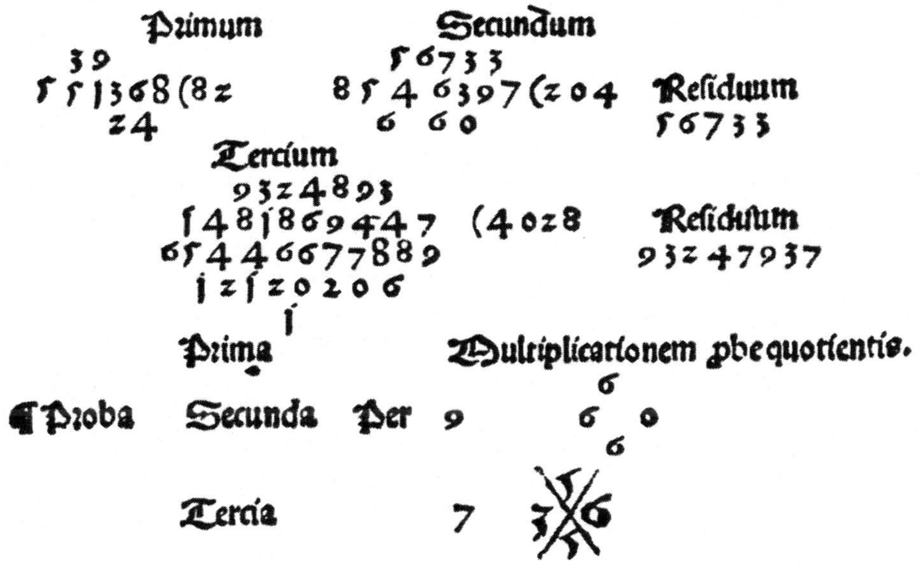

FIGURE 2.5: Husswirt's casting out 9s and 7s. No. 52, 1504, f. 10, r.

These 'due linee oblique descendentes ac se intersecantes per modum crucis hoc modo (quali) ╳' (f. 3, r.) [2.6] are dispensed with in all other cases.

FIGURE 2.6: Husswirt's use of X. No. 52, 1504, f. 3, r.

Husswirt performs subtraction and multiplication by the complementary method (ff. 4, r., 4, v.).[6] In division, the divisor is below the dividend, while the quotient is on the right [2.7].

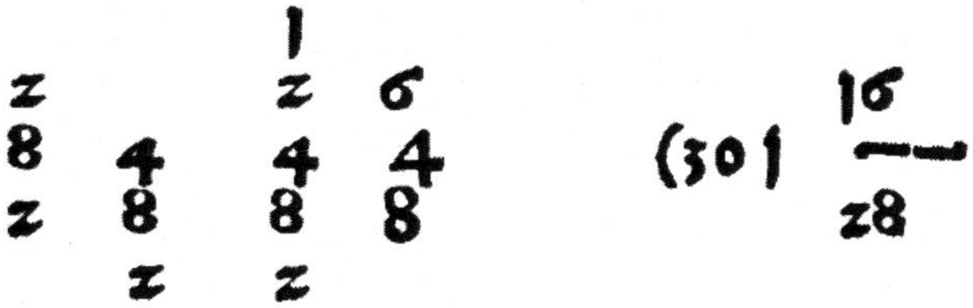

FIGURE 2.7: Husswirt's division.[7] No. 52, 1504, f. 6, v.

Duplation follows multiplication, and mediation, division;[8] but in dealing with counters (*proiectilibus*) and fractions he places them before multiplication, because they are needed there in abacus calculating.

[6]Dr. Susan R. Benedict, *A Comparative Study of the Early Treatises Introducing into Europe the Hindu Art of Reckoning,* p. 75, says that complementary multiplication appears in the works of the thirteenth century.

[7]ED. This division depicts $8444 \div 28 = 301\ 16/28$.

[8]ED. Duplation is the operation which doubles a quantity. Mediation is the operation which halves a quantity. These operations were associated with computing on an abacus.

A dot between the terms of both arithmetic and geometric progressions is his only symbol [2.8].

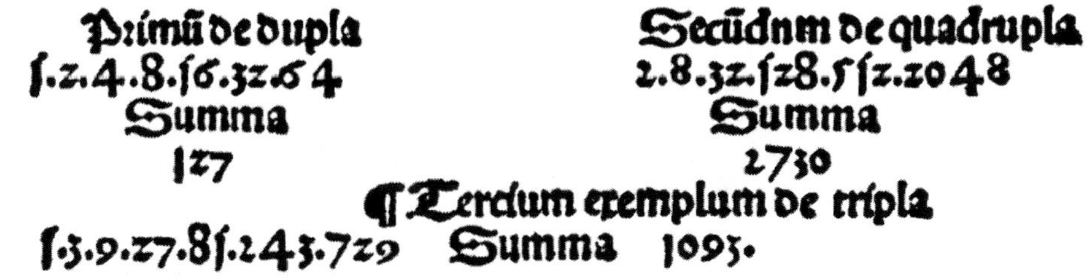

FIGURE 2.8: **Husswirt's dot in progressions. No. 52, 1504, f. 7, v.**

The terms of a fraction are widely separated by a fractional line, broken in the case of two or more digit numbers, as well as general in the early days of printing [2.9].

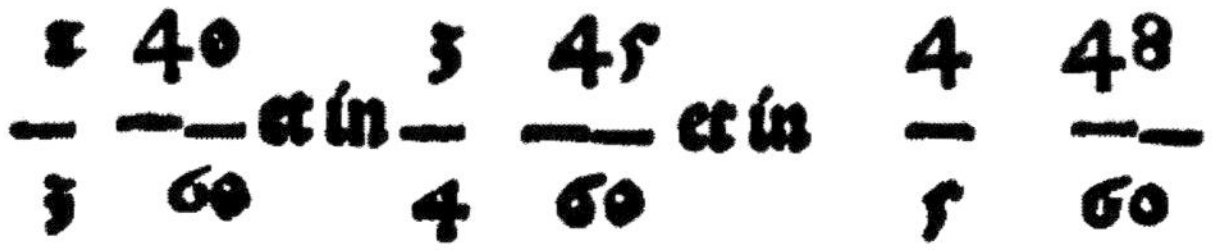

FIGURE 2.9: **Husswirt's fractional line. No. 52, 1504, f. 12, r.**

2.2 German: Johannes Wolphius (No. 100.) (1534)

The *Rudimenta Arithmetices* was first printed in Latin at Nürnberg in 1527; this study concerns the edition of 1534, Frankfurt. The text of Wolphius is very similar to that of Husswirt. Twenty-seven folios are devoted to arithmetic, while the remaining twenty-nine contain elements of geometry by Johannes Vogelin. At the end is the date [2.10]:

FIGURE 2.10: **Wolphius' MDXXXIIII. No. 100, 1534, f. 56, r.**

Wolphius calls 0 *cifra* or *nulla* (f. 3, r.). In grouping digits in numeration he writes [2.11]:

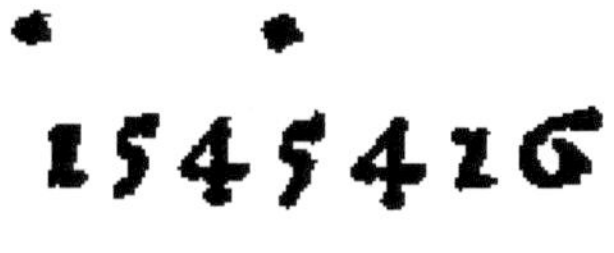

FIGURE 2.11: **Wolphius' grouping of digits. No. 100, 1534, f. 3, v.**

The check by nines is not mentioned by him; seven is used throughout.

Duplation and mediation respectively precede multiplication and division, which order is different from that used by Husswirt. A short method of dividing by 100 is an additional feature; Wolphius divides 63740 by 100 thus [2.12]:

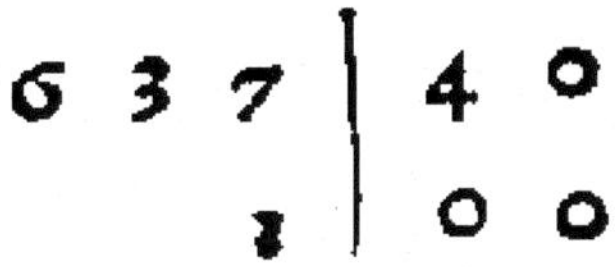

FIGURE 2.12: Wolphius' division. No. 100, 1534, f. 9, v.

When reducing fractions to lowest terms, vertical bars take the place of the equality sign [2.13].

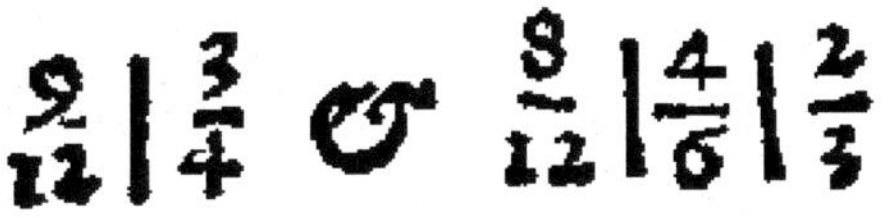

FIGURE 2.13: Wolphius' equality for reduced fractions. No. 100, 1534, f. 16, v.

2.3 Dutch: Gemma, Reinerus Frisius (No. 37.) (1549, 1550)

The *Arithmeticae practicae Methodus Facilis Per Gemmam Frisium, Medicum Ac Mathematicum* is the beginning of the title of an arithmetic the most popular in the Latin schools of the sixteenth century,[9] as evidenced by fifty-nine or more editions in that century cited by Smith besides numerous editions after 1600.[10] The very fact that it was in use for more than a century indicates its degree of appeal to teachers during that period. The arithmetic of Gemma Frisius was prescribed in the curriculum of the seventeenth century.[11]

There is little in Gemma's arithmetic to distinguish it from the early commercial arithmetic of the Italian writers. Gemma calls zero *cyphram,* and says it is written like the letter o, or a circle [2.14].

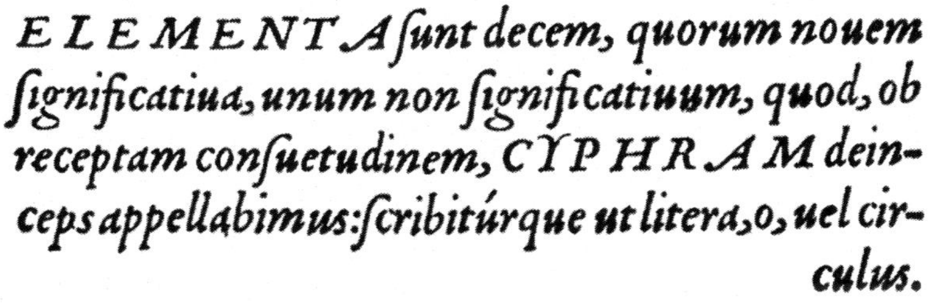
ELEMENTA ſunt decem, quorum nouem
ſignificatiua, unum non ſignificatiuum, quod, ob
receptam conſuetudinem, CYPHRAM dein-
ceps appellabimus: ſcribitúrque ut litera, o, uel cir-
culus.

FIGURE 2.14: Frisius' zero. No. 37, 1550, f. 3, v.

Vertical bars serve to group the digits in numeration, as 23|456|345|678 [2.15].

[9] D. E. Smith, *Rara,* p. 201.

[10] The editions we use, 1549 and 1550, were printed at Paris and are substantially the same. All references are to the 1549 ed.

[11] Friedrich Grundel, *Die Mathematik an den Deutschen Höheren Schulen,* (Berlin, 1928), p. 95, 107. The author singles out the arithmetic of Gemma Frisius as a representative text of this century and discusses it at length, pp. 75–80.

FIGURE 2.15: Frisius' grouping of digits. No. 37, 1550, f. 4, v.

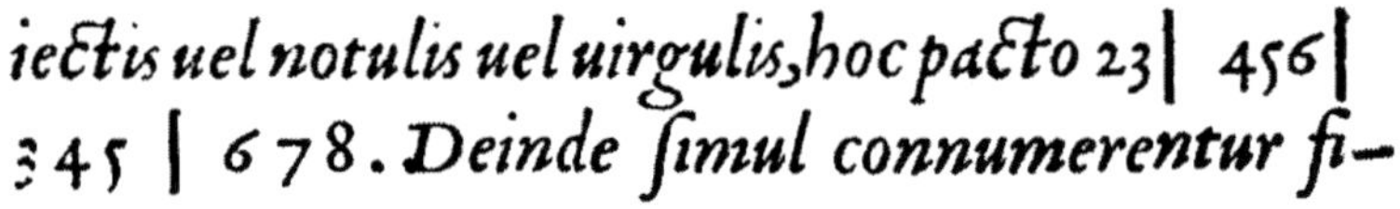

He uses guiding lines in multiplication by the complementary method [2.16], in reduction of fractions to a common denominator (f. 22, v.) [2.17][12] as well as in application of the rule of false position (f. 36, r.) [2.18] but he fails to use them in casting out 9's.

FIGURE 2.16: Frisius' multiplication. No. 37, 1550, f. 9, r.

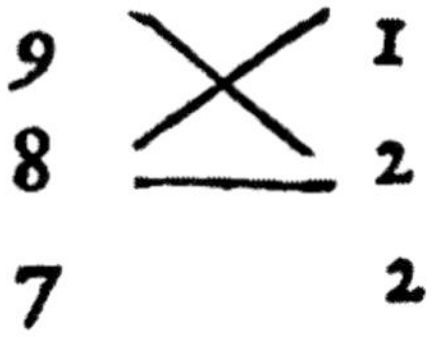

FIGURE 2.17: Frisius' reduction of fractions. No. 37, 1550, f. 22, v.

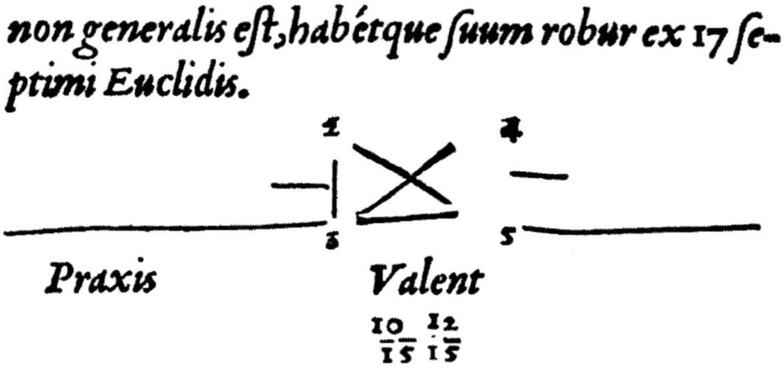

FIGURE 2.18: Frisius' false position. No. 37, 1550, f. 36, r.

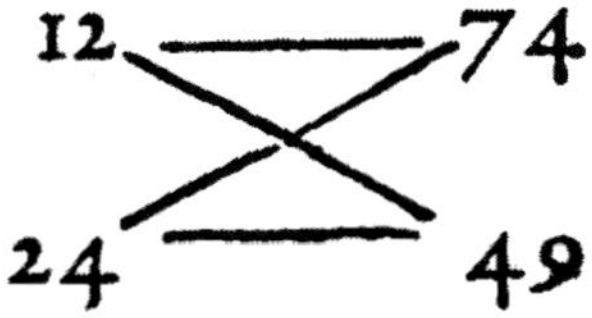

Gemma Frisius finds it convenient in multiplication and division to cancel the digits in the multiplier and dividend as soon as used. There is a section on multiplication by 10 and 100 (f. 11, r.) followed by a discussion on duplation and mediation. Because he does not consider these latter as distinct operations from multiplication and division, they are not treated separately as was customary in the Latin books of that time. The terms of a progression are separated by a comma; progressions, however, are no longer included by Gemma among the fundamental operations. In extracting roots of numbers, grouping is effected by dots below, thus [2.19]:

[12]ED. Gemma makes reference to 'ex 17 septimi Euclidis' (Prop. 17, Book 7 of Euclid's *Elements*).

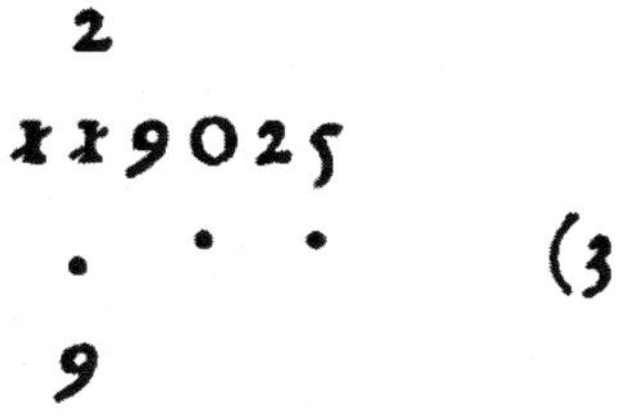

FIGURE 2.19: Frisius' extracting roots. No. 37, 1550, f. 41, r.

2.4 Italian: Girolamo Tagliente (No. 89.) (1520)

An author whose work was so widely used that it probably had thirty-six editions should be important enough to be included among the individual writers listed by Cajori in his *History of Mathematical Notations.* Such was the popular mercantile treatise on arithmetic published in Italian by Tagliente. This textbook was influential in spite of the arrangement of the fundamental operations, in which multiplication and division precede addition and subtraction. Curious also are his forms of multiplication (ff. 9, v., 11, r., 11, v.) [2.20].[13]

FIGURE 2.20: Tagliente's multiplication. No. 89, 1520, Left: f. 11, r. Right: f. 11, v.[14]

[13] Two forms are illustrated in Smith's *Rara,* p. 116.

[14] ED. f. 11, v., on the right, shows the multiplication of 9876 and 6789 yielding the product of 67048164. The light gray lines and numbers in this figure are due to the bleed through of ink from the other side of the folio (which would be f. 11, r. shown on the left).

Tagliente introduces his book with finger symbols (ff. 3, v., 4, r.) also used in a later edition [2.21, 2.22].[15]

FIGURE 2.21: **Tagliente's finger symbols. No. 89, 1520, f. 3, v.**

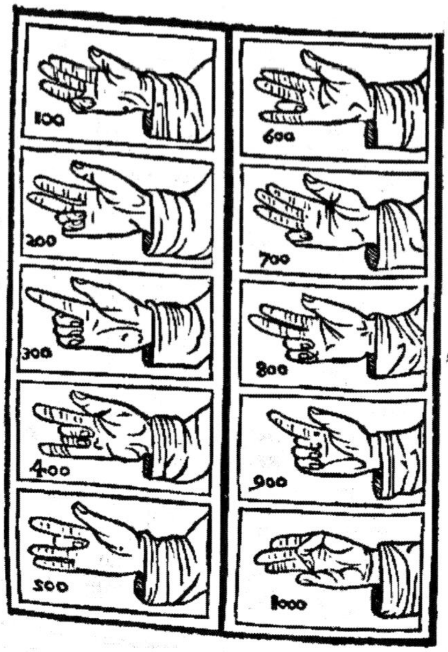

FIGURE 2.22: **Tagliente's finger symbols. No. 89, 1520, f. 4, r.**

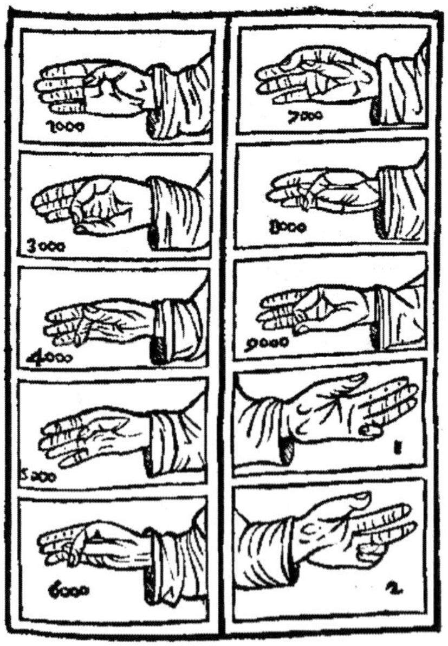

[15] L. L. Jackson, *The Educational Significance of Sixteenth Century Arithmetic,* (New York, 1906), p. 28.

He follows Calandri in accompanying problems with illustrations [2.23].

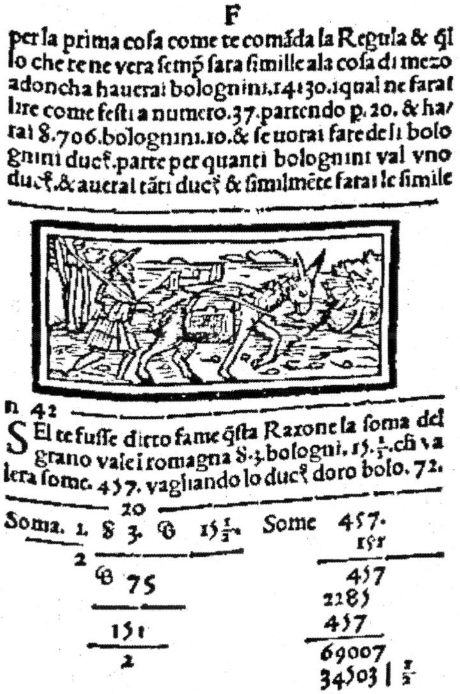
F

per la prima cosa come te comãda la Regula & ql
lo che te ne vera sẽp sara simille ala cosa di mezo
adoncha hauerai bologníni.14130.iqual ne farai
lire come festi a numero.37.partendo p.20.& ha
rai 8.706.bologníni.10.& se uorai fare deli bolo
gnini duc.parte per quanti bologníni val vno
duc.& auerai tãti duc & similmẽte farai le simile

n 42

SE el te fusse ditto fame qsta Raxone la soma del
grano valei romagna 8.3.bologni.15.½.chi va
lera some.457.vagliando lo duc doro bolo.72.

FIGURE 2.23: Tagliente's illustrated problem. No. 89, 1520, f. 25, r.

Instead of the regular form ╳ in the proof of 9s, Tagliente uses the Greek cross (f. 13, v.) [2.24].[16]

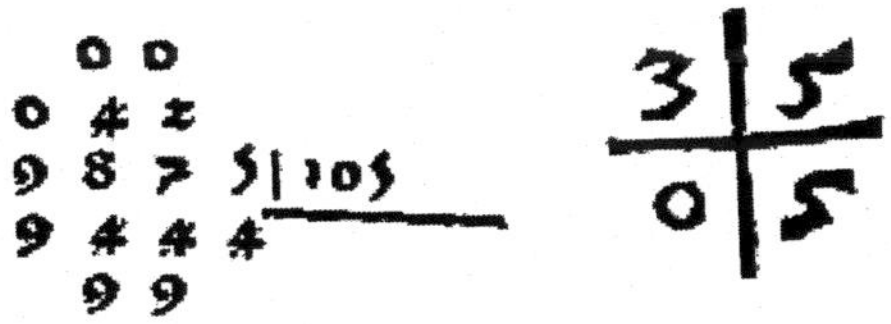

FIGURE 2.24: Tagliente's Greek cross for casting out 9s. No. 89, 1520, f. 14, v.

Division of 3497 by 100 he expresses thus: 34|97 (f. 16, r.) [2.25], as did Wolphius a few years later.

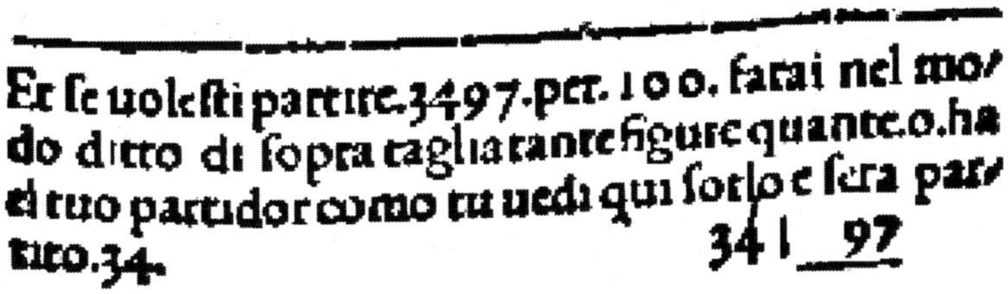
Et se uolesti partire.3497.per.100.farai nel mo
do ditto di sopra taglia tante figure quante.o.ha
el tuo partidor como tu uedi qui sotto e sera par
tito.34. 34|97

FIGURE 2.25: Tagliente's division. No. 89, 1520, f. 16, r.

[16] ED. Schulte's folio information has been corrected.

Tagliente's work is truly a representative Italian arithmetic of the early sixteenth century,[17] for it enters into the various minute details of the operations and commercial rules of the Italian arithmeticians. For example, Tagliente gives several methods for the multiplication of integers, and he deals chiefly with denominate numbers.[18]

2.5 Italian: Pietro Cataneo (No. 19.) (1546, 1559)

Although the work of Cataneo was greatly influenced by his Italian predecessors, it introduces several peculiarities in notation. We draw our conclusions about this work from a very rare edition, his first (Venice, 1546), and further give indication of the use of notations in his second edition (1559). Cataneo uses *Zero* for 0; and following Leonardo of Pisa[19] and other early writers, he arranges the nine digits in decreasing order (f. 2, r.) [2.26].

FIGURE 2.26: Cataneo's digits including zero. No. 19, 1546, f. 2, r.

VEduto che è numero è neceſſario ſaper rileuare più figure, & prima è da cognoſcere le noue figure de gl'Indi quali per Leonardo Piſano ſi manifeſtano eſſere queſte. 9. 8. 7. 6. 5. 4. 3. 2. 1. colle quali & con queſto ſegno o, che in Arabia ſi dice Zero, ogni numero ſi rappreſentu, & detto o per ſe ſolo niſſuna coſa ſignifica, ma accompagnato dallato deſtro con qual uuoi delle altre da maggior ſignificato alla cõpagna, & coſi come le figure ſon.

In his first edition dots enclose all integers in the running text, Whereas in that of 1559 the dot merely follows integers, except in the case of mixed numbers when it is sometimes omitted (f. 15, r.; 1559, f. 21, v.) [2.27, 2.28].

FIGURE 2.27: Cataneo's integers. No. 19, 1546, f. 15, r.

auanza.6.& partito.12.per.6.auanza,0, adunque.6. e la maggior commune miſura di $\frac{1}{3}\frac{8}{0}$ Et ſe partendo il primo auanzo per lo ſecondo alcu-

FIGURE 2.28: Cataneo's integers. No. 19, 1559, f. 21, v.

nea, ſopra laqual ſegnaſti 12.& harai $\frac{12}{2}$ che riducti a interi, partendo 12.per 2.ne uerra 6.interi a punto per lo detto partimento.

Illustrations of problems are found in the margin, as was common in the early texts. There are checks by 9's and 7's.

[17] Dr. Siegmund Günther, *Geschichte der Mathematik,* v. 1 (1908) pp. 518, 519, chooses the work of Tagliente from among the early Italian arithmetics, and gives a detailed account of it.

[18] Ed. Denominate numbers are numbers associated with units of measurement.

[19] Ed. Leonardo of Pisa (1170–1250) is more commonly known today as Fibonacci.

Four methods of multiplication are given, and the short method of dividing by 10, illustrated as follows [2.29]:

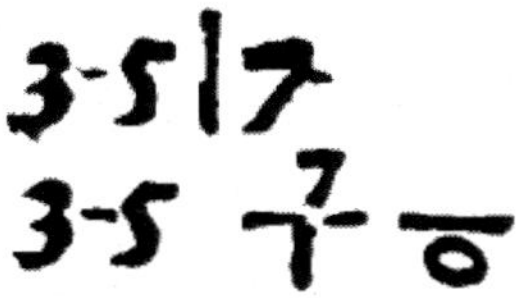

FIGURE 2.29: Cataneo's division by 10. No. 19, 1546, f. 9, v.

It is singular that in the 'a danda' method of division,[20] the divisor is placed above the dividend. He also puts dots above the numbers in the dividend as soon as carried down [2.30].

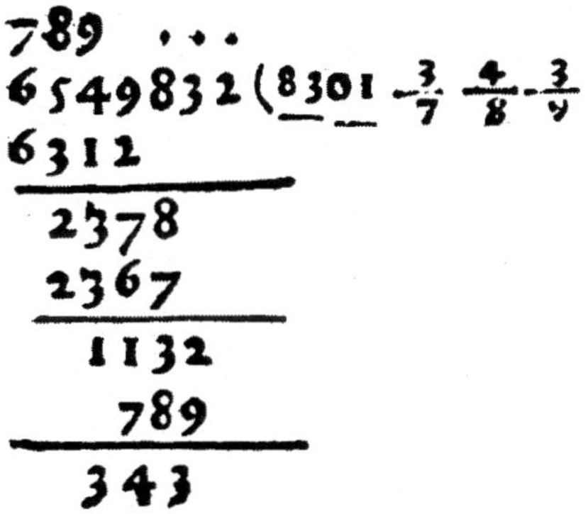

FIGURE 2.30: Cataneo's division. No. 19, 1546, f. 11, v.

Fractional lines are still broken, and guiding lines are used in all the fundamental operations with fractions. His arithmetic is followed by a few folios relating to geometry.

2.6 Italian: Francesco Feliciano (No. 29.) (1550, 1563, 1669, 1692)

In 1526 appeared an Italian arithmetic (No. 29.) which had many subsequent editions, one even as late as 1692. This work, which was in use for over a century and a half after it was first published, is worthy of more than one reference given by Cajori.[21]

The text of Feliciano is replete with historical references, among them, Aristotle, Euclid, Boethius, Scarobosco, Borghi, and Pacioli. Their influence is shown in the first part of the book, which opens with a discussion of the various kinds of numbers and is, in general, commercial in character. The '*Scala grimaldelli*,' [22] [2.31] as it was called,

[20] Ed. The 'a danda' ("by giving") method of division refers to the division algorithm which is the standard division method of today.

[21] Cajori, *Notations*, p. 258, refers to his work but one in connection with the use of X in addition and subtraction of fractions.

[22] Ed. As D.E. Smith states in his *Rara Arithmetica* (4th ed, 1970, Chelsa Publishing Co., p. 148): "The fanciful name, 'Scala grimaldelli', is explained in verse on the title page. Just as it is necessary in attacking a

might, in its day, be considered as a modern, condensed but complete, practical text; one which like Tagliente's arithmetic, would have much influence on the later teaching of elementary mathematics.

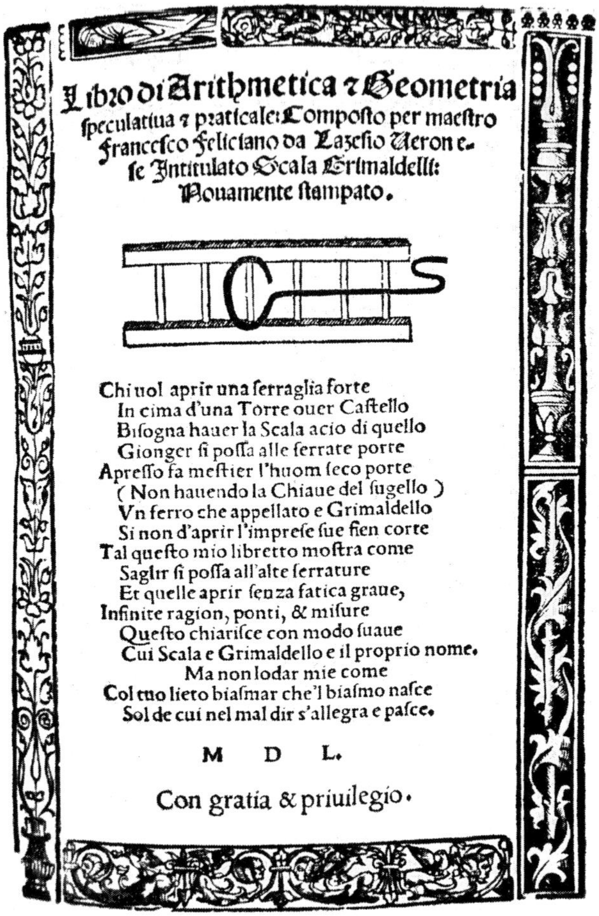
Libro di Arithmetica ⁊ Geometria ſpeculatiua ⁊ praticale: Compoſto per maeſtro Franceſco Feliciano da Lazeſio Veroneſe Intitulato Scala Grimaldelli: Nouamente ſtampato.

Chi uol aprir una ferraglia forte
In cima d'una Torre ouer Caſtello
Biſogna hauer la Scala acio di quello
Gionger ſi poſſa alle ferrate porte
Apreſſo fa meſtier l'huom ſeco porte
(Non hauendo la Chiaue del ſugello)
Vn ferro che appellato e Grimaldello
Si non d'aprir l'impreſe ſue ſien corte
Tal queſto mio libretto moſtra come
Saglir ſi poſſa all'alte ferrature
Et quelle aprir ſenza fatica graue,
Infinite ragion, ponti, & miſure
Queſto chiariſce con modo ſuaue
Cui Scala e Grimaldello e il proprio nome.
Ma non lodar mie come
Col tuo lieto biaſmar che'l biaſmo naſce
Sol de cui nel mal dir s'allegra e paſce.

M D L.

Con gratia & priuilegio.

FIGURE 2.31: **Feliciano's title page of *Libro di Arithmetica e Geometria*.[23] No. 29, 1550, f. 1, r.**

We refer to two editions of each of the two centuries in which it was used. Like his contemporary, Cataneo, in the earlier edition of 1550 Feliciano encloses integers and fractions in the running of the text by dots. He is, however, inconsistent in placing a dot between the integer and fraction of a mixed number, omitting it, or using it, sometimes on the same page (f. 6, r.). Feliciano uses *nulla* or *zero* for 0 (f. 8, v.). The usual X in his proof by 9's and 7's is written as follows (f. 11, r.) [2.32]:

castle to have a ladder (scala) and a skeleton key (grimaldello) to open locks, so in attacking mathematics it is necessary to have a book that answers the same purposes."

[23]ED. From the Plimpton Collection and used with the permission of the Rare Book & Manuscript Library, Columbia University

FIGURE 2.32: Feliciano's proof by 9s. No. 29, 1526, f. 11, r.

Addition and subtraction are performed in the margin with denominate numbers only. Of the six methods of multiplication illustrated, one is an uncommon variation, that of writing the numbers of the multiplicand in reverse order, and then multiplying them from left to right by those of the multiplier from right to left; thus, to multiply 9876 by 6543 (f. 12, r.) [2.33]:

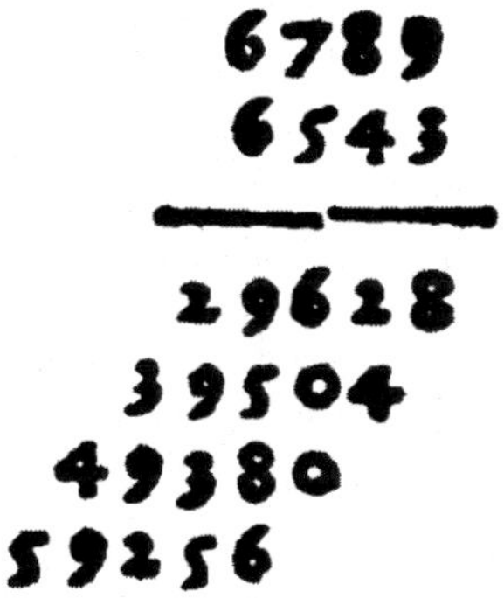

FIGURE 2.33: Feliciano's early multiplication. No. 29, 1526, f. 12, r.

It is curious to note here that in 1669, the editor writes 9876 in the usual order (p. 23.) [2.34].

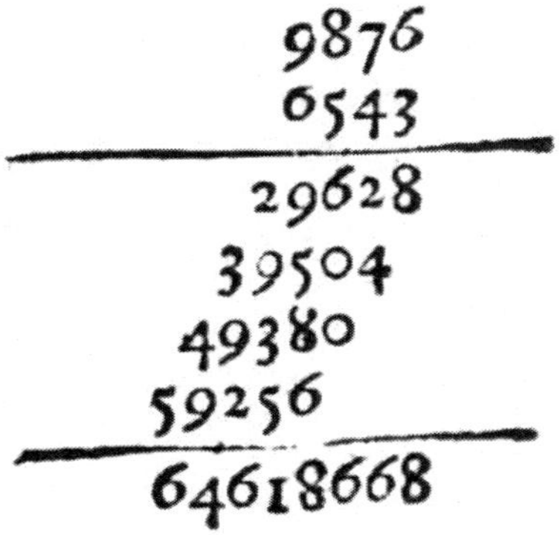

FIGURE 2.34: Feliciano's later multiplication. No. 29, 1669, p. 23

The second part of this arithmetic is devoted mainly to algebra. There Feliciano writes **R** for square root, **R** *cuba* and **R** *relata* for cube and fifth roots respectively (f. 40, v.)[2.35, 2.36, 2.37].

multiplica ℞.4.fia ℞.4 fa ℞. 16.el qual 16

FIGURE 2.35: Feliciano's square root. No. 29, 1563, f. 40, v.

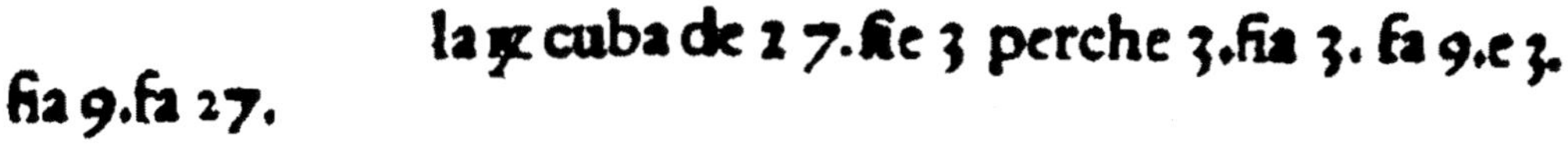
la ℞ cuba de 27.fie 3 perche 3.fia 3. fa 9.e 3. fia 9.fa 27.

FIGURE 2.36: Feliciano's cube root. No. 29, 1563, f. 40, v.

la ℞ relata de 243.fie 3.perche 3.fia 3 fa 9 e 9.fia 9 . fa 81 , e 3.fia 81. fa 243.

FIGURE 2.37: Feliciano's fifth root. No. 29, 1563, f. 40, v.

The usual p̃ and m̃ are symbols for *plus* and *minus* (f. 46, r.) (in 1669, he uses *men* and *più*),[24] as also *co.*, *ce.*, and *cu.* for *cosa*, *censa*, and *cubus* (f. 60 , v.) [2.38].

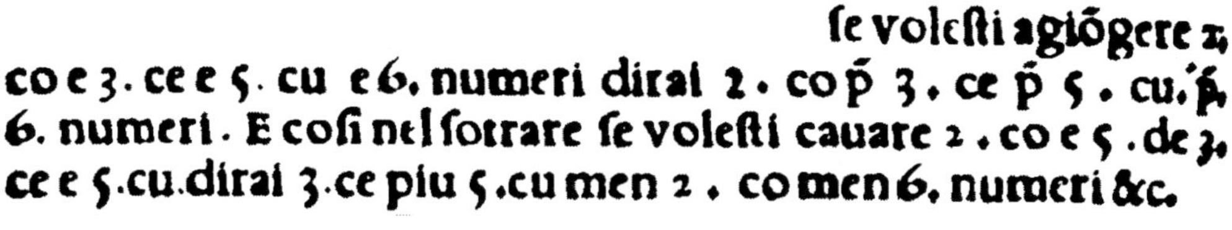
se volesti agiõgere 2. co e 3. ce e 5. cu e 6. numeri dirai 2. co p̃ 3. ce p̃ 5. cu. p̃. 6. numeri. E cosi nel sotrare se volesti cauare 2. co e 5. de 3. ce e 5. cu. dirai 3. ce piu 5. cu men 2. co men 6. numeri &c.

FIGURE 2.38: Feliciano's abbreviations for powers of unknowns. No. 29, 1563, f. 60, v.

An addition of twenty-three problems of algebra is made to the second part of the text in the 1669 edition. It is there that we find plus indicated by the cross of the form ✠, *minus* by a long bar. A and AA are used to represent *cosa* and *censo*, and in the case of more variables, recourse is had to B, C, as far as D (1669, p. 185), following Vieta [2.39].

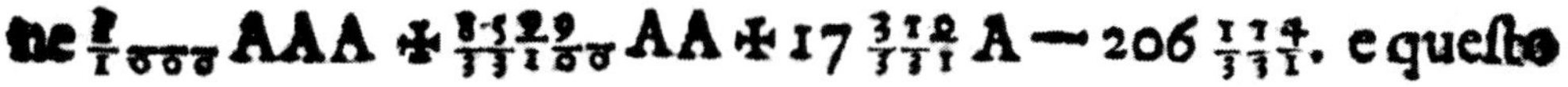
ne $\frac{1}{1000}$ AAA ✠ $\frac{1529}{331000}$ AA ✠ 17 $\frac{310}{331}$ A — 206 $\frac{114}{331}$. e questo

FIGURE 2.39: Feliciano's variables. No. 29, 1669, p. 185

D. E. Smith in his *Rara Arithmetica* does not take note of these additional problems of algebra in this edition of 1669, nor does he describe the 1692 edition, which is es-

[24] ED. This sentence is misleading. In the 1550 edition p̃ and m̃ are used, but so are *piu*, *men* and *mẽ*. Furthermore, in the 1669 edition *men* and *più* are used, but so is m.. Additionally, on the preceding folio (45, v.) p. is used.

sentially the same, except that the eight and the twelve-pointed Lultese cross are used interchangeably (1692, pp. 168, 169) [2.40].

PROBLEMA IX.

Giouanni, Giacomo, Domenico, & Antonio giuocano alle Carte, e d'eſſi Giacomo guadagna 8. ſcudi più di Gionanni; Domenico guadagna 6. ſcudi manco di Giouanni; Antonio guadagna 9. ſcudi più di Domenico: E Franceſco (che ad eſſi taglia al giuoco della Baſſetta) perde ſcudi 5. ſi domanda la riſolutione.

Suppoſto, che Gionanni guadagni 1 A. Dunque ne ſegue neceſſariamente, che Giacomo guadagni 1 A ✠ 8. Domenico 1 A — 6. & Antonio 1 A ✠ 3. il che raccolto tutto inſieme fà 4 A ✠ 5. e queſtò è vguale a 5. ch'è ciò chè perde Franceſco: Leuato dunque 5. da ogni parte, s'haurà 4 A, vguale a nulla, e diuiſo tutto per 4. s'haurà ſimilmente 1 A ; vguale a nulla. Dunque il ſuppoſto, che Giouanni guadagnaſſe 1 A ; era niente ; Giacomo, che s'è poſto, che guadagnaſſe 1 A ✠ 8. e 8. Domenico, che s'è poſto il ſuo guadagno foſſe 1 A — 6. e — 6. cioè perdeua 6. & Antonio, che s'è poſto guadagnaſſe 1 A ✠ 3: e 3. ſi che frà Giacomo, & Antonio vinceuaſi 11. e frà Domenico; e Franceſco perdeuaſi 11. com'è conueniente

1 A	: : : :	0
1 A ✠ 8	. . .	8
1 A — 6	. . .	— 6
1 A ✠ 3	. . .	3
1 A ✠ 5	vguale a	5
5		5
4 A	vguale a	0
1 A	vguale a	0

FIGURE 2.40: Feliciano's eight- and twelve-pointed cross problem. No. 29, 1692, p. 168

2.7 Italian: Giovanni Sfortunati (No. 85.) (1561)

The seven editions of Sfortunati's book *Nuovo Lume. Libro di Arithmetica,* prove that he was a popular writer. In his text he followed along the lines of the Italian writers of business arithmetics of his time.[25] Sfortunati in 1561 grouped the digits in numeration as we do today [2.41].[26]

no le infraſcritte figure cioe 58, 907, 124, 356, 090. Hab 58,907,124,366,

FIGURE 2.41: Sfortunati's grouping of digits. No. 85, 1561, f. 4, r.

[25] D. E. Smith, *Rara,* p. 174, also mentions the influence of Borgi, Feliciano, Pacioli, and Calandri upon the arithmetic of Sfortunati.

[26] Ed. The number on the right side was printed in the margin (by the original printer). Note, there is a typographical error. Namely, the second to last digit showing should be a 5, not a 6. Also, the last 3 digits, 090, are missing entirely.

Notable, too, is the fact that although, as Cajori states (pp . 254, 257, 260.), he used the form X in the process of false position, in proportions involving fractions, and in division of fractions, he does not use it in his check by 9's and 7's. It is to be noted also that dots follow only the integers in the running text.

In division, the divisor is placed above or below the dividend, and dots are placed after the numbers of the dividend when brought down (ff. 18, v., 20, v.).[2.42][27,28]

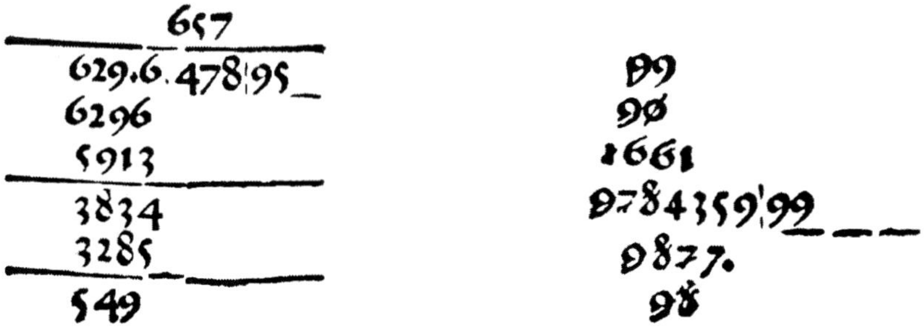

FIGURE 2.42: Sfortunati's division. No. 85, 1561, Left: f. 18, v. Right: f. 20, v.

Division of 84789 by 10 is carried out thus: 8478 | 9 (f. 17, r.)[2.43].

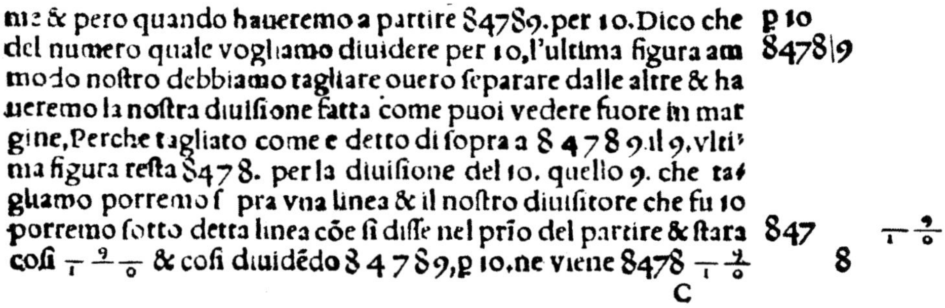

me & pero quando haueremo a partire 84789.per 10.Dico che del numero quale vogliamo diuidere per 10,l'ultima figura am modo noſtro debbiamo tagliare ouero ſeparare dalle altre & haueremo la noſtra diuiſione fatta come puoi vedere fuore in margine,Perche tagliato come e detto di ſopra a 8 4 7 8 9.il 9.vltima figura reſta 8478. per la diuiſione del 10. quello 9. che tagliamo porremo ſ pra vna linea & il noſtro diuiſitore che fu 10 porremo ſotto detta linea cõe ſi diſſe nel prĩo del partire & ſtara coſi 9/10 & coſi diuidẽdo 8 4 7 8 9,p 10.ne viene 8478 9/10

FIGURE 2.43: Sfortunati's division. No. 85, 1561, f. 17, r.

As does Feliciano, he adds a part on geometry, but there is no section on algebra.

2.8 English: Cuthbert Tunstall (No. 91.) (1538, 1544)

The treatise, *De arte supputandi libri quatuor, Cuthberti Tonstalli,* written in Latin, first appeared in London in 1522, while the seven subsequent editions were published in Paris or Strasburg. It was the work of the Englishman Tunstall, a classicist, who included in his work much of the business arithmetic of Italian authors.

It is strange that a book so prolix that, among other things, it included addition, subtraction, and division tables (pp. 21, 29, 45.) [29] should have been in such great favor

[27] Ed. The left of figure 2.42 is the beginning of the computation for 6,296,478 / 657, which equals 9583 149/219. The work in the margin shows the first two digits of the quotient (i.e. 95).

[28] Ed. The right of Figure 2.42 is the beginning of the computation for 9,784,356 / 987, which equals 9913 75/329. The work shows the first two digits of the quotient (i.e. 99).

[29] Kästner, *Geschichte der Mathematik*, v. 1, pp. 94–96 refers to these tables.

among the mercantile class. This may, perhaps, be partly explained by the fact that Tunstall was a prominent man in the Church and State, and that he was greatly influenced by Pacioli.[30] Tunstall uses the term *partitio* (division) and *subductio* (subtraction) (p. 7.), and calls zero *siphram* (p. 9.). He mentions *duplication* (duplation) and *dimidiation* (mediation) (pp. 43, 59.), but does not treat them as separate species. He is most careful to arrange numbers directly under each other with the aid of a vertical dash (p. 55. In the 1544 ed. the dash is slanted.) [2.44].

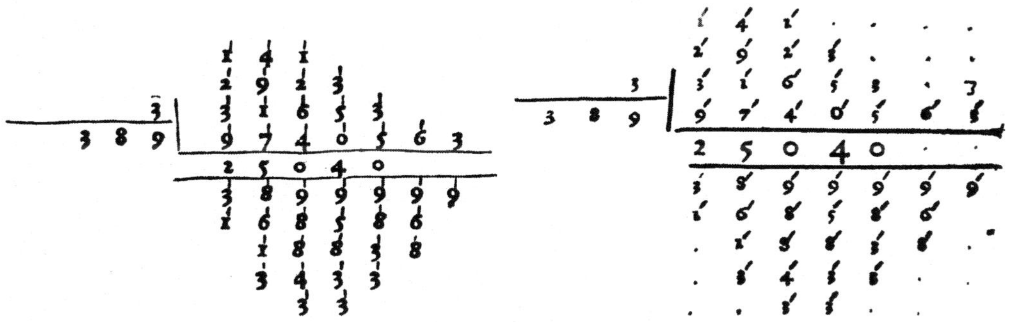

FIGURE 2.44: Tunstall's division of 25040 by 389. Left: No. 91, 1538, p. 55 Right: No. 91, 1544, p. 83

For grouping digits in the extraction of roots he uses dots above [2.45].

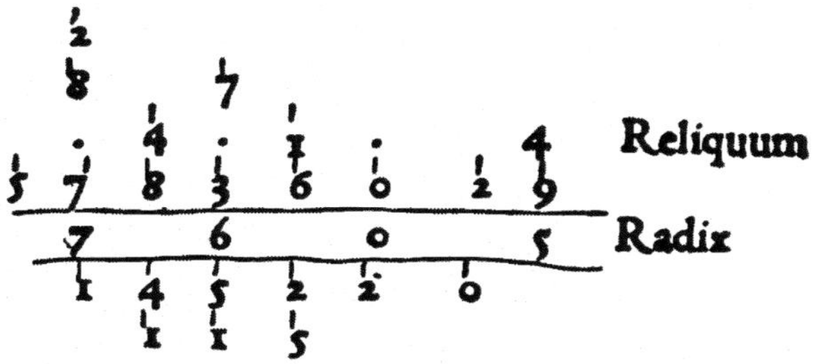

FIGURE 2.45: Tunstall's extraction of roots. No. 91, 1538, p. 65

Our method of subtraction is clearly displayed in the following manner (p. 30.) [2.46]:

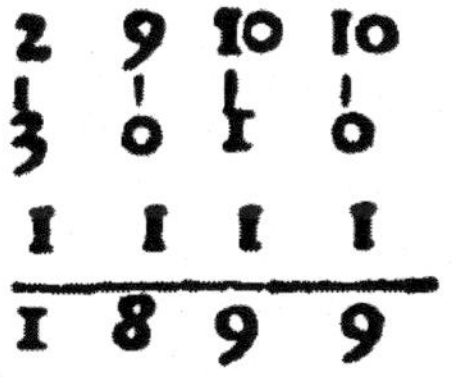

FIGURE 2.46: Tunstall's subtraction. No. 91, 1538, p. 30

There are generally no broken fractional lines; occasionally the lines are omitted altogether after the first fraction (pp. 75, 80.) [2.47].

[30]Tunstall and Recorde are listed as the representative English writers of arithmetic in the sixteenth century by Günther, *Geschichte der Mathematik*, v. 1, pp. 343, 344, and also by Cantor, *Vorlesungen über Geschichte der Mathematik,* v. 2, pp. 475–480.

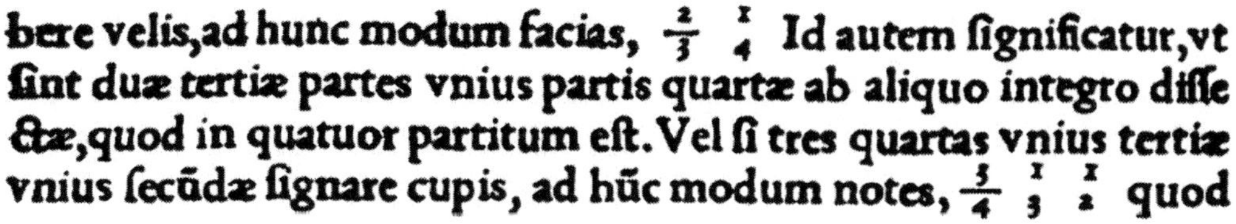
bere velis,ad hunc modum facias, $\frac{2}{3}$ $\frac{1}{4}$ Id autem significatur,vt sint duæ tertiæ partes vnius partis quartæ ab aliquo integro disse ctæ,quod in quatuor partitum est. Vel si tres quartas vnius tertiæ vnius secūdæ signare cupis, ad hūc modum notes, $\frac{3}{4}$ $\frac{1}{3}$ $\frac{1}{2}$ quod

FIGURE 2.47: Tunstall's fractions. No. 91, 1538, p. 75

The 1544 Strasburg edition presents only this minor variation, that the fractional lines are broken throughout (1544, p. 133.) [2.48].

FIGURE 2.48: Tunstall's fractions. Left: No. 91, 1538, p. 83 Right: No. 91, 1544, p. 133

2.9 Austrian: Georg Peurbach (No. 67.) (1539)

The Latin work *Elementa Arithmetices* (Venice, 1539)[31] of the mediaeval writer, Georg Peurbach, is included (ff. 32–67.) in a volume with the *Elementa Geometriae* by Johann Vogelin. It contains merely a summary of the fundamental operations with at most two or three illustrative problems. It is surprising that a book so theoretical in character should have had about seventeen editions, and yet we must remember that it was written by the able mathematician, Peurbach, teacher of Regiomontanus, to explain the Hindu-Arabic notation, and that it embodies many points of superiority over other early texts.[32]

Only integers in the running text are followed by dots, as are also the terms of a progression (f. 39, v.) [2.49].

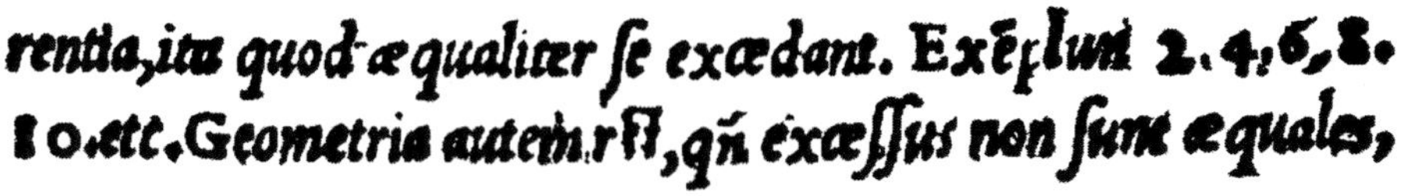
rentia,ita quod æqualiter se excedant. Exēplum 2.4.6.8. 10.etc.Geometria autem [illegible], qñ excessus non sunt æquales,

FIGURE 2.49: Peurbach's integers. No. 67, 1539, f. 39, v.

Fractions are commonly left in improper form, with broken fractional lines (f. 52, r.) [2.50].

[31] Ed. This edition was published more than seven decades after Georg Peurbach's (1423 - 1461) death.

[32] Gerhardt, *Geschichte der Mathematik in Deutschland,* (München, 1877), pp. 9, 10, gives an idea of the importance of Peurbach's work.

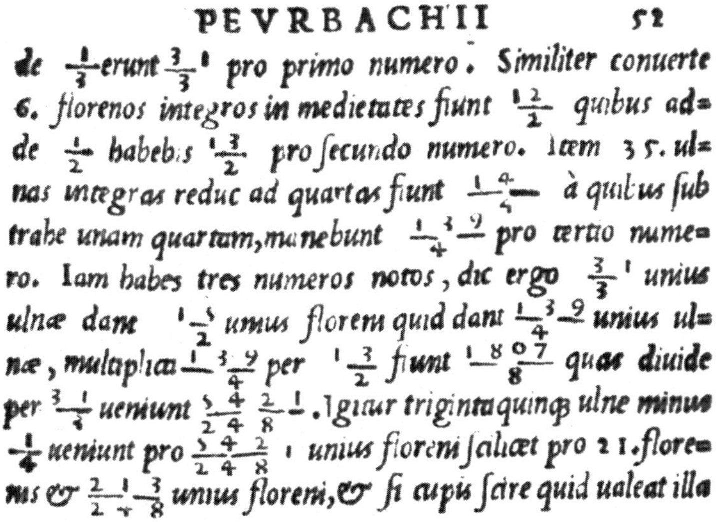
PEVRBACHII 52

de 1/3 erunt 3/3 1 pro primo numero. Similiter conuerte 6. florenos integros in medietates fiunt 12/2 quibus ad=de 1/2 habebis 13/2 pro secundo numero. Item 35. ul=nas integras reduc ad quartas fiunt 14/4 à quibus sub trahe unam quartam, manebunt 139/4 pro tertio nume=ro. Iam habes tres numeros notos, dic ergo 3/3 1 unius ulnæ dant 15/2 unius floreni quid dant 139/4 unius ul=næ, multiplica 139/4 per 13/2 fiunt 1807/8 quas diuide per 31/4 ueniunt 54/24 21/8. Igitur trigintaquinque ulne minus 1/4 ueniunt pro 542/248 1 unius floreni scilicet pro 21. flore=nis & 21/2 3/8 unius floreni, & si cupis scire quid ualeat illa

FIGURE 2.50: **Peurbach's fractions.** **No. 67, 1539, f. 52, r.**

Mediation and duplation strangely both precede multiplication (ff. 35, v., 36, r.). In the extraction of roots, numbers are grouped by placing dots above (f. 41, r.) [2.51].

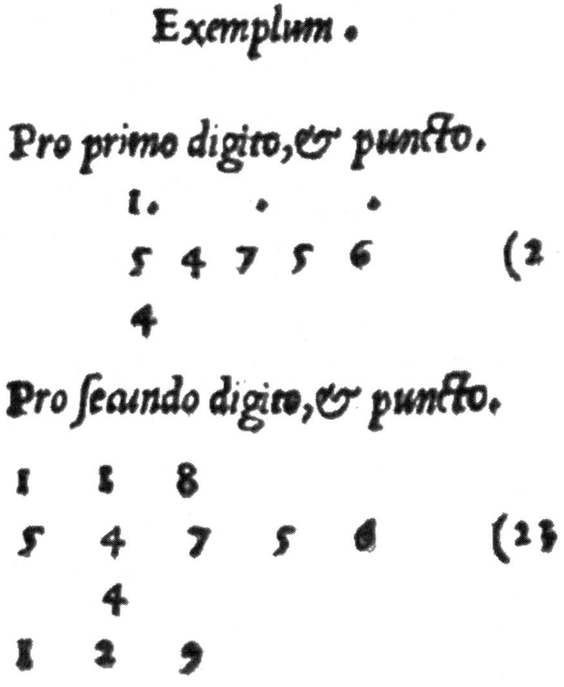
Exemplum.

Pro primo digito, & puncto.

1. . .

5 4 7 5 6 (2

4

Pro secundo digito, & puncto.

1 1 8

5 4 7 5 6 (23

4

1 2 9

FIGURE 2.51: **Peurbach's extraction of roots.** **No. 67, 1539, f. 41, r.**

While this work of Peurbach is noted because of the number of editions, it adds very little as regards the development of notations.

2.10 French: Oronce Finé (No. 33, 34.) (1555, 1587, 1670)

The first work of the French mathematician, Orontius Finaeus, *Orontii Finaei Delphinatis, Promathesis,* was written in Latin in 1532. It was divided into four parts: arithmetic, geometry, cosmography, and horography.[33] This study is based in detail upon the 1555 Latin edition (No. 33.), which contains only the first of these parts, comparing it with the two Italian editions of 1587 and 1670 (No. 34.)

The part on arithmetic deals with integers, common fractions, sexagesimal fractions, and proportion. Zero is called *tziphra* (f. 1, v.). Both subtraction and multiplication are performed by the complementary method.[34]

It is curious that Finé places the quotient between the dividend and divisor; thus, to divide 4761 by 23 (f. 18, r.) [2.52]:

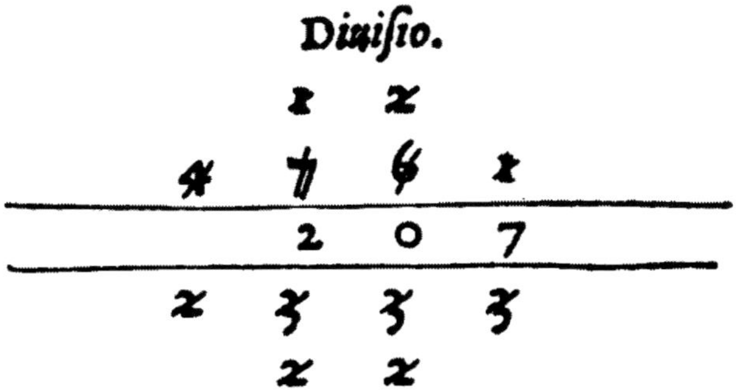

FIGURE 2.52: **Finé's division. No. 33, 1555, f. 18, r.**

Sometimes (f. 9, r.) he writes down the multiples of the divisor before dividing [2.53].

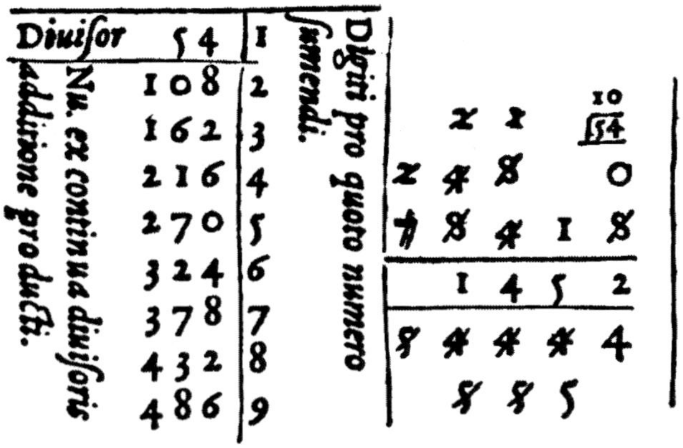

FIGURE 2.53: **Finé's division. No. 33, 1555, f. 9, r.**

[33] Ed. Horography is the study of timekeeping and clocks.

[34] Tropfke, *Geschichte,* v. I, 1921, p. 29 mentions this method used in subtraction in the 1555 ed., p. 28. Cajori refers to his use of complementary multiplication in 1532 (p. 264.).

It is important to note that this author omits the fractional line after the first fraction when multiplying fractions by one another; thus, 3/4 times 1/2 is written (f. 20, v.) [2.54]:

$$\frac{3}{4}\quad\begin{matrix}1\\2\end{matrix}$$

FIGURE 2.54: Finé's multiplication of fractions. No. 33, 1555, f. 20, v.

Guiding lines are used in all operations with fractions.

Finé's interest in astronomy and geometry led him to include a treatment of sexagesimal fractions in his arithmetic [2.55].[35]

Signa.	Gradus.	Minuta.	Secūda.	Tertia.
8 .	25 .	36 .	53 .	30.
5 .	23 .	45 .	00 .	27.
2 .	18 .	50 .	39 .	16.
5 .	8 .	12 .	33 .	13.

FIGURE 2.55: Finé's quasi-sexagesimal based systems of angular measurement. No. 33, 1555, f. 36, r.

Finé devotes many folios to proportions, and makes use of an inclined bar in writing them, somewhat as follows: 8⁄4, 6⁄3, meaning our 8:4::6:3 (f. 59, r.) [2.56].[36]

8/4, 6/3: ut enim 8 ad 4, ſic 6 ad 3

FIGURE 2.56: Finé's fractions. No. 33, 1555, f. 59, r.

He uses vertical bars to group the digits in the extraction of roots, 5|30|84|16 (f. 11.) [2.57].

[35] Ed. This system, for modern readers, has some familiar and unfamiliar conventions. On the familiar side: Gradus. = degrees, Minuta = minutes, Secunda = seconds and Tertia. = 1/60 of a second. On the unfamiliar side: 1 Signa. = 30 Gradus. (i.e. 30 degrees) and 12 Signa. = 1 total revolution (i.e. 360 degrees) - not shown in this chart, though it does account for the 5 in the Signa. column.

[36] Ed. It may appear that Finé's 'inclined bar' is equivalent to our modern slanted fraction bar, but that is not the case. Finé's example of 8/4, 6/3 is what he calls a 'discontinuous proportion'. Earlier, on the same folio, Finé discuss what he calls a 'continuous proportion' using the example 8/4/2/1. It is clear, in this 'continuous' example, that Finé is not discussing a complex fraction, but, rather he is using the 'inclined bar' to indicated which pairs of adjacent numbers share a common ratio. Thus, for Finé, the 'inclined bar' conveys more information than a fraction bar does.

FIGURE 2.57: Finé's grouping of digits. No. 33, 1555, f. 11, v.

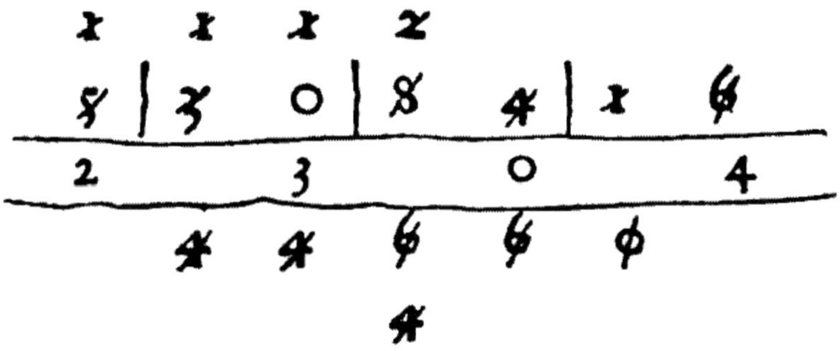

In the Italian editions of 1587 and 1670, which are essentially the same, a few deviations from the notations of the Latin editions may be cited. *Tziphra* becomes *zero* (1587, f. 2, r.) [2.58],

FIGURE 2.58: Finé's zero. No. 34, 1587, f. 2, r.

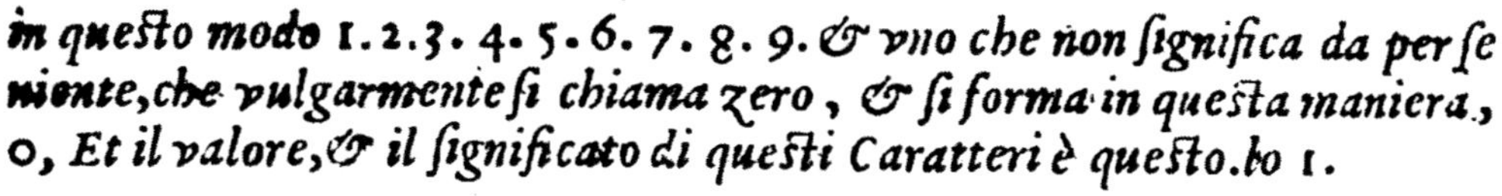
in queſto modo 1.2.3.4.5.6.7.8.9. *& vno che non ſignifica da per ſe niente, che vulgarmente ſi chiama zero, & ſi forma in queſta maniera,* 0, *Et il valore, & il ſignificato di queſti Caratteri è queſto. lo* 1.

and all abstract integers in the running text are followed by dots, as was customary in the early printed Italian texts.

In subtraction he explains the complementary simultaneously with the usual method (*Ibid.,* f. 5, v.). Proportions are now indicated in the following manner (f. 72, r.) [2.59]:

FIGURE 2.59: Finé's proportions. No. 34, 1587, f. 72, r.

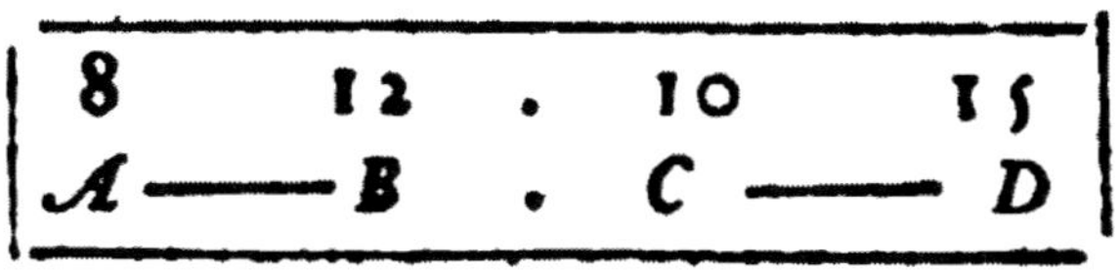

2.11 French: Pierre Savonne (No. 82.) (1571, 1619)

Pierre Savonne, called Talon, wrote a French commercial arithmetic in 1563, which, although not included among the sources of either Cajori or Tropfke, was so popular that an edition appeared as late as 1672. Our study is based upon examination of the editions 1571 and 1619, both printed at Lyons, the commercial center of France. Savonne devotes most of his arithmetic to banking and exchange, and so offers few marked peculiarities in notation. The commercial nature of the book accounts for his introduction of numerous short methods of multiplication and division.

In the 1571 edition he speaks of the form (0) 'qui s'appelle nulle, & entre marchans zero),' whereas in the 1619 edition Savonne merely calls it *zero* (f. 1, r.; 1619, p. 1.) [2.60].

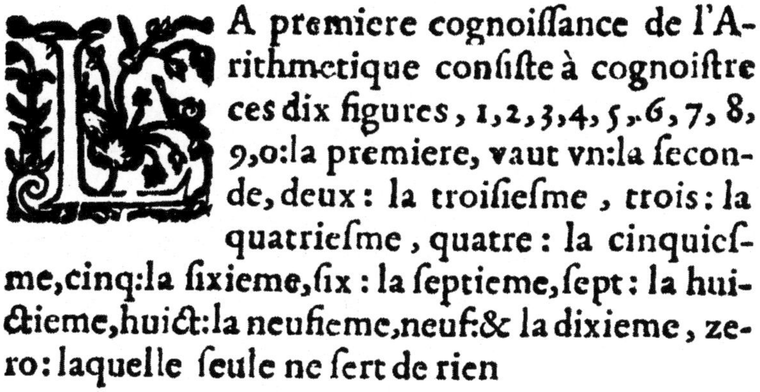
LA premiere cognoiſſance de l'A-
rithmetique conſiſte à cognoiſtre
ces dix figures, 1,2,3,4,5,6,7,8,
9,0: la premiere, vaut vn: la ſecon-
de, deux: la troiſieſme, trois: la
quatrieſme, quatre: la cinquieſ-
me, cinq: la ſixieme, ſix: la ſeptieme, ſept: la hui-
ctieme, huict: la neufieme, neuf: & la dixieme, ze-
ro: laquelle ſeule ne ſert de rien

FIGURE 2.60: Savonne's zero. No. 82, 1619, p. 1

This text uses the Greek and the Latin cross respectively in the check by 9's in the two editions under comparison (*Ibid.,* f. 9, r.; p. 33.) [2.61].

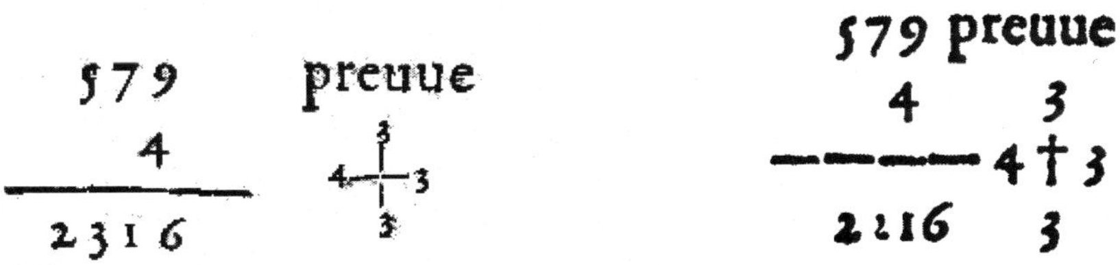

FIGURE 2.61: Savonne's check by 9s. Left: No. 82, 1585, p. 21 Right: No. 82, 1619, p. 33

In both, subtraction is performed by the complementary method and dots are placed to the right of the borrowed numbers, thus (f. 6, r.):

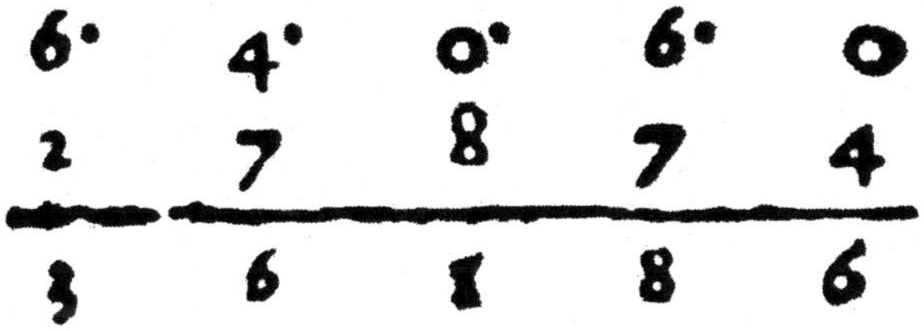

FIGURE 2.62: Savonne's subtraction.[37] No. 82, 1571, f. 6, r.

In division the divisor is placed below the dividend, a survival of the scratch method of division [2.63].

[37]ED. From the Plimpton Collection and used with the permission of the Rare Book & Manuscript Library, Columbia University.

FIGURE 2.63: Savonne's division. No. 82, 1585, p. 24

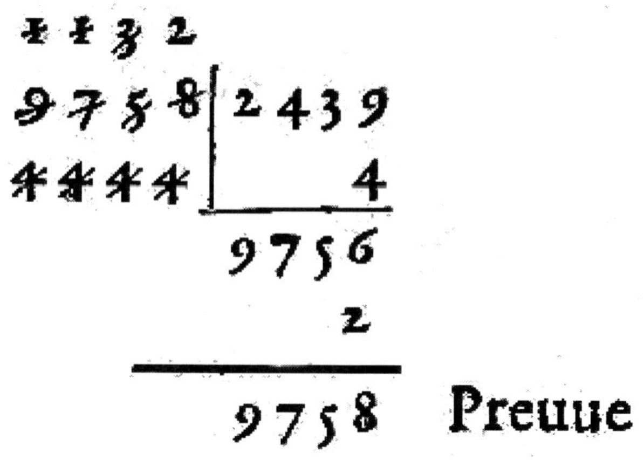

It is strange that we find the use of broken fractional lines in the later edition (1619, p. 48.) [2.64],

FIGURE 2.64: Savonne's fractions. No. 82, 1619, p. 48

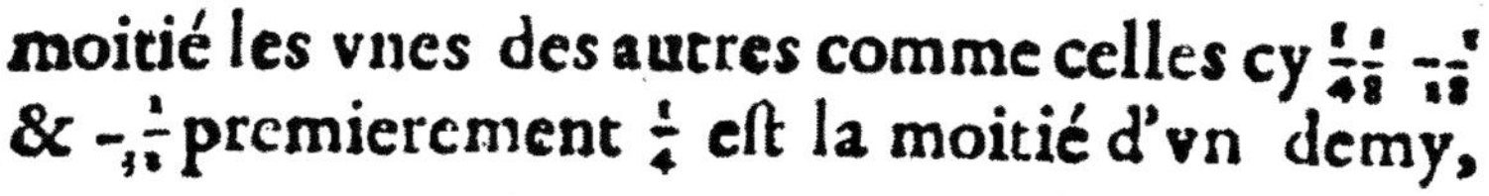
moitié les vnes des autres comme celles cy $\frac{1}{4}$ $\frac{1}{8}$ $\frac{1}{16}$
& $\frac{1}{32}$ premierement $\frac{1}{4}$ eſt la moitié d'vn demy,

and that the fractional lines are quite frequently omitted altogether in the performance of the fundamental operations with fractions (*Ibid.*, p. 54.) [2.65].

FIGURE 2.65: Savonne's fractions. No. 82, 1619, p. 54

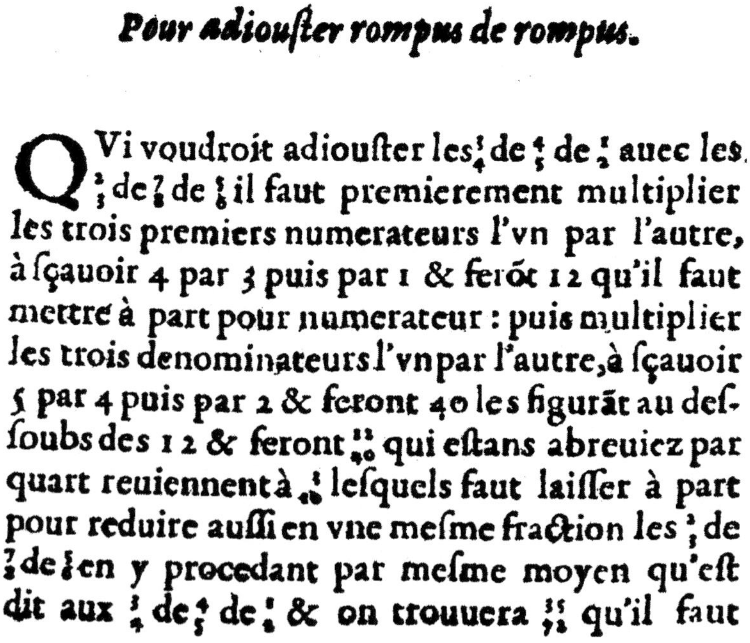
Pour adiouſter rompus de rompus.

QVi voudroit adiouſter les $^{3}_{4}$ de $^{4}_{5}$ de $^{1}_{2}$ auec les $^{2}_{3}$ de $^{7}_{8}$ de $^{5}_{6}$ il faut premierement multiplier les trois premiers numerateurs l'vn par l'autre, à ſçauoir 4 par 3 puis par 1 & ferõt 12 qu'il faut mettre à part pour numerateur : puis multiplier les trois denominateurs l'vn par l'autre, à ſçauoir 5 par 4 puis par 2 & feront 40 les figurãt au deſſoubs des 12 & feront $^{12}_{40}$ qui eſtans abreuiez par quart reuiennent à $^{3}_{10}$ leſquels faut laiſſer à part pour reduire auſſi en vne meſme fraction les $^{2}_{3}$ de $^{7}_{8}$ de $^{5}_{6}$ en y procedant par meſme moyen qu'eſt dit aux $^{3}_{4}$ de $^{4}_{5}$ de $^{1}_{2}$ & on trouuera $^{35}_{72}$ qu'il faut

2.12 French: Pierre de La Ramée (No. 74, 75.) (1599, 1613)

Of the several books on mathematics written by Petrus Ramus, we shall consider his *Arithmeticae Libri Duo,* one with a commentary of Schonerus in 1599 (No. 74.), the other with Snell's[38] commentary of 1613 (No. 75.), and refer only to the peculiarities in symbolism not noted by Cajori or Tropfke.

This work, unlike the arithmetic of Savonne, was too theoretical to meet the needs of the commercial world, and yet it was popular in the Latin schools for about half a century.[39]

The edition by Schonerus (No. 74.) contains the arithmetic (two books) and geometry of Ramus, and it includes treatises by Schonerus on the Greek theory of numbers, and on sexagesimal fractions (as in Finé). The 1613 edition (No. 75) includes Snell's commentary on Ramus' two books of arithmetic with his addition *De re nummeria* (1635).

In Ramus' arithmetic (Schonerus), the digits in numeration are grouped by dots (1599, p. 2) [2.66]; in Snell's, by commas (1613, p. 2.) [2.67].

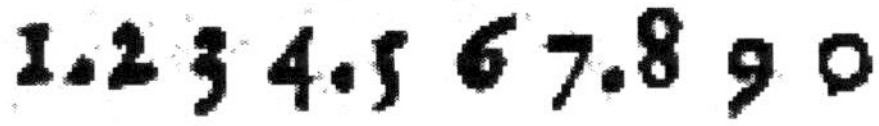

FIGURE 2.66: La Ramée's grouping of digits. No. 74, 1599, p. 2

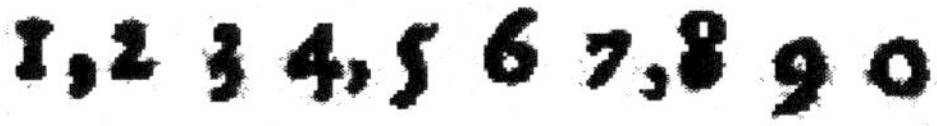

FIGURE 2.67: La Ramée's grouping of digits. No. 75, 1613, p. 2

Zero is referred to as *circulus* in both (1599, p. 1; 1613, p. 2.). His plus sign is a Greek cross with the horizontal stroke much longer than the vertical one, thus (1599, p. 182.) [2.68]:

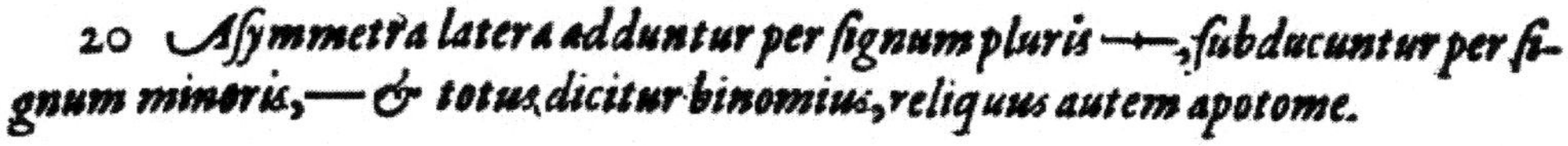
20 *Aſymmetra latera adduntur per ſignum pluris —+—, ſubducuntur per ſignum minoris, — & totus dicitur binomius, reliquus autem apotome.*

FIGURE 2.68: La Ramée's plus sign No. 74, 1599, p. 182

[38] Ed. Willebrord Snell (1580-1626), widely known today for Snell's Law of Refraction.

[39] P. Treutlein, "Das Rechnen im 16. Jahrhundert", *Abhandlungen zur Geschichte der Mathematik,* v. 1, p. 17, in describing this arithmetic, says "in welcher im Ganzen nach der uns heute geläufigen Reihenfolge die Hauptsche des Zahlenrechnens gut dargestellt ist."

Subtraction is performed from left to right, and the remainder is placed above after scratching out the numbers, thus (*Ibid.,* p. 6.) [2.69]:

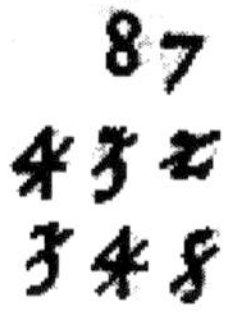

FIGURE 2.69: La Ramée's subtraction. No. 74, 1599, p. 28

Again the divisor is placed below the dividend, and a lunar sign on the right for the quotient, somewhat in this arrangement (*Ibid.,* p. 20.) [2.70]:

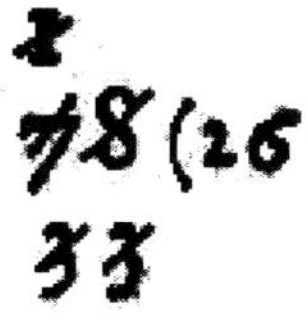

FIGURE 2.70: La Ramée's division. No. 74, 1599, p. 20

Division of 7200 by 100 is performed in this manner (*Ibid.*, p.25.) [2.71]:

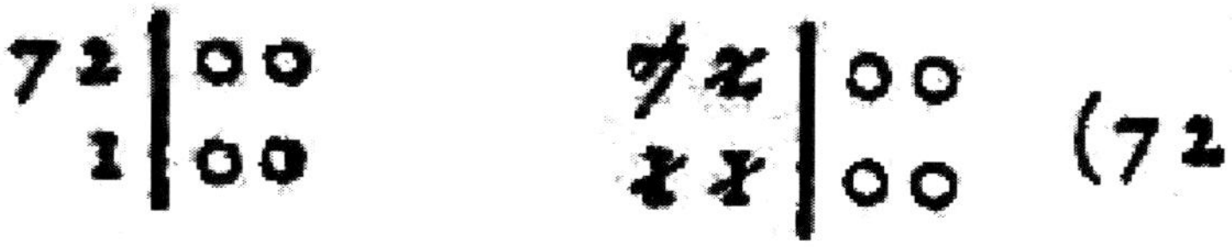

FIGURE 2.71: La Ramée's division. No. 74, 1599, p. 25

When taking a fraction of a fraction, the fractional line is omitted after the first fraction (*Ibid.,* p. 39.) [2.72].

FIGURE 2.72: La Ramée's fractions. No. 74, 1599, p. 39

Ramus uses a dot to group the digits in the extraction of roots (*Ibid.,* p. 154.) [2.73].

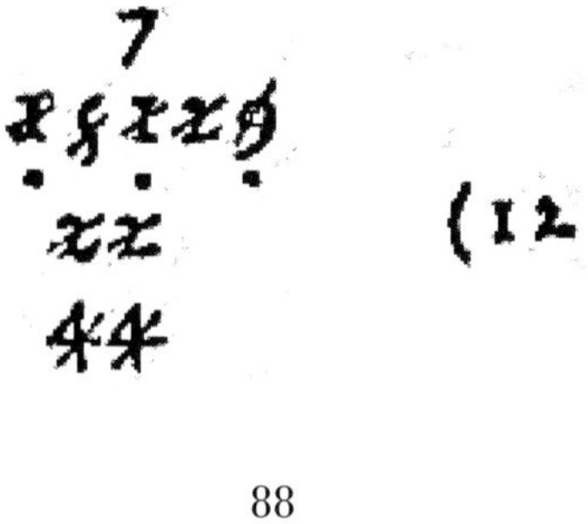

FIGURE 2.73: La Ramée's extraction of roots. No 74, 1599, p. 154

2.13 Italian: Georgio Lapazzaia (No. 56.) (1601)

Again it is desirable to draw attention to a work which was in for use over two centuries. The arithmetic by the abbot Georgio Lapazzaia was first edited in Latin in 1566, and Riccardi records an edition as late is 1784, thus extending over three centuries. This longevity is probably due to the popular style of the work, its commercial nature, and the influence of the Church schools of that time.

The *Libro d'Arithmetica e Geometria,* the particular edition under consideration, published at Naples In 1601, contains very little that is singular in the way of notation. The grouping of digits in numeration is indicated thus (p. 7.) [2.74]

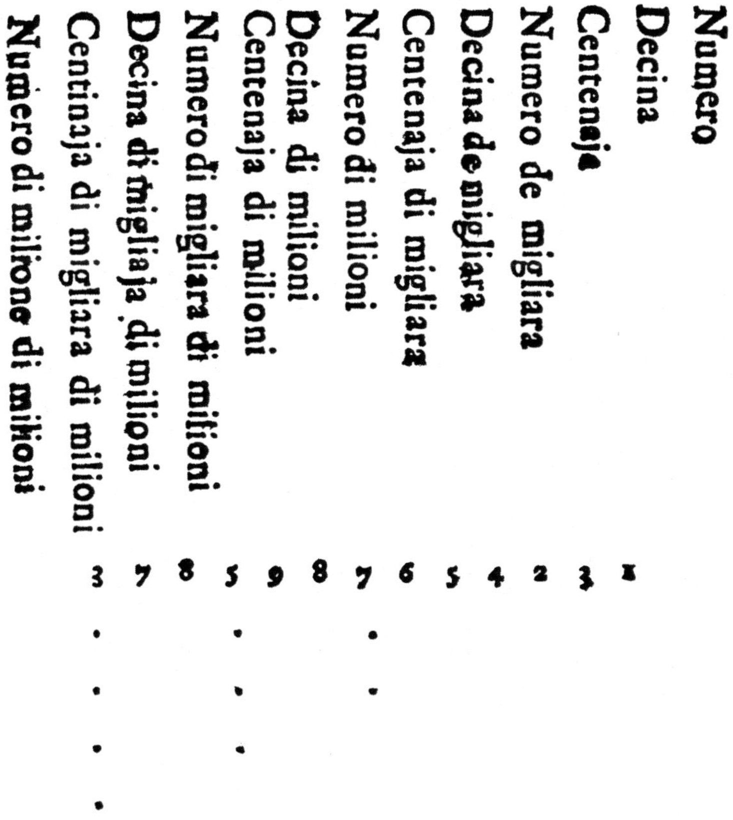

FIGURE 2.74: Lapazzaia's grouping of digits. No. 56, 1727, p. 7

Lapazzaia is consistent in using dots below when grouping numbers for root extraction (p. 117.) [2.75].

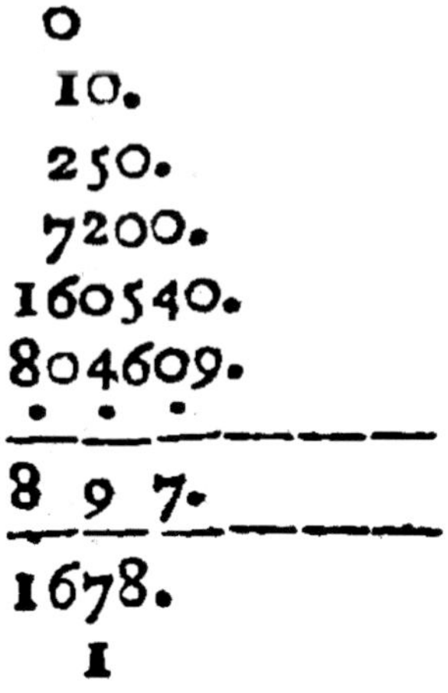

FIGURE 2.75: Lapazzaia's root extraction. No. 56, 1727, p. 117

There are found also the familiar two intersecting lines for results in casting out the 9's and 7's, and the guiding lines in carrying out the fundamental operations with fractions.

Fundamental operations are carried out first with abstract, then with denominate numbers. Subtraction by the complementary method and division by the galley method offer nothing novel in representation. The last part of the work deals with mensuration.

2.14 Italian: Guiseppe Figatelli (No. 32.) (1678, 1699)

Another 'centenarian' which escaped the attention of Cajori is that of the Italian Figatelli, particularly noteworthy because it is rich in notation. Figatelli wrote his first book on arithmetic in 1664, and Riccardi gives these subsequent editions: 1680, 1686, 1692, 1699, a sixth impression without a date, 1737, 1738, 1778, 1791 and 1797. He does not mention the second impression, Venice, 1678, upon which our study is based.

This arithmetic, to which is added algebra, was written not only for merchants, but also for teachers. The Italian symbolism with its p̃ and m̃, his use *nu., co., qu., cu.,* for powers, and the crossed capital R as his basic symbol for roots, his various methods of multiplication (per colonna, baricocolo, ripiego, crosetta, quadrilatero, and piramide) (pp. 6–10.) all show the persistence for centuries of the familiar notations of Pacioli. Figatelli, in both the 1678 and 1699 editions, includes Roman notation, indicating the thousand-fold value of a number by placing a horizontal line above it, thus: $\overline{V}$= 5000 [2.76, 2.77].

Si chè, ſe ſarà ſopra l' $\overline{\text{I}}$. ſignificarà mille; Se ſarà ſopra il $\overline{\text{V}}$. ſignificarà cinquemillà; Sopra il $\overline{\text{X}}$. diecimilla, & ſopra il $\overline{\text{C}}$. Centomilla, &c.

FIGURE 2.76: Figatelli's Roman numeral notation. No. 32, 1678, p. 3

ſe ſarà ſopra l'$\overline{\text{I}}$. ſignificarà mille; Se ſarà ſopra il $\overline{\text{v}}$. ſignificarà cinquemilla; Sopra il $\overline{\text{x}}$. diecimila, e ſopra il $\overline{\text{c}}$. Centomilla, &c.

FIGURE 2.77: Figatelli's Roman numeral notation. No. 32, 1699, p. 4

For the sake of abbreviation he writes 400 as [2.78/Left], 6,000 as [2.78/Right], placing the C and M above the number (pp. 3, 4.).

Left	Right
C.	M.
IIII.	V I.
Quattrocento.	Seimilla.

FIGURE 2.78: Figatelli's Roman numeral notation. Left: No. 32, 1678, p. 4 Right: No. 32, 1678, p. 4

Multiplication 'per crosetta' with the check by nines is displayed as follows (p. 8.) [2.79]:

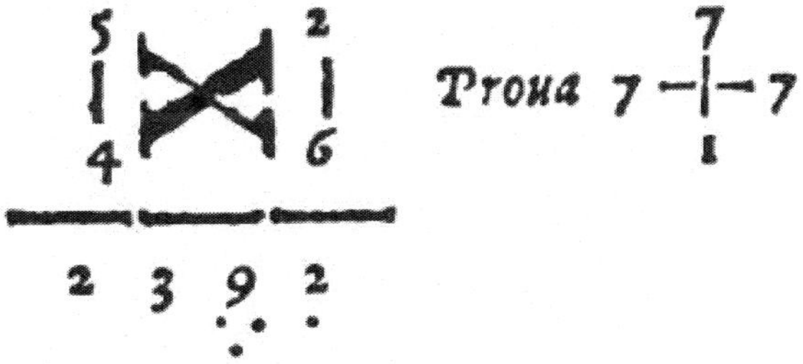

FIGURE 2.79: Figatelli's multiplication. No. 32, 1678, p. 8

Notice that the dots below are the numbers to be carried.

Figatelli's further liberal use of dots for various purposes of convenience is best shows in his illustration of division of denominate numbers, thus (p. 16.) [2.80]:

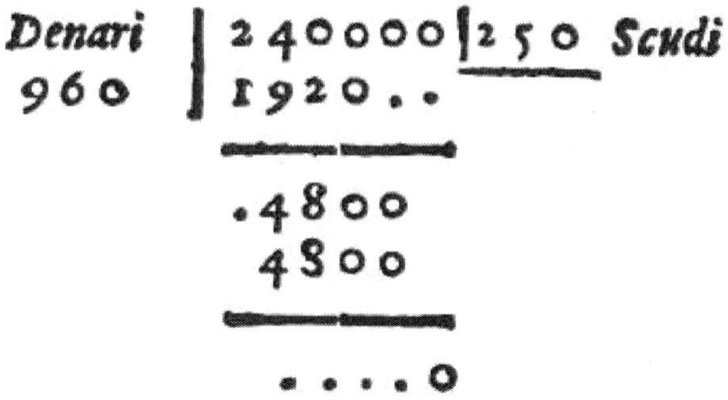

FIGURE 2.80: Figatelli's division. No. 32, 1678, p. 16

In the extraction of roots, dots above are used for grouping; he uses them also in the writing of large numbers, thus (p. 4.) [2.81]:

FIGURE 2.81: Figatelli's extraction of roots. No. 32, 1678, p. 4

Figatelli introduces the sign * for indicating commensurable (cummunicanti) radicals (p. 183.); thus there occurs this problem in addition of radicals (p. 184.) [2.82]:

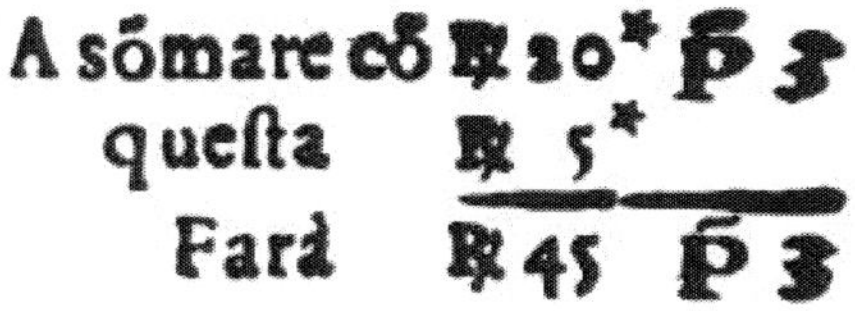

FIGURE 2.82: Figatelli's addition of radicals. No. 32, 1678, p. 184

The notation for proportion is very clumsy; for example (p. 200.) [2.83]:

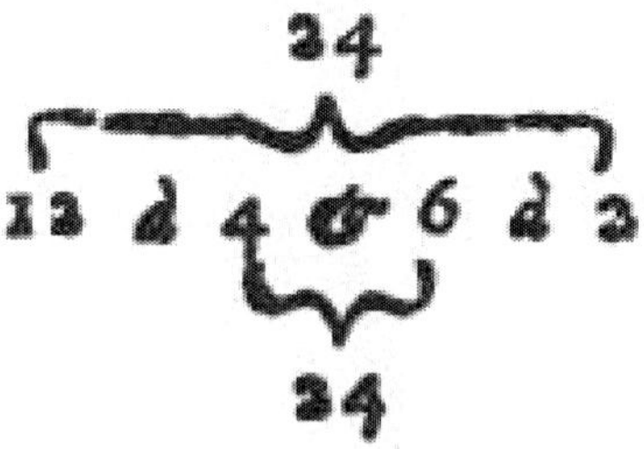

FIGURE 2.83: Figatelli's proportions. No. 32, 1678, p. 200

2.15 German: Simon Jacob (No. 54.)(1612)

Jacob's two arithmetical works must have met with favor between 1550 and 1600, for there were nine editions during that time. These were arithmetics of the Adam Riese type, commercial in nature, extensive in applied problems, and still devoting a few pages to reckoning 'auff den Linien'.[40] Jacob was the most important German reckoning master of the last half of the sixteenth century.[41]

The particular edition of *Ein New und Wolgegrundt Rechenbuch* used was printed at Frankfurt in 1612. The preface to the reader is dated 1552, and the colophon gives the date 1600. (The first edition listed by Smith is of 1560).

There are several things of note as regards symbolism. All integers of the running text are followed by dots. Zero is called *nulla* (f. 1, r.). The digits of large numbers are grouped in this manner (f. 1, v.) [2.84]:

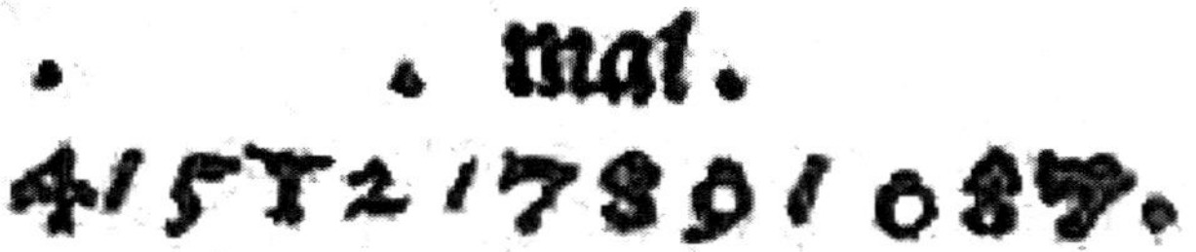

FIGURE 2.84: **Jacob's grouping of digits. No. 54, 1612, f. 1, v.**

Jacob makes use of dots above as a separatrix for the extraction of roots. There are explanations of *proba* by 9's, 3's, 7's, 8's, 11's, etc.

In connection with the reduction of fractions to lowest terms, the author explains divisibility by numbers below 10 (ff. 20, 21.). Tropfke[42] refers to Jacob's test of divisibility by 7 as that used by Ibn Albannā![43] Jacob uses the words of Rudolff for powers of the unknown, namely, *rad.*, *zens*, *cub*, *zensdezens*, etc.; for roots, *radix quadrata*, *radix cubica*, *radix zensdezensico*, but he has no symbols for them (f. 89 ff).

Outside of his use of the plus sign in the form of a Maltese cross in the rule of false position, Jacob uses it only in writing the powers of a binomial. He writes the result of $(x+1)^2$ thus (f. 106, r.): 1. zens ✠ 2. rad. ✠ 1. [2.85]

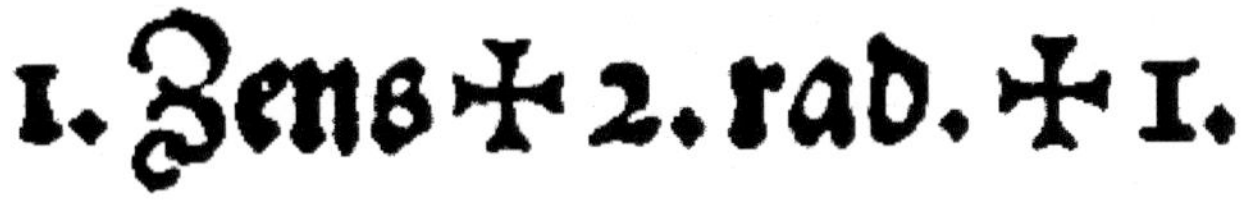

FIGURE 2.85: **Jacob's powers of a binomial. No. 54, 1612, f. 106, r.**

2.16 German: Johann Hemeling (No. 47, 48.) (1667, 1678)

The *Historische Rechne-kunst* (No. 47.) of Hemeling contains one hundred historical problems; interspersed we find couplets that are religious in character. Hemeling uses

[40] Ed. 'auff den Linien' is German for 'on the Lines'.

[41] L. L. Jackson, *The Educational Significance of Sixteenth-Century Arithmetic,* (New York, 1906), p. 28.

[42] Tropfke, *Geschichte,* v. I, 1921, p. 106.

[43] Ed. al-Marrakushi ibn al-Banna (1256–1321) was an Islamic mathematician who spent most of his life in Morocco. He wrote at least eighty-two books on mathematics.

one, two, three, or more dots superimposed to mark the groups of digits in numeration, thus (p. 4.) 10000000, and also to indicate borrowing in subtraction (p. 8.):

```
8 2 4 8 0 0
4.3.5 4 0 0
-----------
3 8 4 4 0 0
```

It is interesting that he should write the partial products of 1 2 3 4 5 by 1 4 3 2 thus (p. 11.):

```
1 2 3 4 5
  4 9 3 8 0
    3 7 0 3 5
      2 4 6 9 0
---------------
1 7 6 7 8 0 4 0
```

This order has been proposed very often as more useful in lending itself to cutting off non-significant digits.

We find Hemeling using the ordinary plus sign, and ÷[44] for minus (p. 177.) [2.86].

32. Gekaufft in Hildesheim 1 ℞ Wahr umb 13 thl 1
6 Pf / wieviel demnach umb 114 thl ÷ 5 Pf ? Antw.
55 lb.

FIGURE 2.86: Hemeling's plus sign. No. 48, 1678, p. 177

1 R is his symbol for the unknown X (p. 174.). Hemeling wrote various other works, but our study is based upon the *Selbstlehrende Rechneschul* or *Selbstlehrendes Rechnebuch* of 1678, of which there is a revised edition by E. A. Talgener as late as 1817, an indication of the probability that it was still used in many schools.

In this work the dots used for grouping in numeration are slightly changed in position, thus (1678, p. 21.) [2.87]:

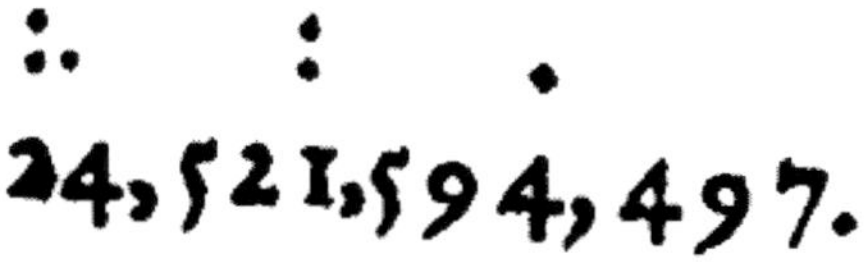

FIGURE 2.87: Hemeling's grouping of digits. No. 48, 1678, p. 21

The otherwise impossible subtraction is performed by the addition of one to the next highest number in the subtrahend, but now Hemeling indicates this addition by placing the dot below the numbers.

[44] ED. This is the symbol that Sister Leontius published. It is possible that it is a typographical error. As figure 2.86 depicts the two dots, though not centered, are directly above/below each other.

There are addition and multiplication tables, and the additions are performed downwards. Hemeling checks by the inverse operations, and does not recommend the check by nines.

Again Hemeling uses ÷ for minus,[45] and we still find this redundant symbol of subtraction in the nineteenth-century revised edition. Occasionally we find him using the Maltese cross for plus (pp. 381, 478, 483.), but the usual sign is +. Hemeling deals extensively with denominate numbers, and his book is largely commercial in character.

2.17 English: Jonas Moore (No. 63.) (1660, 1681)

Moor's Arithmetick, as it is called, had three editions. All of them contain vulgar arithmetic, including the use of Napier's rods, logarithms, and decimals, besides algebra. Moore also wrote several other mathematical works, but our discussion is largely confined to *Moor's Arithmetick in Two Books* (London, 1660), which abounds in notation much like the works of Oughtred, his contemporary. Besides the seven references made by Cajori to Moore's 1660 and 1688 editions, many other symbols are worthy of note. Moore treats the fundamental operations in integers, decimals, and denominate numbers simultaneously. After proportion, fractions, and algebra, follow two treatises, the *Ellipsis* and *Conical Sections,* and finally a table of the squares and cubes of numbers under 100, the fourth powers of numbers under 300, and the fifth and sixth powers under 200. Zero is called *cypher* (p. 2.)[2.88],

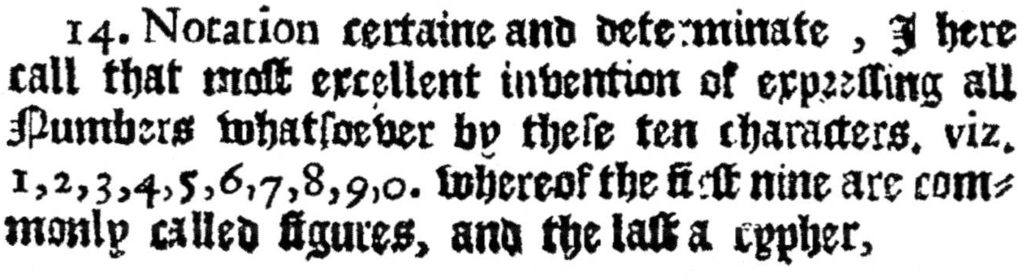
14. Notation certaine and determinate, I here
call that moſt excellent invention of expreſſing all
Numbers whatſoever by theſe ten characters, viz.
1,2,3,4,5,6,7,8,9,0. whereof the firſt nine are com-
monly called figures, and the laſt a cypher,

FIGURE 2.88: Moore's zero. No. 63, 1660, v. 1, p. 2

and large numbers are divided into groups of three by a colon, thus: 1:758:527:821 (p. 4.) [2.89].

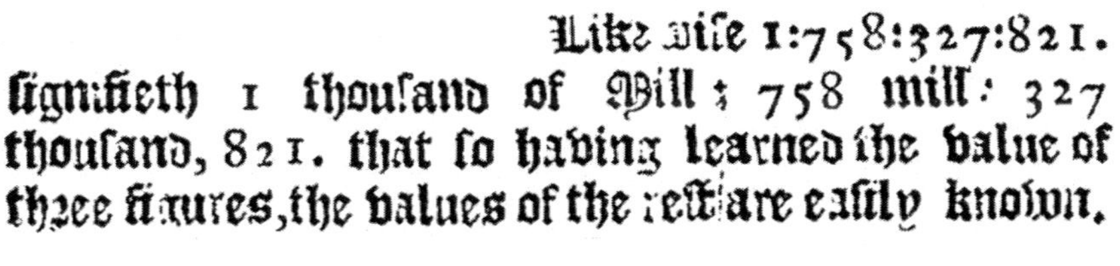
Like wiſe 1:758:327:821.
ſignifieth 1 thouſand of Mill; 758 mill: 327
thouſand, 821. that ſo having learned the value of
three figures, the values of the reſt are eaſily known.

FIGURE 2.89: Moore's grouping of digits. No. 63, 1660, v. 1, p. 4

[45] See Cajori, *Notations,* p. 242.

Moore briefly speaks of casting out 9's as a check for addition. Various ways of writing the decimal fraction 245.1234 are given by Moore (pp. 10, 11.) [2.90]:

FIGURE 2.90: Moore's decimal fractions. No. 63, 1660, v. 1, p. 10

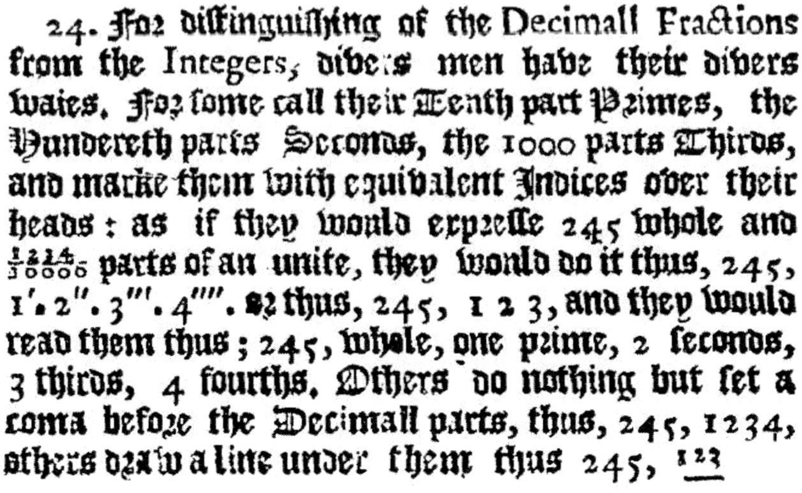
24. For distinguishing of the Decimall Fractions from the Integers, divers men have their divers waies. For some call their Tenth part Primes, the Hundereth parts Seconds, the 1000 parts Thirds, and marke them with equivalent Indices over their heads: as if they would expresse 245 whole and $\frac{1234}{10000}$ parts of an unite, they would do it thus, 245, 1'. 2''. 3'''. 4''''. or thus, 245, 1 2 3, and they would read them thus; 245, whole, one prime, 2 seconds, 3 thirds, 4 fourths. Others do nothing but set a comma before the Decimall parts, thus, 245, 1234, others draw a line under them thus 245, 123

but the 'best and most distinct way' is by means of a 'separatrix, a separating line', thus [2.91]:

FIGURE 2.91: Moore's separatrix. No. 63, 1660, v. 1, p. 12

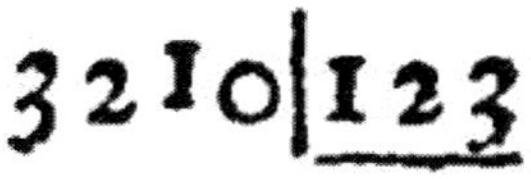

In some of the work following, however, Moore uses an elevated and inverted comma instead (p. 20.) [2.92],

FIGURE 2.92: Moore's grouping of digits. No. 63, 1660, v. 1, p. 20

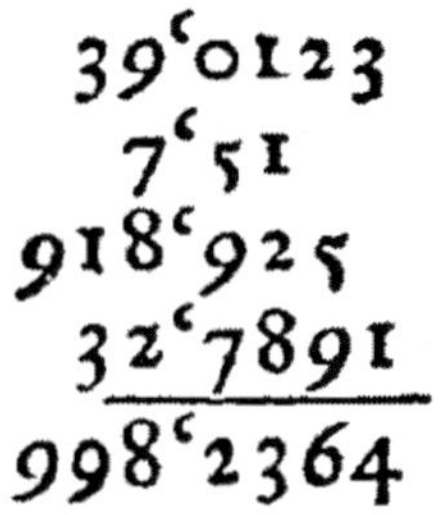

which is the only decimal symbol which Cajori attributes to him. He gives the sign of addition as St. George's cross, elevated and small (p. 16.), thus [2.93]:

36. The signe of Addition or Affirmation is + (or pl. for plus more) as, + A, whereby is signified that any magnitude or number noted by A, is affirmed to be, and to be added, so + 34 more by 34.

FIGURE 2.93: Moore's addition. No. 63, 1660, v. 1, p. 16

The sign of multiplication is referred to as St. Andrew's cross, but it is sometimes made much like the plus sign, whereas throughout most of his work Moore correctly uses the small elevated $^{+}$, thus [2.94]:

39. The signe of Multiplication (if any be used) is this +, the difference of this signe from the signe of Addition is, that of Addition is St. Georges Crosse, this is St. Andrews, but most commonly letters are joyned together by Multiplication without a signe.

FIGURE 2.94: Moore's multiplication. No. 63, 1660, v. 1, p. 16

Cajori (p. 197.) refers to this edition (p. 108.) for his use of the letter x for multiplication, but Moore uses no symbol whatever for multiplication on the indicated page.

The symbol for subtraction is first represented as a very small horizontal line, sometimes much longer, and at other times as two short successive dashes (pp. 16, 84, 85.) [2.95].

$$100 \times 2 = 200 -\!- 2^{+}2 = 200 \times 100 = \frac{20000}{2} = 10000.$$

FIGURE 2.95: Moore's subtraction. No. 63, 1660, v. 1, p. 85

Moore also uses ± for *plus* and *minus* (p. 191.) [2.96].

$$\frac{2}{5} \pm \frac{1}{5} = \frac{3}{5} \text{ or } \frac{1}{5}.\ \frac{2}{7} + \frac{4}{7} + \frac{6}{7} = \frac{12}{7} \text{ or } \frac{5}{7}.\ \frac{54}{60} \pm \frac{13}{60} = \frac{67}{60}.$$

FIGURE 2.96: Moore's plus and minus. No. 63, 1660, v. 1, p. 191

Bypartition (mediation) is demonstrated (p. 26.) from left to right [2.97],

consist of uneven Numbers, if you begin from the left hand, taking the half of it, augment the figure following by 10. if the figure before were odde, and so go on.

$$\frac{4862}{2431} \qquad \frac{47892}{23946} \qquad \frac{9011780}{4505890} \qquad \frac{2350}{1175}$$

FIGURE 2.97: Moore's mediation. No. 63, 1660, v. 1, p. 26

but the author states (p. 27.) that 'it will be of most speed to do it from the right hand.' [2.98]

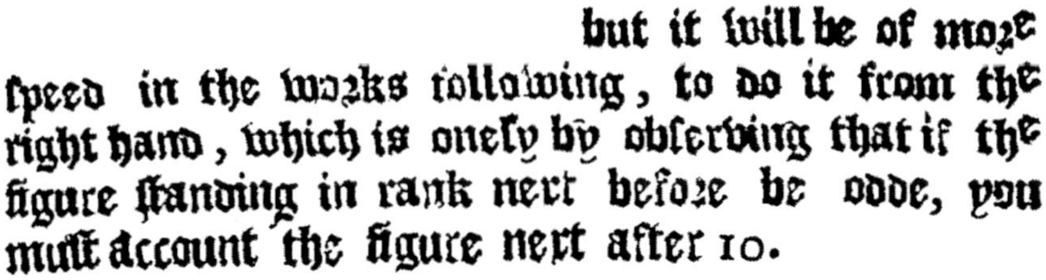
but it will be of more ſpeed in the works following, to do it from the right hand, which is onely by obſerving that if the figure ſtanding in rank next before be odde, you muſt account the figure next after 10.

FIGURE 2.98: Moore's mediation. No. 63, 1660, v. 1, p. 27

There are illustrations of *duplication*, *triplication*, and *reduplication* (pp. 26, 27, 28.). Moore relates that he taught his pupils to remember multiplication by 5, 6, 7, 8, and 9, as follows (p. 32.) [2.99]:

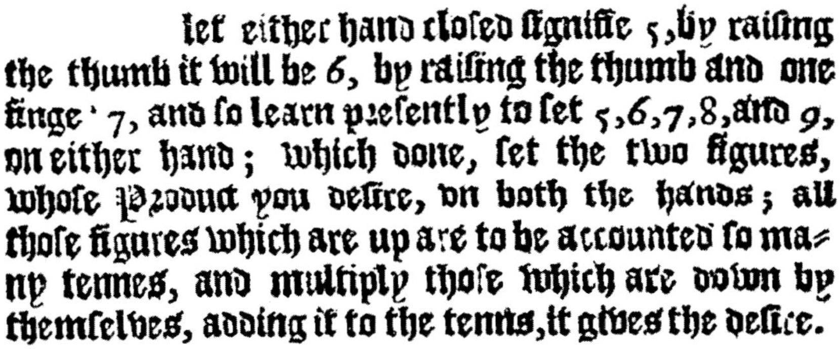
let either hand cloſed ſignifie 5, by raiſing the thumb it will be 6, by raiſing the thumb and one finge' 7, and ſo learn preſently to ſet 5, 6, 7, 8, and 9, on either hand; which done, ſet the two figures, whoſe Product you deſire, on both the hands; all thoſe figures which are up are to be accounted ſo many tennes, and multiply thoſe which are down by themſelves, adding it to the tenns, it gives the deſire.

FIGURE 2.99: Moore's multiplication. No. 63, 1660, v. 1, p. 32

> "let either hand closed signifie 5, by raising the thumb it will be 6, by raising the thumb and one finger 7, ...; which done, set the two figures, whose Product you desire, on both the hands; all those figures which are up are to be accounted so many tennes, and multiply those which are down by themselves, adding it to the tennes, it gives the desire."

This is a form of complementary multiplication still practiced by the Wallachians at the end of the nineteenth century.[46]

In this work dots are placed below [2.100/Left] the numbers in the dividend as soon as they are carried down, but dots above [2.100/Right] are used to mark the periods for square root (pp. 48, 65.).

[46] P. Treutlein, "Das Rechnen im 16. Jahrhundert", Abhandlungen zur Geschichte der Mathematik, v. 1, p. 48, refers to an article by Dr. Pick (1874) in which similar instructions are given for multiplication. This process may also be found in the treatises of the thirteenth century.

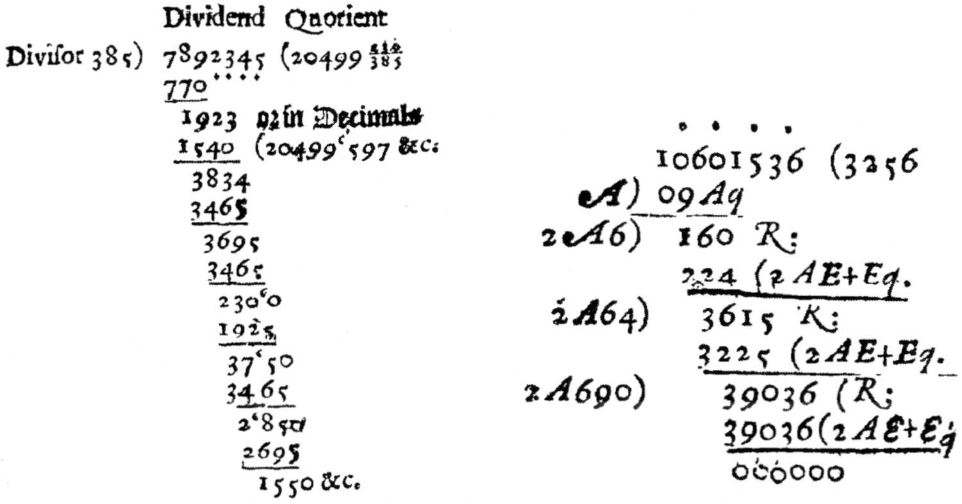

FIGURE 2.100: Moore's dots used in division and finding square roots. No. 63, 1660, Left: v. 1, p. 48 Right: v. 2, p. 65

This edition (p. 2 of the 2nd. part) also uses a ∺ symbol for the continual proportion, thus [2.101]:

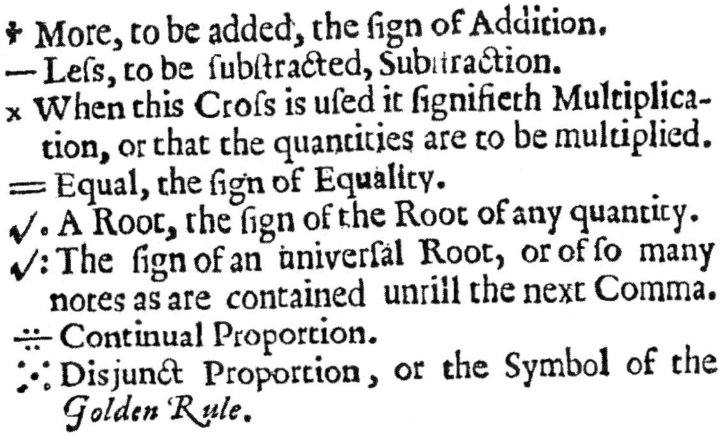
+ More, to be added, the ſign of Addition.
— Leſs, to be ſubſtracted, Subtraction.
× When this Croſs is uſed it ſignifieth Multiplication, or that the quantities are to be multiplied.
= Equal, the ſign of Equality.
√. A Root, the ſign of the Root of any quantity.
√: The ſign of an univerſal Root, or of ſo many notes as are contained untill the next Comma.
∺ Continual Proportion.
∷ Disjunct Proportion, or the Symbol of the *Golden Rule*.

FIGURE 2.101: Moore's symbols. No. 63, 1660, v. 2, p. 2

The powers of the root are represented in various ways (p. 7.) [2.102],

FIGURE 2.102: Moore's symbols. No. 63, 1660, v. 2, p. 7

8.	7.	6.	5.	4.	3.	2.	1.	0	$\overline{1}$.	$\overline{2}$.	$\overline{3}$.	$\overline{4}$.	$\overline{5}$.	$\overline{6}$.	$\overline{7}$.	$\overline{8}$.	1
11^{VIII}.	75^{VII}.	89^{VI}.	32^{V}.	$91''''$.	$75'''$.	$15''$.	$12'$.	1	$12'$.	$15''$.	$75'''$.	$91''''$.	32^{V}.	89^{VI}.	75^{VII}.	11^{VIII}.	2
12^{VIII}.	18^{VII}.	9^{VI}.	12^{V}.	$38''''$.	$56'''$.	$39''$.	$15'$.	1	$15'$.	$39''$.	$56'''$.	$38''''$.	12^{V}.	9^{VI}.	18^{VII}.	12^{VIII}.	3
256.	128.	64.	32.	16.	8.	4.	2.	0	$\frac{1}{2}$.	$\frac{1}{4}$.	$\frac{1}{8}$.	$\frac{1}{16}$.	$\frac{1}{32}$.	$\frac{1}{64}$.	$\frac{1}{128}$.	$\frac{1}{256}$.	4
ſquared ſq.ſquare	2. ſurſolid	ſquare cube	ſurſolid	ſquared ſquare	cube	ſquare	root		root	ſquare	cube	ſq.ſquare	ſurſolid	ſq.Cube	2. ſurſolid	ſq.ſq.ſquare	5
QQQ.	2S.	QC.	S.	QQ.	C.	Q.	R	0	R.	Q.	C.	QQ.	S.	QC.	2S.	QQQ.	6
Aqcc.	Aqqc.	Acc.	Aqc.	Aqq.	Ac.	Aq.	A.	0	A.	Aq.	Ac.	Aqq.	Aqc.	Acc.	Aqqc.	Aqcc.	7
	&c.	aaaaaa.	aaaaa.	aaaa.	aaa.	aa.	a.	0	a.	aa.	aaa.	aaaa.	&c.				8
a^8.	a^7.	a^6.	a^5.	a^4.	a^3.	aa.	a.	0	a,	ca.	a^3.	a_4.	a^5.	a^6.	a^7.	a^8.	9

by means of the great letters, the lesser characters, and by exponents, according to the way of Oughtred, of Harriot, and of Descartes, respectively; for example, the third power of A is written Ac, aaa, and a^3. The square root of any quantity is expressed by $\surd$, cube root $\surd$c, fourth root $\surd$qq, etc., according to Oughtred (p. 11.). Moore gives a list (p. 12.) of the letters appointed by him as symbols for proposed quantities, as follows [2.103]:

FIGURE 2.103: Moore's symbols. No. 63, 1660, v. 1, p. 12

As if *A* and *E* ſignifie two numbers: let *Z* ſignifie the ſumme of theſe numbers, *X* the difference, *Æ* the rectangle or multiplication $\frac{r}{s}$ the proportion, *K*, the ſumme of the ſquares, *L* the difference of the ſquares, *M* the ſumm of the cubes, and *N* the difference of the cubes of them.

Later (p. 72.) he gives almost similar symbols in the lesser characters [2.104].

THe general Varieties go upon theſe Queſtions, that having any two of the numbers mentioned *Chap.* 1. *Sect.* 16. thereof, *a*, *b*, *z*, *x*, *p*, $\frac{r}{s}$ *m*, *n*, to find out any of the reſt.

Queſt. Having two nunbers given, whereof *a* the greater and *b* the leſſer, to find their ſumme, difference, rectangle, proportion, ſumme of ſquares, and differences of ſquares, *z*, *x*, *p*, $\frac{r}{s}$ *aa*, *bb*, *m* & *n*.

FIGURE 2.104: Moore's symbols. No. 63, 1660, v. 2, p. 72

A few of the more important variations in another work of Moore (1681) are well worth noting. In his *A New Systeme of the Mathematicks,* in which is included 'Arithmetick, as well Natural and Decimal, as in Species, or the Principles of Algebra', Moore arranges the nine digits in decreasing order. In this work the dot displaces the colon in grouping the digits in numeration. More noteworthy is the fact that in this later work Moore consistently uses the dot as a decimal symbol. The comma is used only in the case of logarithms. These additional signs are listed: [2.105] (division), [2.106] (involution), and [2.106] (evolution), all used in Brancker's translation of Rahn (1668).

Diviſion in Single whole Quantities.

Its *Sign* is [$-\!\!\!\div$] *i. e.* divided by.

FIGURE 2.105: Rahn's division. No. 73, 1668, p. 7

Involution in Single whole Quantities.

Its *Sign* is [$\circledast$]

FIGURE 2.106: Rahn's involution. No. 73, 1668, p. 8

Evolution, or Extraction of Roots out of Single whole Quantities.

The *Sign* of this Operation is [ω]

FIGURE 2.107: Rahn's evolution. No. 73, 1668, p. 8

2.18 Swiss: John Alexander (No. 1) (1693)

The *Synposis Algebraica,* a posthumous work in Latin by John Alexander, appeared at London in 1693; it was translated into English by Sam. Cobb sixteen years later. Like the works of Vieta and Oughtred, to whom he refers (pp. 3, 23.), this book is important to us here because of its notations. Alexander uses + and −, but not × or ÷. He gives = or Descartes' sign ∞ as symbols for equality (p. 2.), but uses the former throughout his work [2.108].

FIGURE 2.108: Alexander's equality. No. 1, 1693, p. 2

Pro denotanda autem æqualitate [quæ in Algebra utramque paginam facit] adhiberi ſolent hæc ſigna; = vel ∞; ut $2b=3a$; hoc eſt, $2b$ æqualia $3a$, &c.

Cajori[47] says that he used ·· for therefore, whereas we find it was ·.· . Probably an omission in the printing (p. 7.) accounts for this error [2.109].

FIGURE 2.109: Alexander's 'therefore'. No. 1, 1693, p. 7

ex $4a-2c$	L. ſit $a=9$	∵ $4a-2c=32$
ſub. $2a-6c$	$c=2$	& $2a-6c=6$
reſt. $2a+4c$		&, $2a+4c=26$

This author writes the proportion a.b::b.x or a ∺ b ∺ c (p. 16.) [2.110],

FIGURE 2.110: Alexander's proportion. No. 1, 1693, p. 16

	(1)		(2)		(2)		(3)
	a.		b::		b.		X
hoc eſt, ut	a	ad	b:	ita	b	ad	X

	(1)		(2)		(3)		(4)
	a,		b::		c.		X
ut	a	ad	b:	ita	c	ad	X

Vel etiam ſic;

a ∺ b: b ∺ X

a ∺ b: c ∺ X

[47] Cajori, *Notations,* v. II, p. 282.

although Cajori states that he expressed geometrical ratio by a dot, a.b, and also by 4 a ∺ b.[48] In other places, however, Cajori refers to the correct symbol ∹ (*Ibid.,* pp. 287, 289.).

Whereas others designate involution and evolution with the Archimedean spiral and the double epsilon (⊚ and ϵϵ), Alexander says he uses exponents (p. 19.) [2.111].

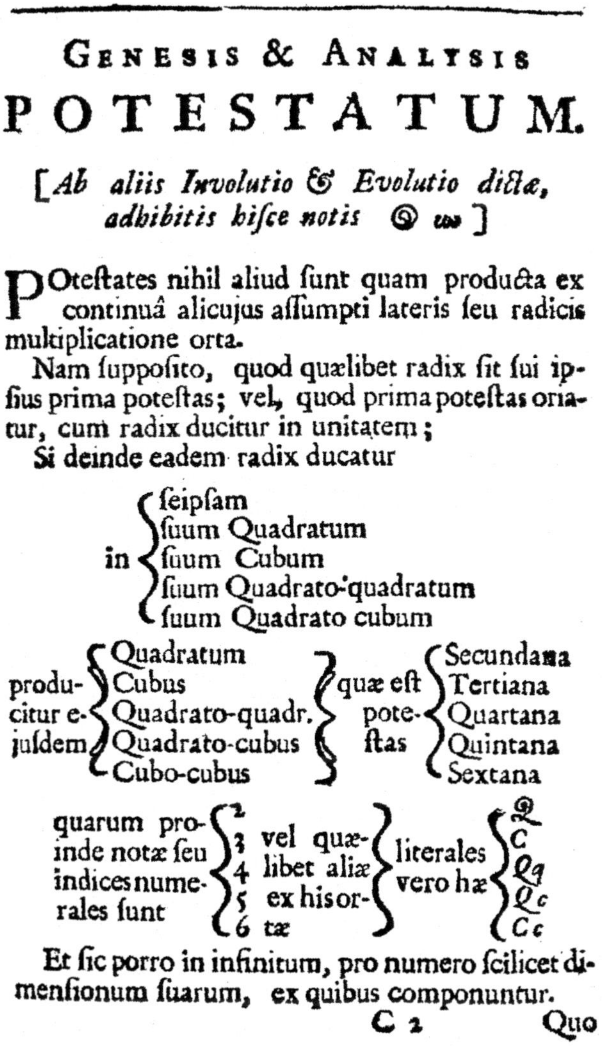

GENESIS & ANALYSIS
POTESTATUM.

[*Ab aliis Involutio & Evolutio dictæ, adhibitis hiſce notis* ⊚ ϵϵ]

POteſtates nihil aliud ſunt quam producta ex continuâ alicujus aſſumpti lateris ſeu radicis multiplicatione orta.

Nam ſuppoſito, quod quælibet radix ſit ſui ipſius prima poteſtas; vel, quod prima poteſtas oriatur, cum radix ducitur in unitatem;

Si deinde eadem radix ducatur

in { ſeipſam / ſuum Quadratum / ſuum Cubum / ſuum Quadrato-quadratum / ſuum Quadrato cubum }

producitur eiuſdem { Quadratum / Cubus / Quadrato-quadr. / Quadrato-cubus / Cubo-cubus } quæ eſt poteſtas { Secundana / Tertiana / Quartana / Quintana / Sextana }

quarum proinde notæ ſeu indices numerales ſunt { 2 / 3 / 4 / 5 / 6 } vel quælibet aliæ ex his ortæ } literales vero hæ { Q / C / Qq / Qc / Cc }

Et ſic porro in infinitum, pro numero ſcilicet dimenſionum ſuarum, ex quibus componuntur.

C 2 Quo

FIGURE 2.111: Alexander's symbols. No. 1, 1693, p. 19

[48]Cajori, *Notations,* v. I, p. 276.

However, square, cube, and fourth roots are written $\sqrt{}$q., $\sqrt{}$c., and $\sqrt{}$qq., respectively (pp. 24, 28, 32.) [2.112, 2.113, 2.114].

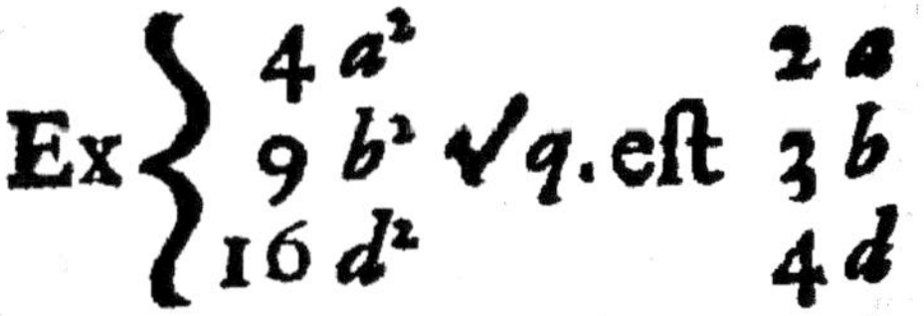

FIGURE 2.112: Alexander's square root symbol. No. 1, 1693, p. 24

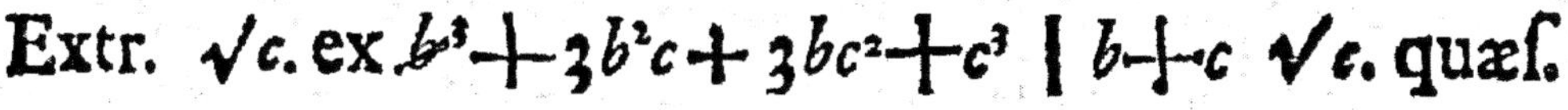

FIGURE 2.113: Alexander's cube root symbol. No. 1, 1693, p. 28

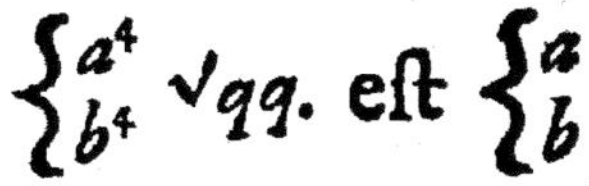

FIGURE 2.114: Alexander's fourth root symbol. No. 1, 1693, p. 32

To express the square root of a quantity he makes use of the vinculum with the radical sign, in this manner (p. 35.): $\sqrt{a^2 + b^2}$ [2.115].

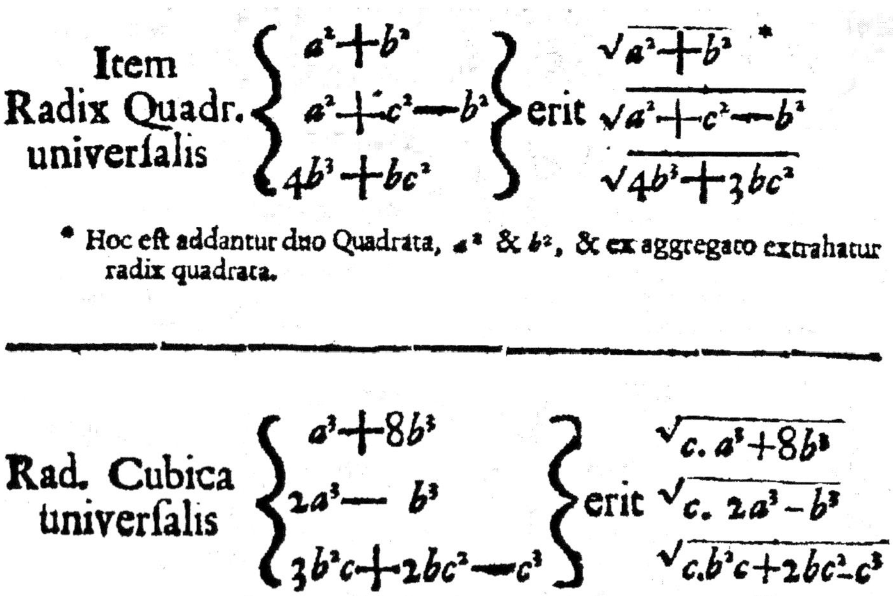

FIGURE 2.115: Alexander's radical sign with vinculum. No. 1, 1693, p. 35

Chapter 3

Topical Survey of the Use of Notations

3.1 Notation and Numeration

3.1.1 Finger and counter reckoning

The finger symbolism of number is found in the works of Baeda, 'The Venerable Bede' (c. 673–735), was practiced in both the East and West during the Middle Ages, and is mentioned by several sixteenth-century arithmeticians. The 1491 Calandri (No. 14.) [3.1] was probably the first printed work to illustrate finger symbolism [3.2] (not the 1494 Pacioli as given in Smith's *Rara,* p. 56).

FIGURE 3.1: Calandri's title page of *Pictagoras arithmetrice introductor.* No. 14, 1491, f. 1, v.

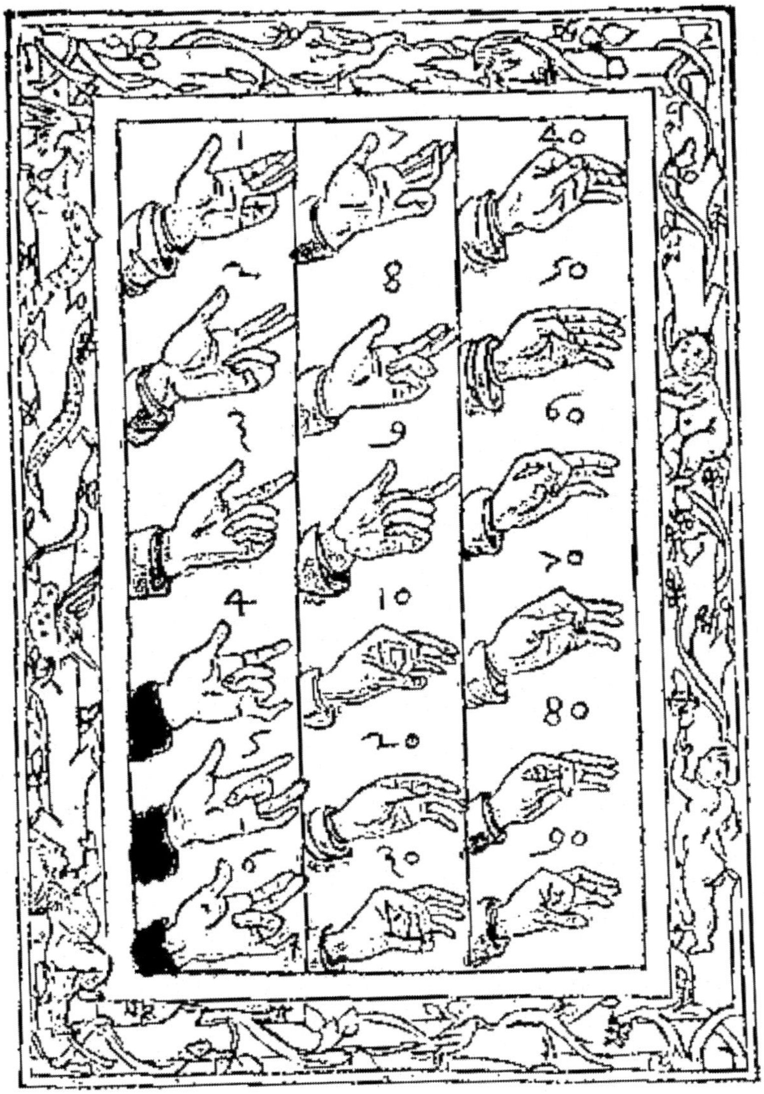

FIGURE 3.2: Calandri's finger reckoning. No. 14, 1491, f. 3, v.

Aventinus wrote a most complete treatise (*Abacus vetustissima veterum latinorum per digitos* . . . , Regensburg, 1522.) on numerical finger symbolism using both fingers and arms. (See P. Treutlein, "Das Rechnen im 16. Jahrhundert", *Abhandlungen zur Geschichte der Mathematik,* v. 1, p. 22, for a discussion of this work.)

The treatment of finger symbols by Tagliente (1520) has been noted above.[1] Another writer who discusses this subject is Jan Bronckhorst of Holland, called Noviomagus[2] from his birthplace. In his *De numeris libri II* (No. 65.) the first book is 'de Logistice' or 'de arte computandi.' There is an illustration of finger reckoning (f. 19, v.) [3.3],

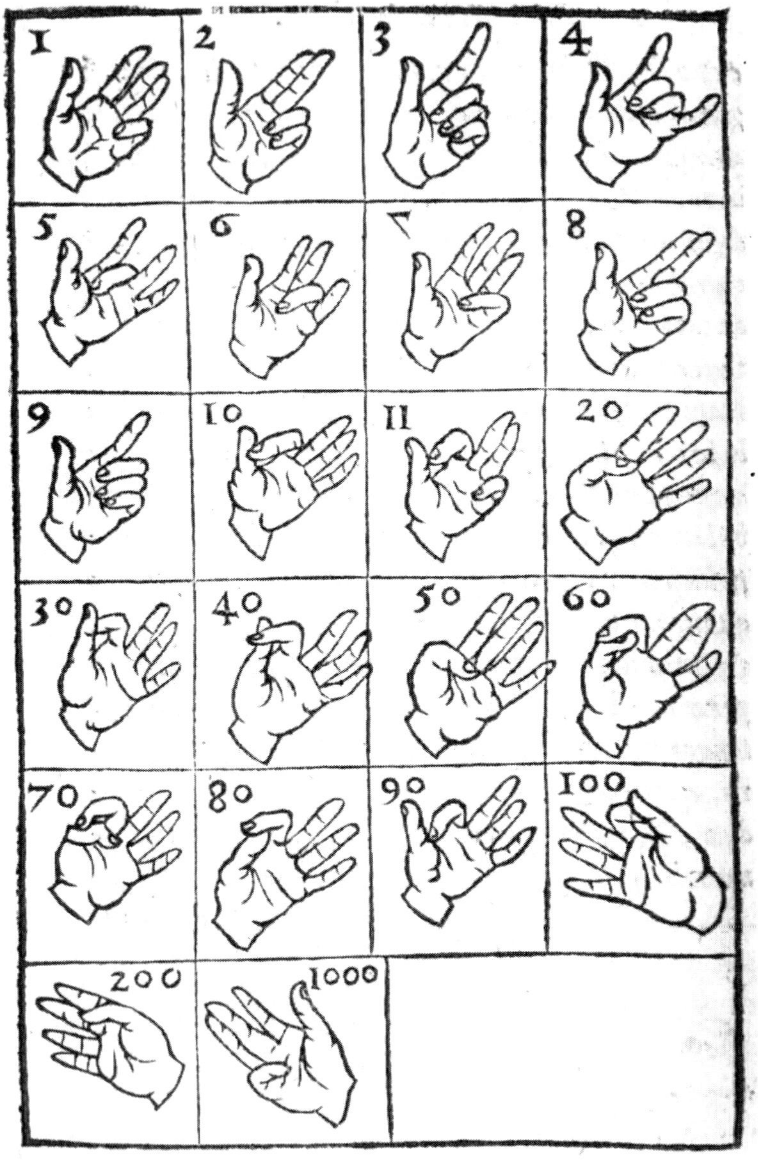

FIGURE 3.3: Noviomagus' finger reckoning. No. 65, 1544, f. 19, v.

practically the same as that described by Pacioli, with the exception of the symbols for 100, 200, and 1000.

[1] ED. See Figures 2.21 and 2.22 on page 70 as well as Section 2.4 on page 69.

[2] ED. Jan van Bronckhorst (1494–1570), Neomagus or Noviomagus is the Dutch city Nijmegen.

As to counter or abacus reckoning, not only Husswirt (1501) (mentioned above[3]) but also Reisch (No. 77.) [3.4], Blasius (No. 8.), and Feme (No. 31.) give early treatments.

FIGURE 3.4: **Reisch's counter reckoning. No. 77, 1503, f. 79, v.**

[3]ED. See section 2.1 on page 63.

The titles such as *Rechenbiechlin auf den linien* by Köbel (1514) [3.5], and *Rechnung auff der Linien und Federn* by Riese (1522) indicate the work on the lines.[4] It was not uncommon to find the title-page bearing a woodcut showing computing with counters, as in Köbel's *Rechenbiechlin* (1514) [3.5], and Recorde's *Ground of Artes* (1558) [3.6].

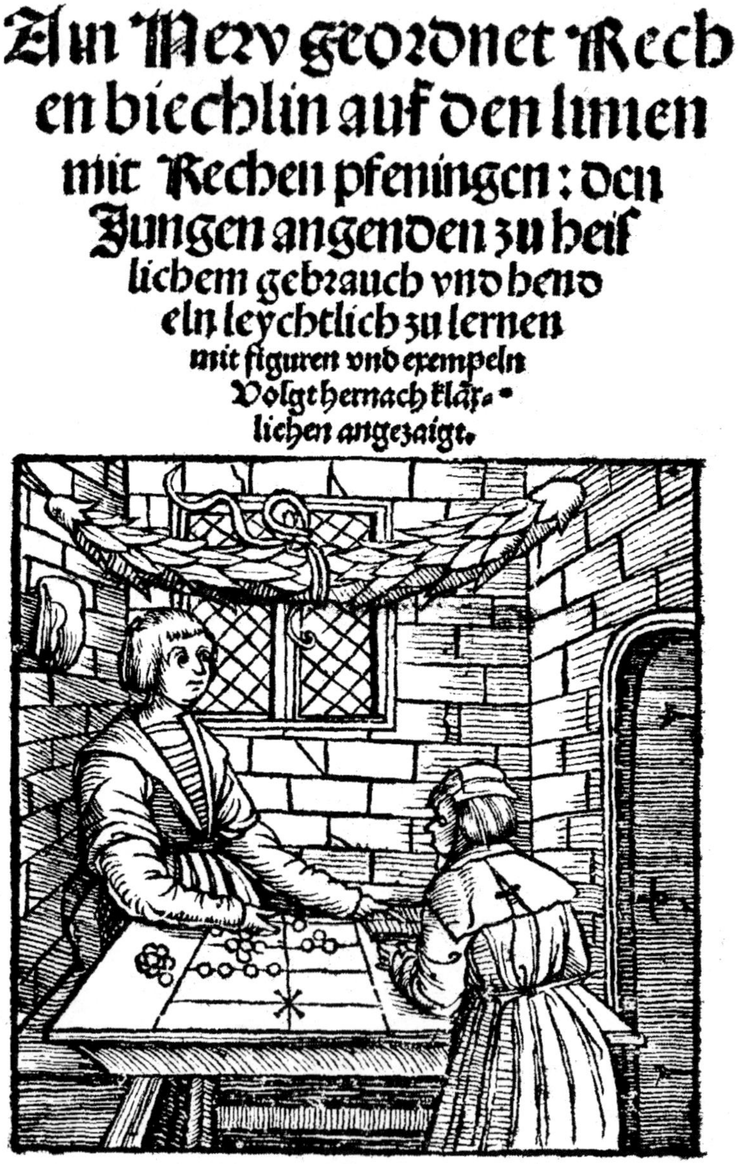
Ain New geordnet Rech
en biechlin auf den linien
mit Rechen pfeningen: den
Jungen angenden zu heiſ
lichem gebrauch vnd hend
eln leychtlich zu lernen
mit figuren vnd exempeln
Volgt hernach klär=
lichen angezaigt.

FIGURE 3.5: Köbel's title page of *Rechenbiechlin,* 1514, Augsburg.

[4] ED. 'work on the lines' refers to calculating (reckoning) on a table marked with lines. The lines represent the various, non-negative, powers of ten. This table is the counter or reckoning abacus mentioned above.

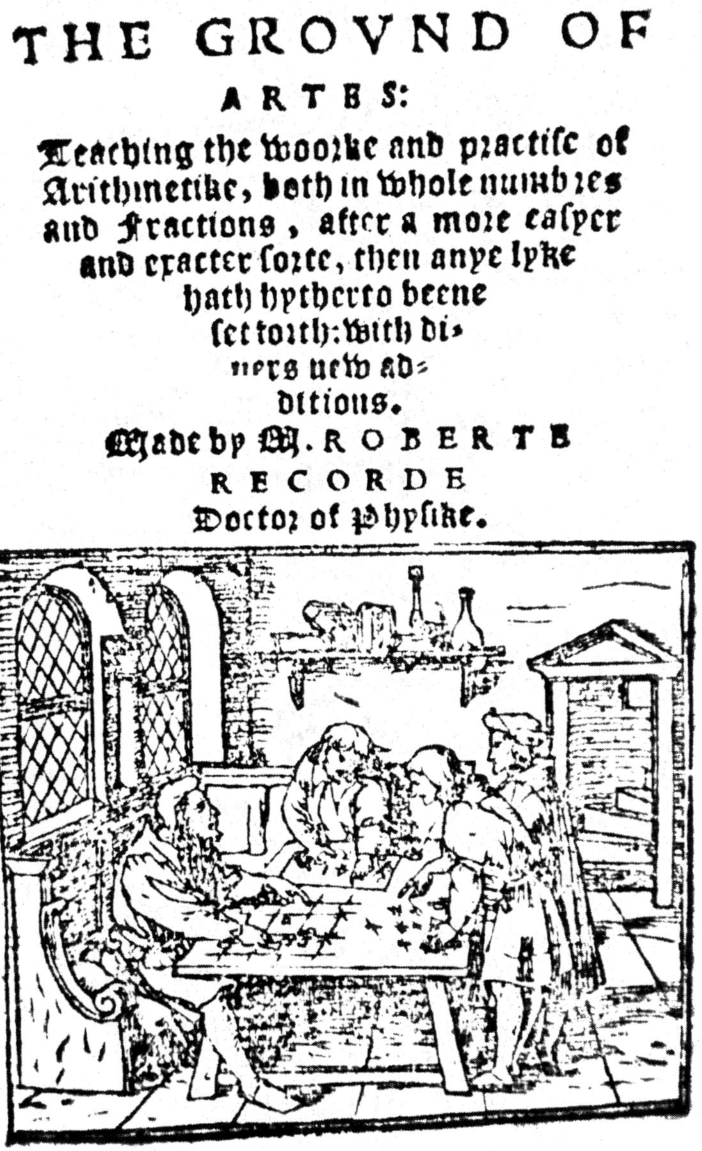
THE GROVND OF
ARTES:
Teaching the worke and practise of
Arithmetike, both in whole numbres
and Fractions, after a more easyer
and exacter sorte, then anye lyke
hath hytherto beene
set forth: with di-
uers new ad-
ditions.
Made by M. ROBERTE
RECORDE
Doctor of Physike.

FIGURE 3.6: Recorde's title page of *The Ground of Artes,* 1558, London.[5]

[5] ED. From the Plimpton Collection and used with the permission of the Rare Book & Manuscript Library, Columbia University.

Noviomagus (No. 65, ff. 32, 36.) [3.7] also briefly performs all fundamental operations upon the line abacus.

FIGURE 3.7: Noviomagus' counters. No. 65, 1544, f. 36, r.

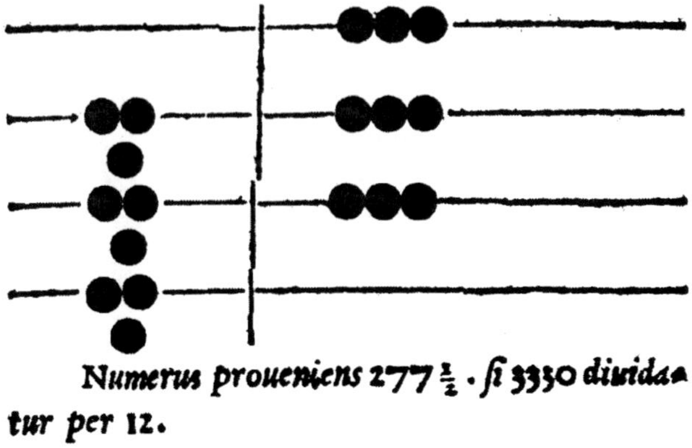

The German Ulricus Regius (No. 76, ff. 94–100.) closes his book with several folios on the use of counters [3.8].

FIGURE 3.8: Regius' counters. No. 76, 1536, f. 91, r.

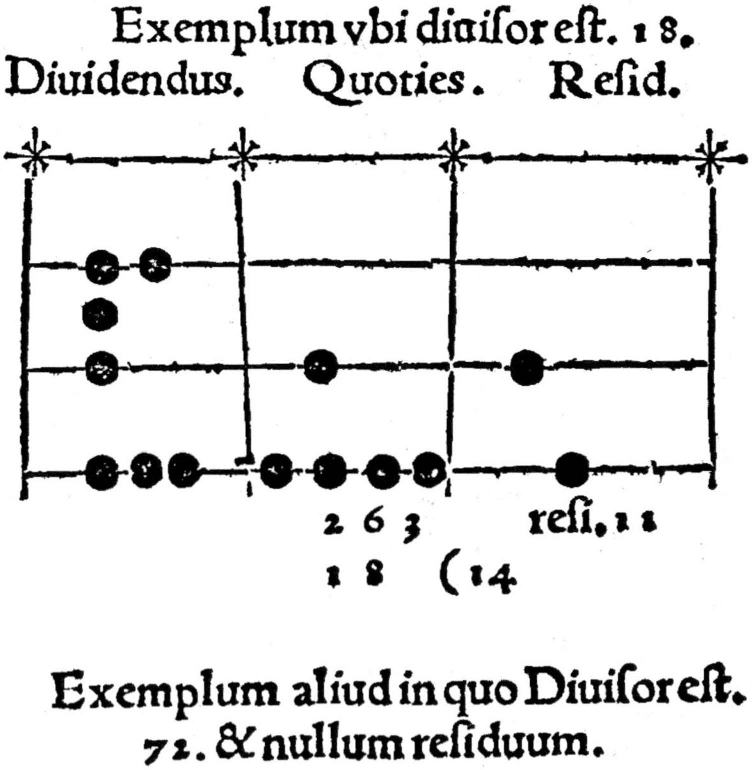

Johann Scheubel (No. 83, pp. 13, 14, 25, 31, 43–45, 54, 55.) speaks of numeration and of the fundamental operations 'per calculos.' [3.9]

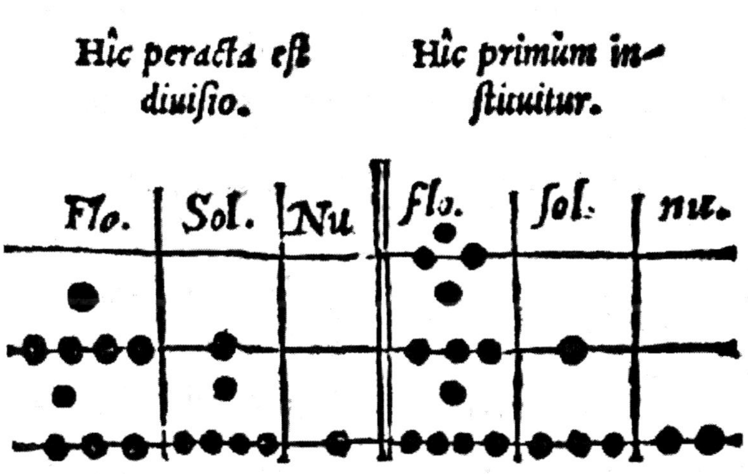

FIGURE 3.9: Scheubel's counters. No. 83, 1560, p. 55

In the early seventeenth century Sesen (No. 84, p. 5.) gives this topic extensive treatment.

From such works we see that line reckoning was common in the first half of the sixteenth century, especially in Germany and Holland. One of its last serious advocates is Lucas à S. Edmundo, who as late as 1697, devotes a chapter (No. 58 , pp. 214–220) to it, entitled 'Arithmetica Calculatoria', with illustrations of counter reckoning [3.10].

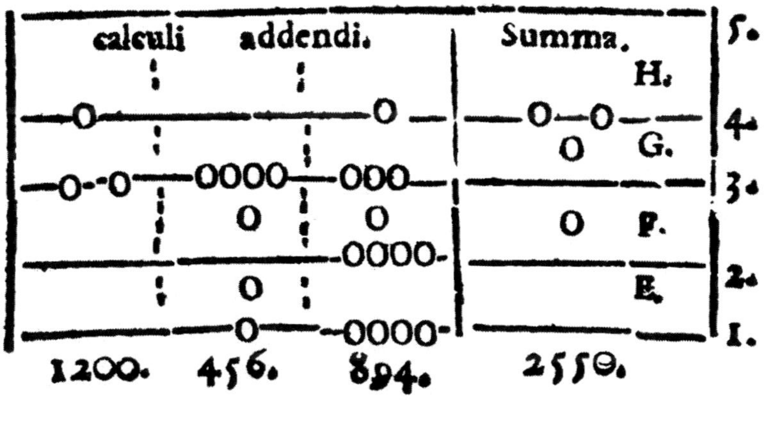

FIGURE 3.10: Lucas' counter reckoning for addition. Note the misplaced counter in the column above 894. No. 58, 1697, p. 216

A detailed account of the method of casting with counters (jettons) as found in Legendre's work of 1753 (No. 57.)

3.1.2 Roman numeral symbols

Blondel (No.9, Tome Second, pp. 56–59.), in his chapter on numeration of the Romans, speaks of the probable origin of the characters of the Roman notation, X, C, M, V,

L, and D. X, he says, was formed by crossing the line 1 when, in counting, people came to the end of their fingers. V is given as the upper part of a bisection of X; and L as the lower half of 100 made in this manner ⊏ (seen on ancient inscriptions); and D was the right half of the symbol for 1000 [3.11].

FIGURE 3.11: **Blondel's Roman numerals. No. 9, 1683, v. 2, p. 56**

CIↃ

There is frequent use of the principle of subtraction in the case of IX, rarely of IV. We find the use of IX and IIII in Regius (No. 76, f. 49, v.) and Steinmetz (No. 86, f. 11, v.). The same notation appears in Noviomagus (No. 65, f. 20, r.), Vuelpius (No. 98, f. 6, r.), and as late as 1695 in Behme (No. 7, p. 3.) [3.18]. The latter writer gives, however, 34 as XXXIV [3.18]. The use of this principle is extended to 8 as IIX and 80 as XXC by Waser (No. 99, f. 9.) [3.12] and Lucas (No. 58, f. 8, r.) [3.13], respectively, to 400 as CD, and to 900 as C∞ by Waser (No. 99, f. 9.) [3.12].

FIGURE 3.12: **Waser's Roman numerals. No. 99, 1603, f. 9, v.**

IV *Quatuor.*
IIX *Octo.*
IX *Novem.*
XL *Quadraginta.*
VL *Quadraginta quinque.*
XXC *Octoginta.*
XC *Nonaginta.*
CD *Quadringenta.*
C∞ *Noningenta.*

FIGURE 3.13: **Lucas' Roman numerals. No. 58, 1697, f. 8, r.**

PROLUSIO II.

Variæ notæ, & expreſſiones numerorum.

Notæ vulgares Arithmeticæ ſunt: 1. 2. 3. 4. 5. 6 7. 8. 9. 0. Veteres Latini utebantur literis I. V. X. L. C. D. M. vocantur autem numeri Eccleſiaſtici.

I. Unum, II. duo, III. tria, IV. quatuor, V. quinque; VI. ſex, VII ſeptem, VIII. octo, IX. novem, X. decem, XX. viginti, XXX. triginta, XL. quadraginta, L. quinquaginta, LX. ſexaginta, LXX. ſeptuaginta, LXXX. vel XXC: octuaginta, XC. nonaginta, C. centum, D. ſeu IↃ. quingenta, M. vel CIↃ. mille, CM. centum millia, M̄. millio.

Blondel (No. 9, ibid.) uses both IIII and IV, VIII and IIX, VIIII and IX, XXXX and XL, and the forms LXXXX and XC. [3.14]

FIGURE 3.14: **Blondel's Roman numerals. No. 9, 1683, v. 2, p. 56**

CHAPITRE III.

Numeration des Romains.

CHAP. III. Numeration des Romains.

LEs Romains avoient encore d'autres caracteres pour marquer leurs nombres dont l'uſage eſt demeuré à la Chambre des Comptes & aux Finances du Roy. Les voici en lettres Majuſcules.

I	1	X	10	C ou Ɔ	100	M ou CIƆ	1000.
II	2	XX	20	CC ou IIC	200	XM	10000.
III	3	XXX	30	CCC ou IIIC	300	L M	50000.
IIII ou IV.	4	XXXX ou XL	40	CCCC ou IIIIC	400	C M	100000.
V	5	L	50	D	500	D M	500000.
VI	6	LX	60	DC	600		
VII	7	LXX	70	DCC	700		
VIII ou IIX	8	LXXX	80	DCCC	800		
VIIII ou IX	9	LXXXX ou XC	90	DCCCC	900		

He also mentions examples in ancient inscriptions where the subtractive principle is used in writing 8 as IIX and 18 as XIIX, but states that this is very rare.

The principle of multiplication, according to which IIM stands for 2,000, occurs in Noviomagus (No. 65, f. 20, v.), Vuelpius (No. 98, f. 7.), Steinmetz (No. 86, f. 12.) [3.15, 3.16], Lucas (No. 58, f. 8, r.) and in Blondel (No. 9, *ibid.*).

FIGURE 3.15: Steinmetz's Roman numerals. No. 86, 1575, f. 12, r.

ARITHMETICAE.

	Notæ Latinæ	Indicæ	Græcæ	
Centum	C	100	ρ	Numeri ἑκατονταδικοί.
Ducenta	CC	200	σ	
Trecenta	CCC	300	τ	
Quadringenta	CCCC	400	υ	
Quingenta	D	500	φ	
Sexcenta	DC	600	χ	
Septingenta	DCC	700	ψ	
Octingenta	DCCC	800	ω	
Nongenta	DCCCC	900 ϡ *uel* πι		
Mille	M	1000	Α	χιλιαδικοί
Duo millia	MM *uel* IIM	2000	ʹΒ	
Tria millia	MMM *uel* IIIM	3000	ʹΓ	
Quatuor millia	MMMM	4000	ʹδ	
Quinq; millia	VM	5000	ʹε	
Sex millia	VIM	6000	ʹϛ	
Septem millia	VIIM	7000	ʹζ	
Octo millia	VIIIM	8000	ʹη	
Nouem millia	IXM	9000	ʹθ	
Decem millia	XM	10000	α̈	Primus μυριονταδικῶν ἁπλῶν
Viginti millia	XXM	20000	β̈	
Triginta millia	XXXM	30000	γ̈	
Quadraginta millia	XLM	40000	δ̈	
Quinquaginta millia	LM	50000	ε̈	
Sexaginta millia	LXM	60000	ϛ̈	
Septuaginta millia	LXXM	70000	ζ̈	
Octuaginta millia	LXXXM	80000	η̈	

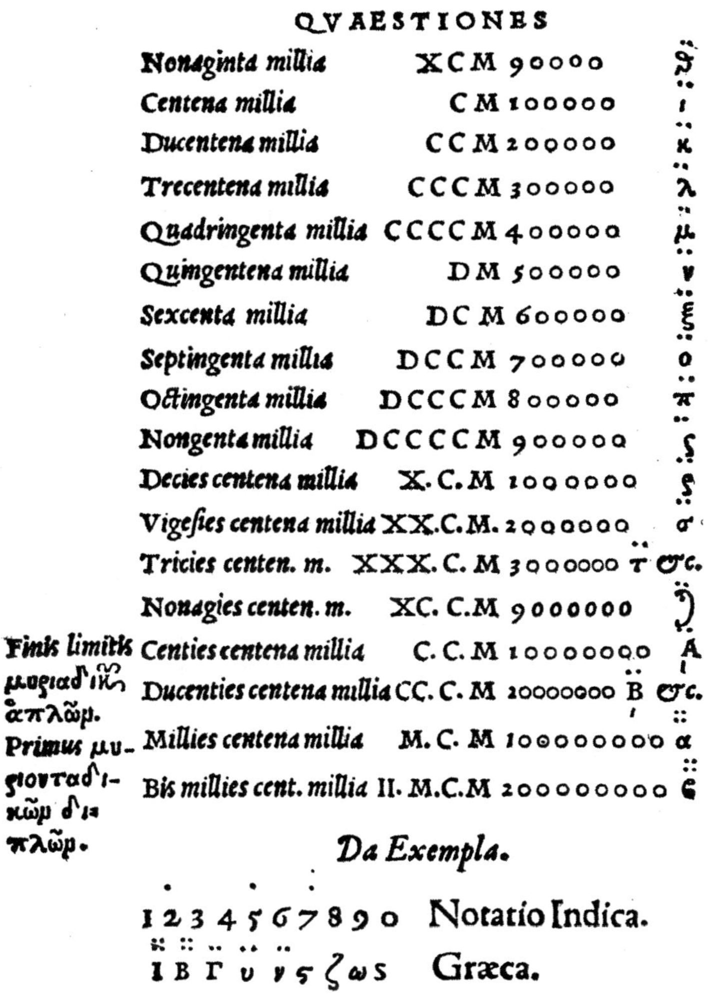

QVAESTIONES

Nonaginta millia	XCM	90000	ϑ
Centena millia	CM	100000	ι
Ducentena millia	CCM	200000	κ
Trecentena millia	CCCM	300000	λ
Quadringenta millia	CCCCM	400000	μ
Quingentena millia	DM	500000	ν
Sexcenta millia	DCM	600000	ξ
Septingenta millia	DCCM	700000	ο
Octingenta millia	DCCCM	800000	π
Nongenta millia	DCCCCM	900000	ϛ
Decies centena millia	X.C.M	1000000	ϛ
Vigesies centena millia	XX.C.M.	2000000	σ
Tricies centen. m.	XXX.C.M	3000000	τ &c.
Nonagies centen. m.	XC.C.M	9000000	ϡ
Centies centena millia	C.C.M	10000000	A
Ducenties centena millia	CC.C.M	20000000	B &c.
Millies centena millia	M.C.M	100000000	α
Bis millies cent. millia	II.M.C.M	200000000	ϐ

Finis limitis μυριαδικῶν ἁπλῶν. Primus μυριοντάδικῶν διπλῶν.

Da Exempla.

1 2 3 4 5 6 7 8 9 0 Notatio Indica.

Ι Β Γ υ ν ϛ ζ ω S Græca.

FIGURE 3.16: Steinmetz's Roman numerals. No. 86, 1575, f. 12, v.

The indication of the thousand-fold value of a number by a dash placed above has been noted previously[6] in the case of Figatelli. Monier (No. 62, pp. 3.) gives these two representations for 10,000, XM and $\overline{\text{X}}$ [3.17].

[6] ED. See Figures 2.76 and 2.77 on page 91.

FIGURE 3.17: Monier's Roman numerals. No. 62, 1693, p. 3

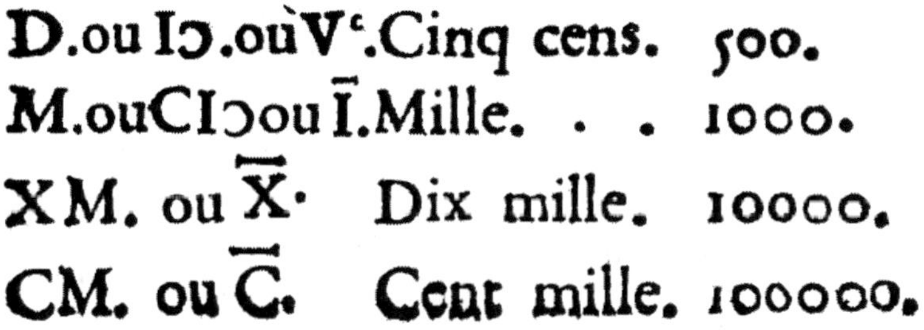

D.ou IↃ.ou V^{c}.Cinq cens. 500.
M.ou CIↃ ou $\bar{I}$.Mille. . . 1000.
XM. ou $\bar{X}$. Dix mille. 10000.
CM. ou $\bar{C}$. Cent mille. 100000.

Figatelli writes $\overset{\text{M}}{\text{VI}}$ for 6,000,[7] and Behme (No. 7, p. 3.) I^{m} for 1,000 [3.18].

FIGURE 3.18: Behme's Roman numerals. No. 7, 1695, p. 3.

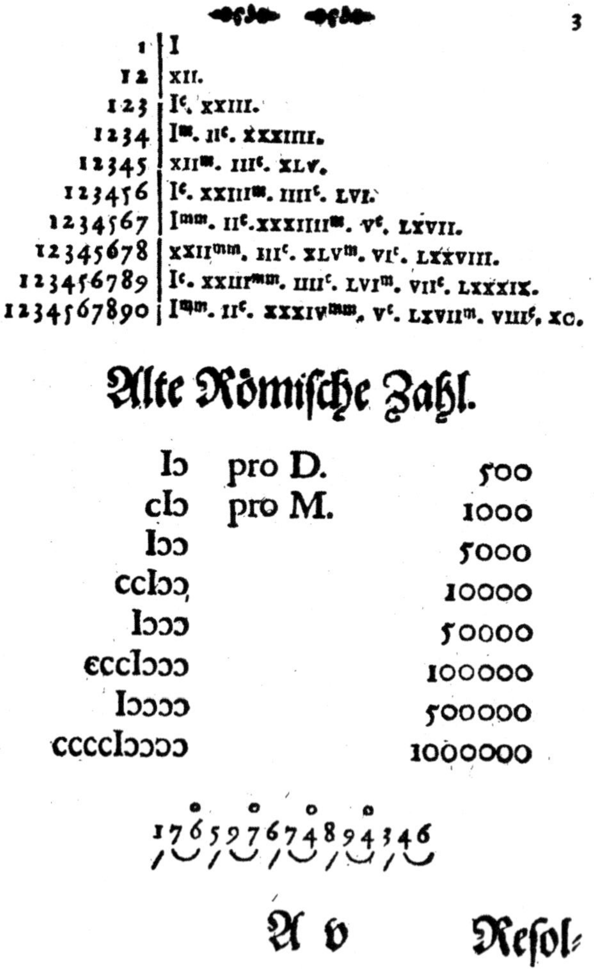

3

1	I
12	XII.
123	I^{c}. XXIII.
1234	I^{m}. II^{c}. XXXIIII.
12345	XII^{m}. III^{c}. XLV.
123456	I^{c}. XXIII^{m}. IIII^{c}. LVI.
1234567	I^{mm}. II^{c}.XXXIIII^{m}. V^{c}. LXVII.
12345678	XXII^{mm}. III^{c}. XLV^{m}. VI^{c}. LXXVIII.
123456789	I^{c}. XXIII^{mm}. IIII^{c}. LVI^{m}. VII^{c}. LXXXIX.
1234567890	I^{mm}. II^{c}. XXXIV^{mm}. V^{c}. LXVII^{m}. VIII^{c}. XC.

Alte Römische Zahl.

IↃ	pro D.	500
cIↃ	pro M.	1000
IↃↃ		5000
ccIↃↃ		10000
IↃↃↃ		50000
cccIↃↃↃ		100000
IↃↃↃↃ		500000
ccccIↃↃↃↃ		1000000

1765976748943 46

A v Resol-

[7]Ed. See Figure 2.78 on page 91.

There are various symbols for 1,000 due to the fact that for this the printer had to make up a separate character. Lucas (No. 58, f. 8, r.) and Blondel (No. 9, ibid.) use M or ⊂|⊃to these Monier (No. 62, p. 3.) adds $\bar{\text{I}}$. Behme (No. 7, p. 3.) gives I^{m} and ⊂|⊃, and Regius (No. 76, f. 49, v.) $\bar{\text{M}}$. Unicorno (No. 93, f. 4, v.) writes M, ⊂|⊃, or ∞, and Waser (No. 99, f. 9.) gives these three symbols, ⊂|⊃, ∞, and ⊂X⊃[3.19];

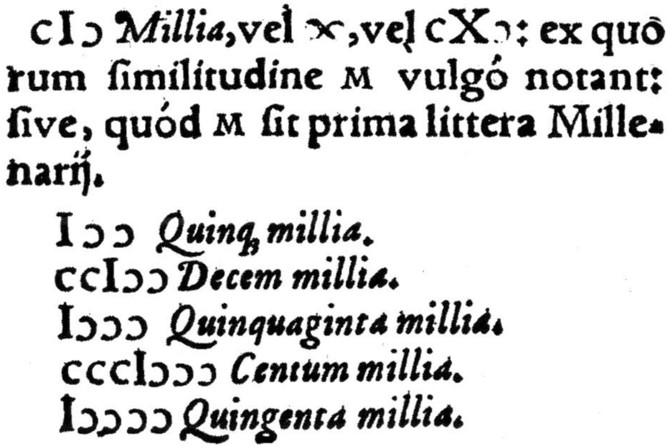
CIↃ *Millia*, vel ∞, vel CXↃ: ex quò
rum similitudine M vulgó notant:
sive, quód M sit prima littera Mille-
narij.
IↃↃ *Quinq; millia.*
CCIↃↃ *Decem millia.*
IↃↃↃ *Quinquaginta millia.*
CCCIↃↃↃ *Centum millia.*
IↃↃↃↃ *Quingenta millia.*

FIGURE 3.19: Waser's Roman numeral symbols for 1000 and higher. No. 99, 1603, f. 9, r.

in fact we find the symbol ∞ [3.20] used in writing the date 1603 as on the title-page of the latter's book.

FIGURE 3.20:
Waser's title page of *Institutio arithmetica:* ∞ used for 1000 in date.
No. 99, 1603, f. 1, r.

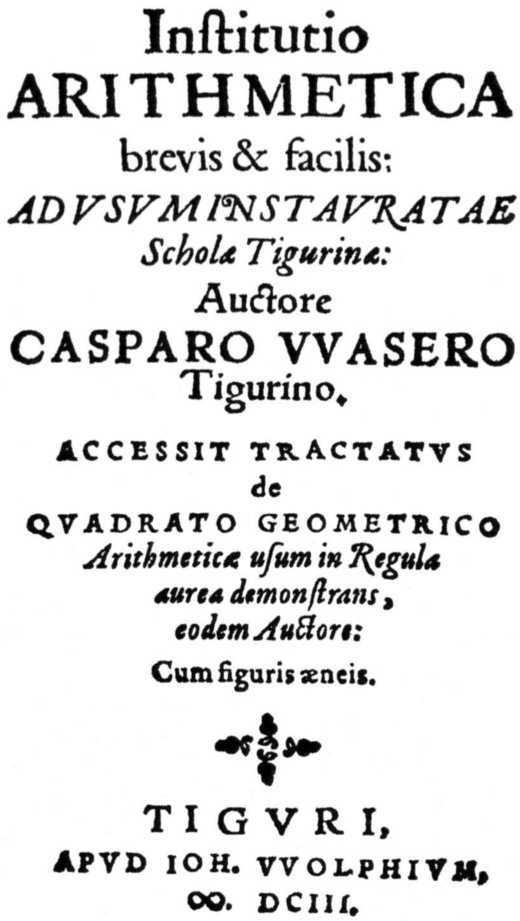

Inſtitutio
ARITHMETICA
brevis & facilis:
AD VSVM INSTAVRATAE
Scholæ Tigurinæ:
Auctore
CASPARO VVASERO
Tigurino.

ACCESSIT TRACTATVS
de
QVADRATO GEOMETRICO
Arithmeticæ uſum in Regula
aurea demonſtrans,
eodem Auctore:
Cum figuris æneis.

TIGVRI,
APVD IOH. VVOLPHIVM,
∞. DCIII.

The principle of multiplication is used by Monier (No. 62, p. 4.) in writing 1,000,000 as MM, whereas Lucas (No. 58, f. 8, r.) uses the dash, $\overline{\text{M}}$. Waser (No. 99, f. 9.) [3.21] and Behme (No. 7, p. 3.) [3.18] give the following old Roman symbol for one million:

FIGURE 3.21:
Wasser's one million in Roman numerals.
No. 99, 1603, f. 9, v.

ccccIↄↄↄↄ *Decies centena millia.*

There are several instances of the prolongation of the last I in the units for the purpose of preventing the subsequent additions of a further unit, as iiij for 4, and ij for 2, in Vuelpius (No. 98, f. 6, v.) and Blondel (No. 9, ibid.) respectively. Feme (No. 31.) and Sesen (No. 84, p. 6.) write iiij for 4, using printed i's. (See account of the *Treviso Arithmetic* (14–78) in D. E. Smith' s *Source Book in Mathematics* (New York, 1929), p. 3, wherein i's are also used for ones.)

3.1.3 Hindu-Arabic numerals

In grouping digits in numeration, vertical bars, commas, dots, numbers, strokes, arcs, and combinations of these occur in the arithmetics under examination.

Cajori says (p. 58.) that Gemma Frisius wrote 24 456 345 678 (without the use of any separatrix) in his 1540 edition; it is noteworthy that he uses the vertical bar as a separatrix of this same number in his 1549 and 1550 editions, as has been mentioned previously[8]. Sfortunati (1561) and Irson (No. 53, p. 11.) mark a number as we do today, while Forestani (No. 35.) and Hodder (No. 51.) employ the dot in place of the comma. Isaac Greenwood's book of 1729, the earliest native American arithmetic, used both of these latter forms (See D. E. Smith, *History of Mathematics,* v. 2, p. 87.).

Dots below as a sign of separation are used by Lapazzaia, as illustrated before [See Section 2.13]. Tartaglia (No. 90, f. 7, r.) in his *General trattato di numeri* of 1556 writes dots above, thus [3.22]:

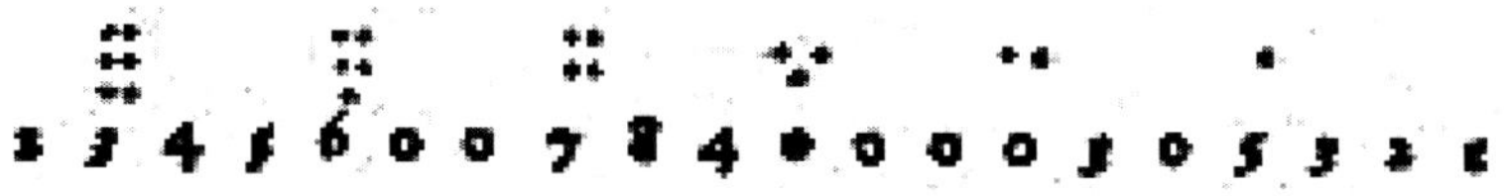

FIGURE 3.22: **Tartaglia's separation. No. 90, 1556, f. 7, r.**

Steinmetz (No. 86, f. 10, r.) marks every third digit with a dot, thus [3.23]:

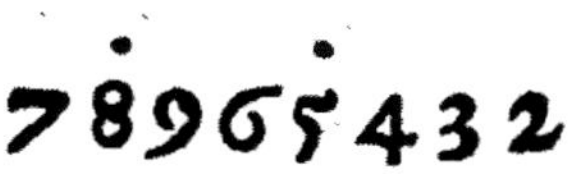

FIGURE 3.23: **Steinmetz's separation. No. 86, 1575, f. 10, r.**

as also do Graffenried (No. 42.) and Sesen (No. 84.). We find it interesting to note the various markings in Clavius (No. 24, pp. 10–12; No. 25, 1586, pp. 13, 14; No. 25, 1618, pp. 2–4; No. 25, 1686, pp. 10, 11.). In the 1585, 1586, and 1618 editions he employs dots and numbers in several ways, thus [3.24, 3.25, 3.26]:

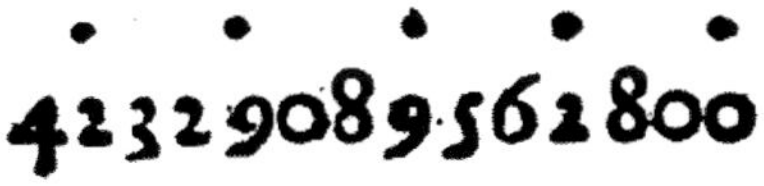

FIGURE 3.24: **Clavius' separation. No. 24, 1585, p. 10**

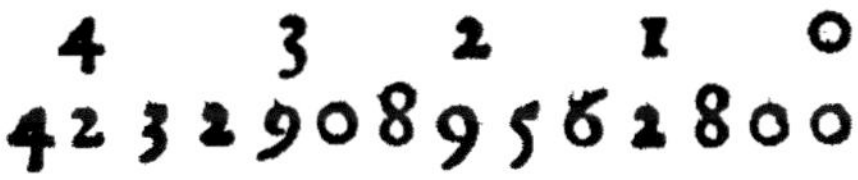

FIGURE 3.25: **Clavius' separation. No. 24, 1585, p. 11**

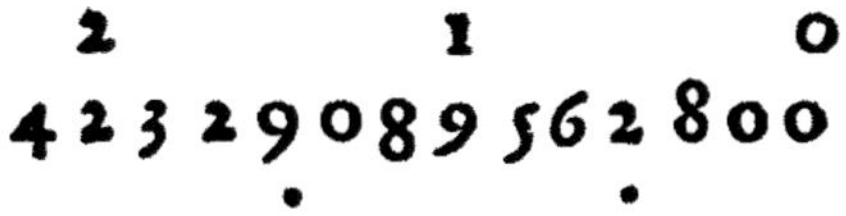

FIGURE 3.26: **Clavius' separation. No. 24, 1585, p. 12**

[8] ED. See section 2.3 on page 67.

In 1686 he uses only numbers, as in [3.27, 3.28].

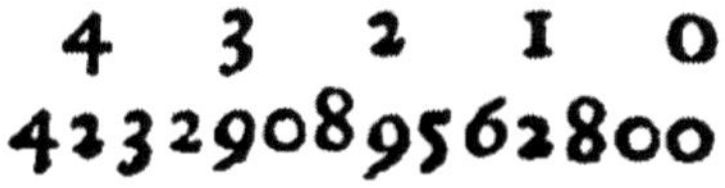

FIGURE 3.27: Clavius' separation. No. 25, 1686, p. 10

FIGURE 3.28: Clavius' separation. No. 25, 1686, p. 11

Cajori (p. 59.) mentions only one method of marking a large number in the 1583 edition, namely $\overset{4}{42}\overset{3}{329}\overset{2}{089}\overset{1}{562}\overset{0}{800}$ whereas the 1586 edition, which is a translation of that of 1583, gives the three symbols for grouping mentioned above.[9] Hemeling, similarly, uses dots above in his work of 1667, but dots and commas in 1678. Lucas (No. 58, p. 2.) uses dots above and below with strokes, as in [3.29].

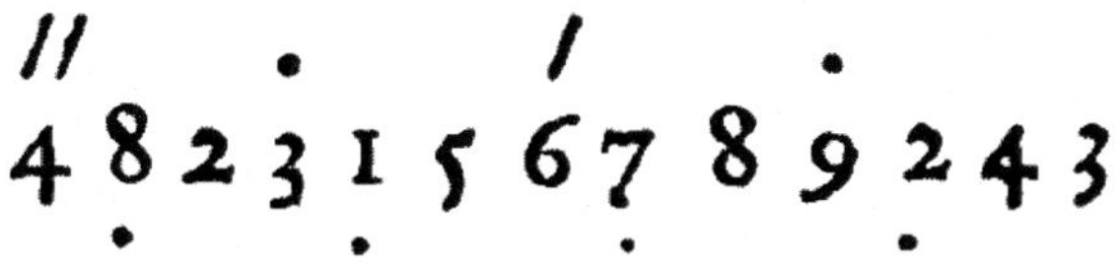

FIGURE 3.29: Lucas' separation. No. 58, 1697, p. 2.

Figatelli, as noted before[10], has dots and numbers, while Behme (No. 7, p. 3.) separates the digits in this particularly peculiar and profuse manner [3.30]:

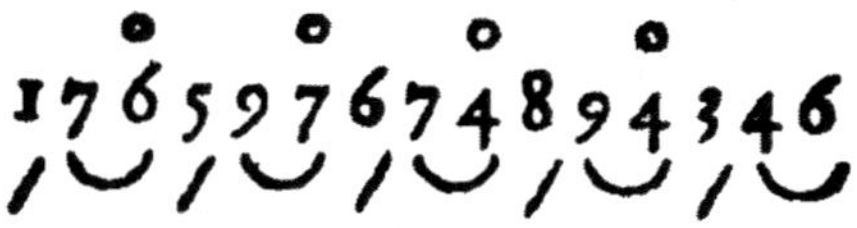

FIGURE 3.30: Behme's separation. No. 7, 1695, p. 3.

Vertical bars and dots occur most frequently as symbols for separating numbers into periods of two digits in the process of extracting square root. Besides Finé, whom we have mentioned earlier,[11] Blondel (No. 9, p. 123.) uses the vertical bar as separatrix somewhat in this manner [3.31]:

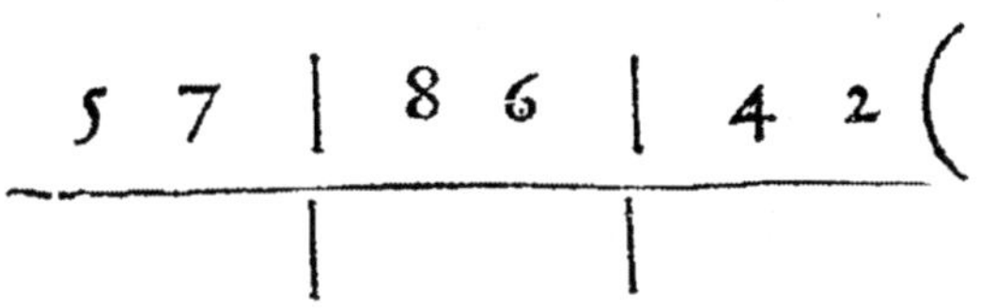

FIGURE 3.31: Blondel's separation. No. 9, 1683, v. 2, p. 123.

[9] Ed. Clavius' 1583 edition (pp. 6–8) includes the same three grouping notations (i.e. both numbers and dots) that are used in the 1585, 1586 and 1618 editions and depicted in Figures 3.24, 3.25, and 3.26.

[10] Ed. See section 2.14 on page 90.

[11] Ed. See Figure 2.52 on page 82.

There are several adherents of dots below the digits, as Gemma, already cited[12], Aurel (No. 3, f. 41, v.), Clavius (No. 24, p. 309; No. 25, 1586, p. 274; No. 25, 1618, p. 268; No. 25, 1586, p. 306.) in all his editions, Peverone (No. 68, p. 55.), and Alsted (No. 2, p. 73.). For the use of dots above we have referred to Peurbach (1539) [2.51], Jacob (1612) [2.84], and Figatelli (1678)[13]. Regius (No. 76, f. 77.), Scheubel (No. 83, f. 77.), Steinmetz (No. 86, f. 58.), Bombelli (No. 11, p. 31.), Unicorno (No. 93, p. 86.), Napier (No. 64, p. 67.), and Harriot (No. 45, p. 130.) also use them. Harriot makes his dots very large and heavy [3.32].

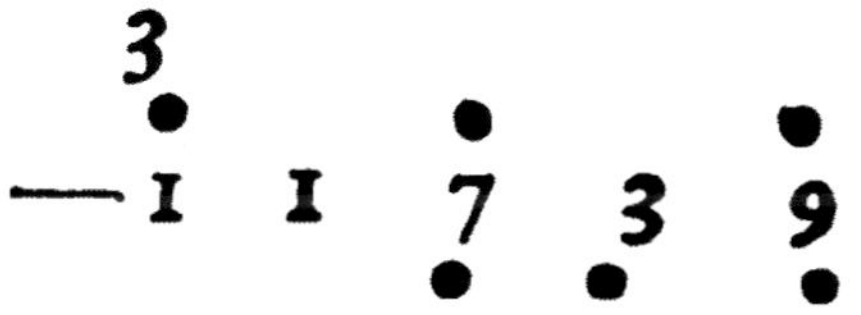

FIGURE 3.32: **Harriot's separation.** No. 45, 1631, p. 130.

We find that Micyllus (No. 60, p. 69.) uses dots above in the extraction of cube roots, but for square roots he does not mark his numbers into periods in any way [3.33].

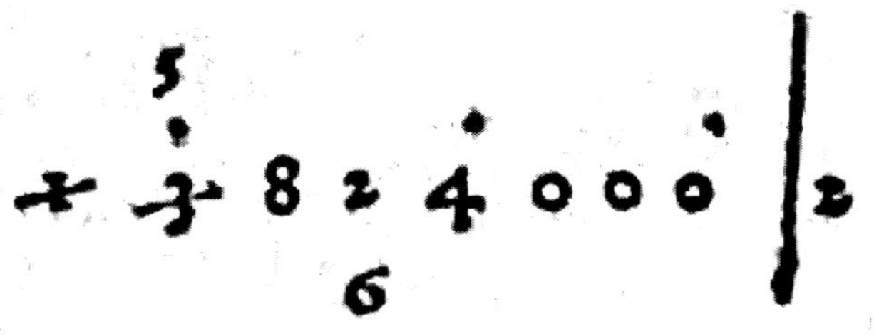

FIGURE 3.33: **Mycillus' dots for extracting cube roots.** No. 60, 1555, p. 69.

In many of the arithmetics of these two centuries we still find the integers of the running text enclosed by dots, as in Cataneo (1546) [2.27], or merely followed by dots as in Peurbach (1539) [2.49], Cataneo (1559) [2.29], and Jacob (1612) [2.84], as noted before. Tartaglia (1556) also places dots after integers, Tagliente (1520) after both integers and fractions, while Noviomagus (1544) uses them only after the integers in mixed numbers. Clavius again shows his fondness for dots, for he uses them after both integers and fractions.

3.2 Signs for Addition and Subtraction

3.2.1 Early symbols for addition and subtraction

The Italian symbolism with its p̃ or p. for plus and m̃ or m. for minus, was used extensively by sixteenth-century and early seventeenth-century mathematicians in Italy. Unicorno (No. 93, p. 118.) uses p. and m. in his 1598 arithmetic, and as late as 1671 we find these same symbols in a work of the Spanish Puig (No. 72, p. 374.). Cataldi (No. 17, p. 8; No. 18, p. 50.), of whom Riccardi lists at least thirty works, uses *più*

[12] Ed. See section 2.3 on page 67.
[13] Ed. See section 2.14 on page 90.

and *meno*, sometimes p̃ and m̃, according to the crowded condition of the particular line. Feliciano's editions testify to the waning strength of the Italian symbols in the latter part of the seventeenth century, for p̃ and m̃ are used in his 1550 edition, but the Maltese cross in 1669, and the eight and twelve-pointed Maltese cross interchangeably in 1692, as we have seen earlier.[14] However, Figatelli's works still adhere to p̃ and m̃ as late as 1678. (A lengthy account of the early history of the signs + and − is given by J. W. L. Glaisher in *Messenger of Mathematics,* v. 51, 1921, pp. 1–149. In it Glaisher credits Widman (1489) with the introduction of these signs; and shows that they had an arithmetical origin, not a commercial origin, as De Morgan (*Camb. Phil. Trans.,* v. 11, 1871, pp. 203–212) believed).

3.2.2 Use of plus and minus in the rule of false position

Several writers, in solving problems by false position, prefix plus or minus in various shapes to the error, as do Vuelpius (No. 98. On the last few folios we find + and − used to denote excess and deficiency in the error.), Lambeck (No. 55, p. 286. Lambeck also uses plus and minus in the solution of algebraic problems in the appendix, p. 300.), Jacob (1612. We have found the use of his plus sign (Maltese cross) only once outside of its application in the rule of false.), and Blondel (No. 9, p. 211, uses. ⁻+ and -.). Gemma Frisius (1549, f. 36, r.) uses + and − in combination with guiding lines in his application of the rule of false position, somewhat in this manner [3.34];

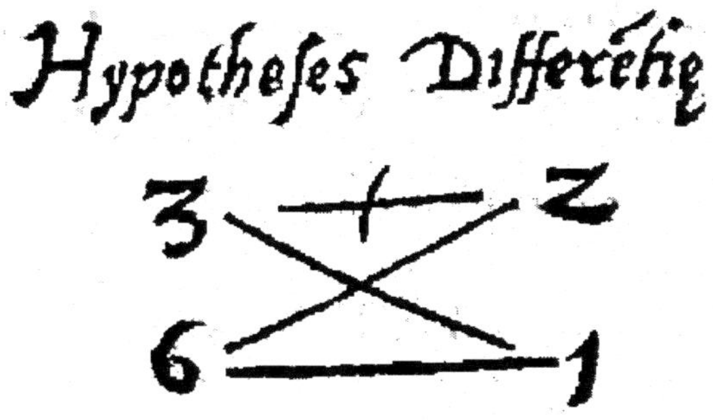

FIGURE 3.34: Frisius's false position. Book 37, 1540, f. 21, r.

Cajori (p. 254.) only mentions the use of the guiding lines in the 1569 edition of Gemma Frisius.

3.2.3 Shapes of the plus sign

The Greek cross was usually used by Descartes (No. 28, p. 15.), as mentioned by Cajori (p. 237.). However, it is interesting to add that he also uses + the Greek cross with each of its four arms terminating in a line drawn across it, but he uses this form only in the terms of a fraction [3.35].

[14] See Figure 2.40 on page 77.

$y + \frac{bx}{z}$, C D futura ſit $\frac{cy}{z} + \frac{bcx}{zz}$. Poſtea, quia lineæ A B, A D, & E F poſitione datæ ſunt, data quoque erit diſtantia puncti A à puncto E, quæ ſi nominetur K, habebitur E B æqualis $k + x$; foret autem ipſa $k - x$, ſi punctum B caderet inter E & A; at verò $-k + x$, ſi E caderet inter A & B. Rurſus, quoniam anguli trianguli E S B omnes dantur, dabitur quoque ratio lateris B E ad B S, quam ſi ponam eſſe ut z ad d, B S fiet $\frac{dk + dx}{z}$, C S verò $\frac{zy + dk + dx}{z}$; quæ quidem foret $\frac{zy - dk - dx}{z}$, ſi punctum S caderet inter B & C; at verò $\frac{-zy + dk + dx}{z}$, ſi C caderet inter B & S. Porrò dantur

FIGURE 3.35: Descartes' plus sign. No. 28, 1649, p. 15

This is probably due to the fact that the form + was too large to use in this case. In the same way the Spanish writer Ventura de Avila (No. 4, pp. 41, 69, 85.), uses a symbol other than the usual Greek cross when necessary in the numerator and denominator of a fraction, namely -+ and +-.

The Latin cross placed horizontally, -+, was used by Vieta (No. 95, p. 107; No. 96, p. 154.) in 1591, according to Cajori (p. 237.). This is the case in the 1630 algebra [3.36],

D -+ A

FIGURE 3.36: Viète's plus sign. No. 95, 1630, p. 107

but throughout his work of 1646, the writer finds that he uses -+ and + interchangeably [3.37].

SI A cubus -+ B in A quadr. 3 -+ D plano in A, æquetur B cubo 2 — D plano in B. A quad. -+ B in A 2, æquabitur B quad. 2 — D plano.

Quoniam enim A quadr. + B in A 2, æquatur B quadr. 2 — D plano. Ductis igitur omnibus in A. A cubus + B in A quad. 2, æquabitur B quad. in A 2 — D plano in A.

FIGURE 3.37: Viète's other plus sign. No. 96, 1630, p. 154

Prestet (No. 70, p. 2.) also gives -+ as the symbol for plus, and we have said that Blondel used it in the rule of false position.[15]

The Maltese cross in its simple form, ✠, we have seen, may be found occasionally in Hemeling.[16] Lucas (No. 58, pp. 165, 286.) uses a form of it as an algebraic sign [3.38]

Signa Algebraica, ut alia prætermittamus, ſunt ✠ plus, & —. minus.

FIGURE 3.38: Lucas' plus sign. No. 58, 1697, p. 286

[15] Ed. See section 3.2.2 on page 124.

[16] Ed. See section 2.16 on page 93.

as well as in his rule of false. Feliciano (1692), we have seen previously,[17] used an eight and twelve-pointed Maltese cross. The latter was also employed by Ceva in his work (No. 20, p. 9.) [3.39].

FIGURE 3.39: Ceva's plus sign. No. 20, 1682, p. 9

A very fanciful twelve-pointed cross was used by Renaldini (No. 78, p. 80.) [3.40].

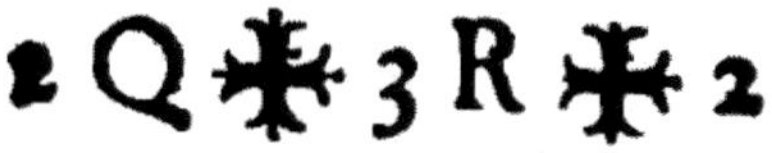

FIGURE 3.40: Renaldi's plus sign. No. 78, 1665, p. 80

Cajori states (p. 238.) that he uses this form on page 80 and throughout the volume. We find three forms of the plus sign used by him on this same page [3.41], and throughout the volume these symbols appear with about equal frequency.

In a recent article (*Isis,* v. 17, Jan. 1932, p. 259.) by J. Pelsener, note is made of the successive use of three different signs to indicate addition, ×, ⨯, and ⟶+, in the treatise, *De Numeris Geometricis* of Schonerus, translated into English by Thom. Bedwell, London, 1614. The author of the article cites this as a unique case of the use of three different notations for addition in the course of a single work. The example of Renaldini given above is another such instance. The use of different symbols for the same operation is not uncommon.

3.2.4 Varieties of minus signs

The sign of subtraction used by Vuelpius (1544), Feliciano (1669), and Blondel (1699) was somewhat longer than the ordinary symbol —. A slight modification of it, ⟝, was employed by Lucas (No. 58, p. 286.) [3.38] and Renaldini (No. 78, p. 21.). Vieta makes this same symbol very heavy so that it renders the minus signs on a page very conspicuous (No. 95, p. 107.) [3.42].

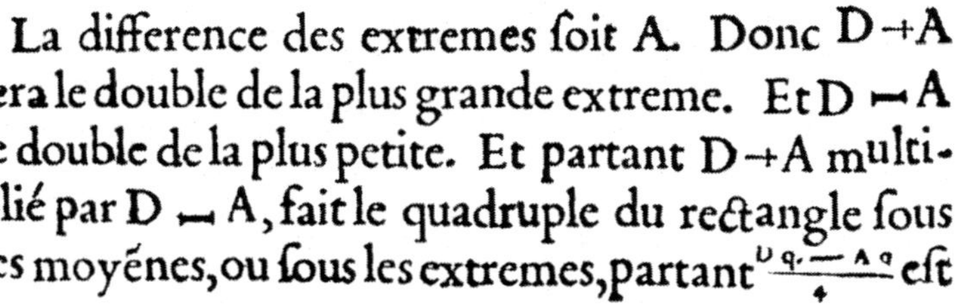
La difference des extremes ſoit A. Donc D+A
ſera le double de la plus grande extreme. Et D — A
le double de la plus petite. Et partant D+A multi-
plié par D — A, fait le quadruple du rectangle ſous
les moyẽnes, ou ſous les extremes, partant $\frac{D q. - A q.}{4}$ eſt

FIGURE 3.42: Viète's minus sign. No. 95, 1630, p. 107

A short dash, sometimes longer, and again, two successive dashes were Moore's varieties, as noted above.[18] Renaldini (No. 78, p. 40.) occasionally uses the latter form

[17] Ed. See Figure 2.40 on page 77.
[18] Ed. See Figure 2.95 on page 97.

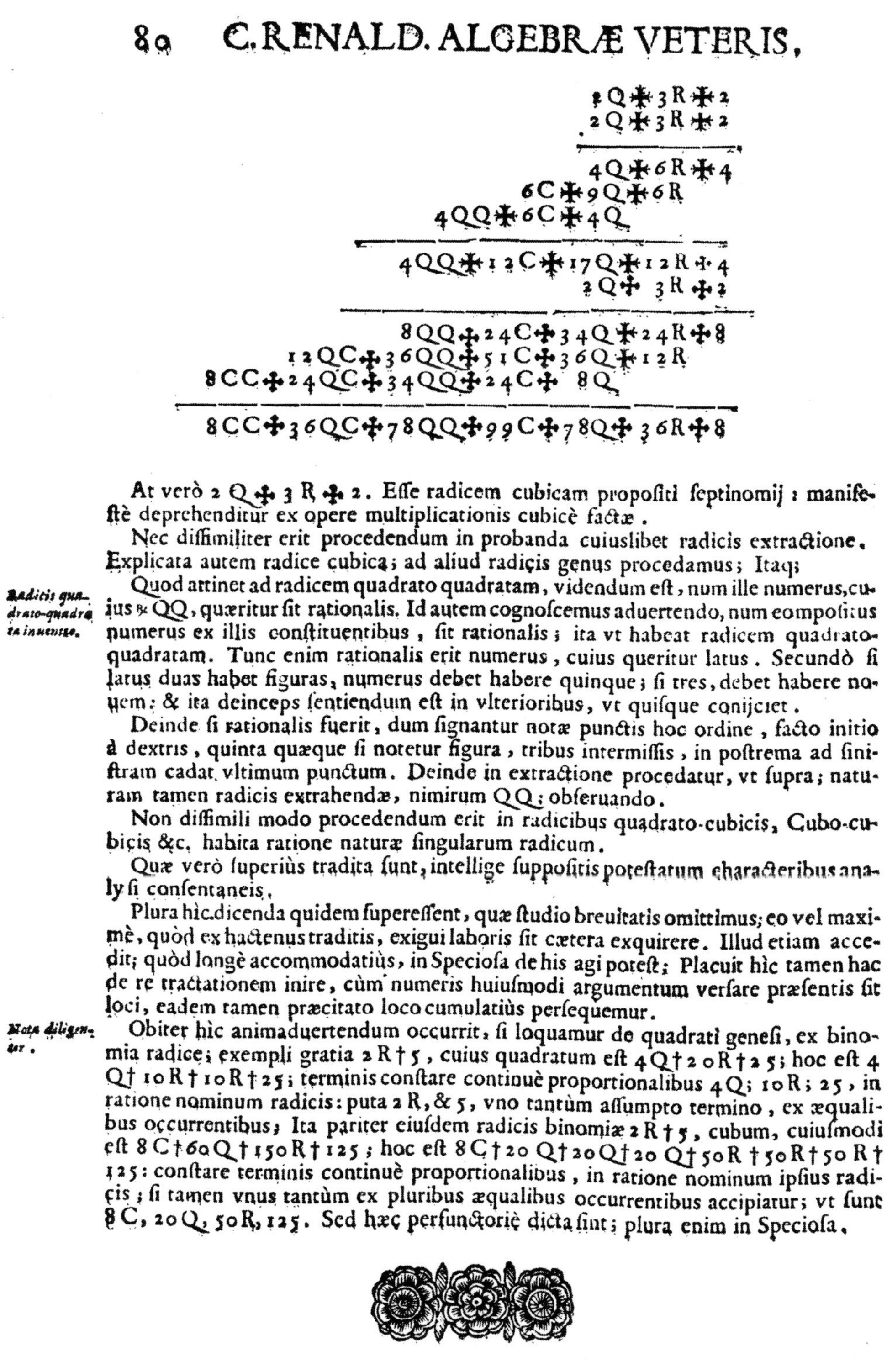

80 C. RENALD. ALGEBRÆ VETERIS,

2Q + 3R + 2
2Q + 3R + 2

4Q + 6R + 4
6C + 9Q + 6R
4QQ + 6C + 4Q

4QQ + 12C + 17Q + 12R + 4
2Q + 3R + 2

8QQ + 24C + 34Q + 24R + 8
12QC + 36QQ + 51C + 36Q + 12R
8CC + 24QC + 34QQ + 24C + 8Q

8CC + 36QC + 78QQ + 99C + 78Q + 36R + 8

At verò 2 Q + 3 R + 2. Esse radicem cubicam propositi septinomij: manifestè deprehenditur ex opere multiplicationis cubicè factæ.

Nec dissimiliter erit procedendum in probanda cuiuslibet radicis extractione. Explicata autem radice cubica; ad aliud radicis genus procedamus; Itaq;

Radicis quadrato-quadratæ inuentio. Quod attinet ad radicem quadrato quadratam, videndum est, num ille numerus, cuius ℞ QQ, quæritur sit rationalis. Id autem cognoscemus aduertendo, num compositus numerus ex illis constituentibus, sit rationalis; ita vt habeat radicem quadrato-quadratam. Tunc enim rationalis erit numerus, cuius queritur latus. Secundò si latus duas habet figuras, numerus debet habere quinque; si tres, debet habere nouem: & ita deinceps sentiendum est in vlterioribus, vt quisque conijciet.

Deinde si rationalis fuerit, dum signantur notæ punctis hoc ordine, facto initio à dextris, quinta quæque si notetur figura, tribus intermissis, in postrema ad sinistram cadat vltimum punctum. Deinde in extractione procedatur, vt supra; naturam tamen radicis extrahendæ, nimirum QQ; obseruando.

Non dissimili modo procedendum erit in radicibus quadrato-cubicis, Cubo-cubicis &c. habita ratione naturæ singularum radicum.

Quæ verò superiùs tradita sunt, intellige suppositis potestatum characteribus analysi consentaneis.

Plura hìc dicenda quidem superessent, quæ studio breuitatis omittimus; eo vel maximè, quòd ex hactenus traditis, exigui laboris sit cætera exquirere. Illud etiam accedit; quòd longè accommodatiùs, in Speciosa de his agi potest; Placuit hìc tamen hac de re tractationem inire, cùm numeris huiusmodi argumentum versare præsentis sit loci, eadem tamen præcitato loco cumulatiùs persequemur.

Nota diligenter. Obiter hìc animaduertendum occurrit, si loquamur de quadrati genesi, ex binomia radice; exempli gratia 2 R † 5, cuius quadratum est 4 Q † 20 R † 25; hoc est 4 Q † 10 R † 10 R † 25; terminis constare continuè proportionalibus 4 Q; 10 R; 25, in ratione nominum radicis: puta 2 R, & 5, vno tantùm assumpto termino, ex æqualibus occurrentibus; Ita pariter eiusdem radicis binomiæ 2 R † 5, cubum, cuiusmodi est 8 C † 60 Q † 150 R † 125; hoc est 8 C † 20 Q † 20 Q † 20 Q † 50 R † 50 R † 50 R † 125: constare terminis continuè proportionalibus, in ratione nominum ipsius radicis; si tamen vnus tantùm ex pluribus æqualibus occurrentibus accipiatur; vt sunt 8 C, 20 Q, 50 R, 125. Sed hæc perfunctoriè dicta sint; plura enim in Speciosa.

FIGURE 3.41: Renaldi's three forms of a plus sign.
No. 78, 1665, p. 80

also. Lambeck (No. 55, p. 300.) and Hemeling (1667) are two adherents of our present symbol for division as their symbol for subtraction [3.43].

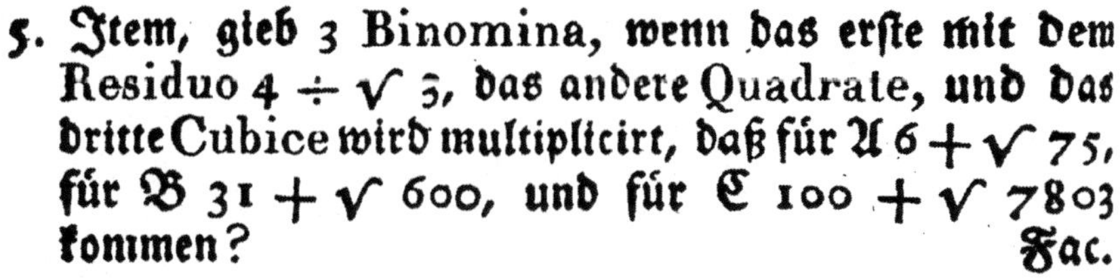
5. Item, gieb 3 Binomina, wenn das erste mit dem Residuo 4 ÷ √ 3, das andere Quadrate, und das dritte Cubice wird multiplicirt, daß für A 6 + √ 75, für B 31 + √ 600, und für C 100 + √ 7803 kommen? Fac.

FIGURE 3.43: Lambeck's minus sign. No. 55, n.d., p. 300

Cajori (p. 242.) says the latter wrote − for minus, but in the example cited he uses ÷, probably a typographical error. The form ∸ [3.44] is, however, used by Curtius (No. 27.) as an operational symbol,

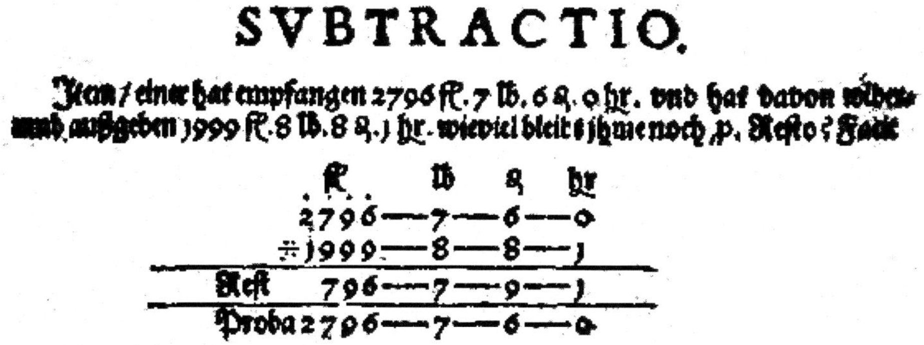
SVBTRACTIO.

Item / einer hat empfangen 2796 fl. 7 lb. 6 ß. 0 hr. vnd hat davon widerumb außgeben 1999 fl. 8 lb. 8 ß. 1 hr. wieviel bleibt jhme noch p. Resto? Facit

	fl.	lb.	ß.	hr.
	2796	7	6	0
	÷1999	8	8	1
Rest	796	7	9	1
Proba	2796	7	6	0

FIGURE 3.44: Curtius' subtraction symbol in *Ein Newes Rechenbuch,* 1604, Nuremberg, p. 32

whereas Graffenried (No. 42, pp. 701 ff.) uses ÷ also with a qualitative meaning [3.45].

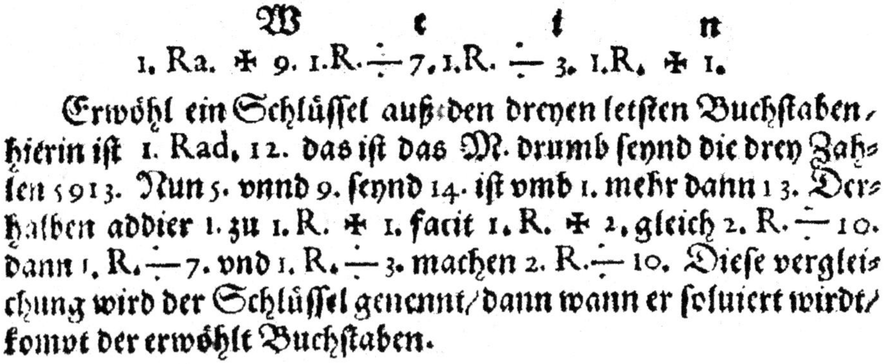
W e i n

1. Ra. ✢ 9. 1.R. ÷ 7. 1.R. ÷ 3. 1.R. ✢ 1.

Erwöhl ein Schlüssel auß den dreyen letsten Buchstaben, hierin ist 1. Rad. 12. das ist das M. drumb seynd die drey Zahlen 5 9 13. Nun 5. vnnd 9. seynd 14. ist vmb 1. mehr dann 13. Derhalben addier 1. zu 1.R. ✢ 1. facit 1.R. ✢ 2. gleich 2.R. ÷ 10. dann 1.R. ÷ 7. vnd 1.R. ÷ 3. machen 2.R. ÷ 10. Diese vergleichung wird der Schlüssel genennt / dann wann er soluiert wirdt / kompt der erwöhlt Buchstaben.

FIGURE 3.45: Graffenried's minus sign. No. 42, 1619, p. 702

Thus we have found that in the early books the forms of type for plus and minus were not standardized. In many cases the symbol was adjusted in size to the space to be filled, and very frequently various forms were displayed on the same page.

3.3 Signs of Multiplication

3.3.1 Early Uses of the St. Andrew's cross, but not as a symbol of multiplication of two numbers. The process of two false positions.

In the early use of St. Andrew's cross X in this process, each bar connects the two numbers which are multiplied, whereas later the two bars are considered as one symbol.

Reference has been made[19] to the use of these guiding lines by Gemma Frisius in solving problems by double false position. We find similar settings of numbers in Micyllus (No. 60, p. 129.), who writes [3.46]

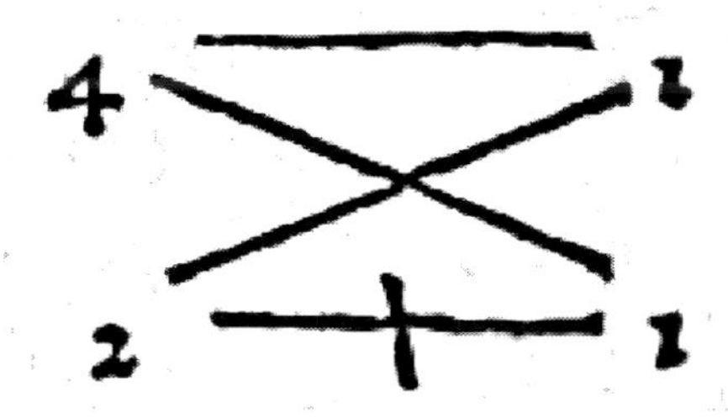

FIGURE 3.46: Micyllus' multiplication. No. 60, 1555, p. 129

Clavius, in his 1585 and 1586 editions, uses ⅹ [3.47/Left] for this purpose, but a large capital letter X [3.47/Right] in the later editions of 1618 and 1686.

FIGURE 3.47: Clavius' early multiplication sign. Left: No. 24, 1585, p. 235 Right: No. 25, 1618, p. 196

A slightly different scheme is found in Bassi (No. 5, p. 176.) who uses

FIGURE 3.48: Bassi's multiplication sign. No. 5, 1645, p. 176

[19] Ed. See section 2.3 on page 67.

Blondel (No. 9, p. 213.) [3.49] uses the same scheme as Gemma and Micyllus.

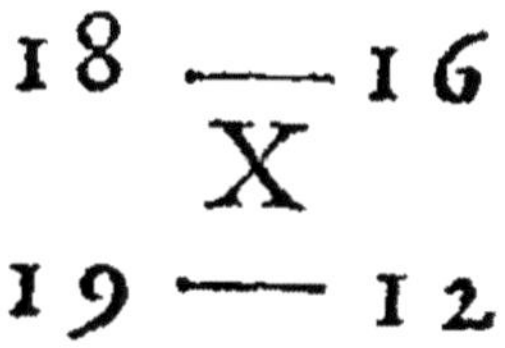

FIGURE 3.49: Blondel's multiplication. No. 9, 1683, v. 2, p. 213

3.3.2 Proportions involving fractions

There were various schemes used in the early arithmetics to serve as aids to computation. In the solution of proportions involving fractions we find the use of $\times$, single bars, horizontal lines, arcs, braces, and other devices. Bassi (No. 5, p. 76.) solves the proportion $\frac{2}{3} : \frac{3}{4} :: \frac{1}{2} : x$ thus [3.50]:

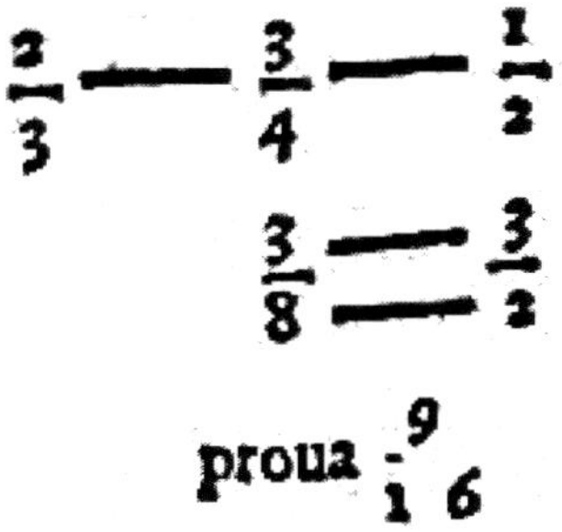

FIGURE 3.50: Bassi's proportions. No. 5, 1645, p. 76

3.3.3 Addition and subtraction of fractions

The cross $\times$ is used by Regius (No. 76, f. 85, v.) in reduction of fractions to equivalent fractions, thus [3.51]:

FIGURE 3.51: Regius' fractions. No. 76, 1536, f. 77. v

Bassi (No. 5, pp. 72, 73.) and Manelli (No. 59, pp. 25, 28.) use the large capital X on its side [3.52],

FIGURE 3.52: **Bassi's fractions. No. 5, 1645, p. 72**

in the addition and subtraction of fractions. Zuchetta (No. 101, pp. 2, 9.) uses dotted lines for this same purpose in the following suggestive manner [3.53]:

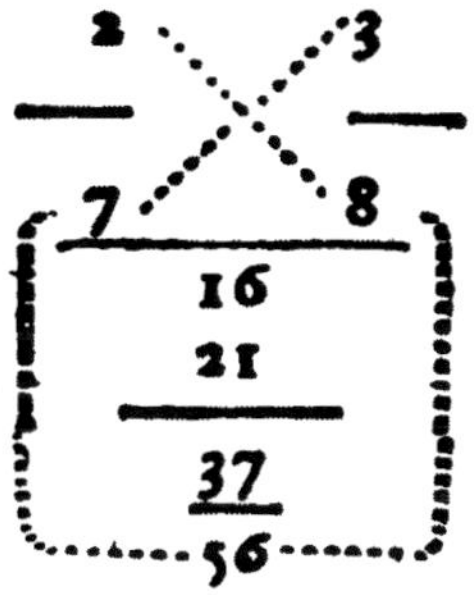

FIGURE 3.53: **Zuchetta's fractions. No. 101, 1600, p. 8**

3.3.4 Multiplication and division of fractions

Zuchetta (No. 101, pp. 10, 13.) uses his dotted lines In a less suggestive way in the multiplication of $\frac{2}{3}$ by $\frac{4}{5}$ as follows [3.54]:

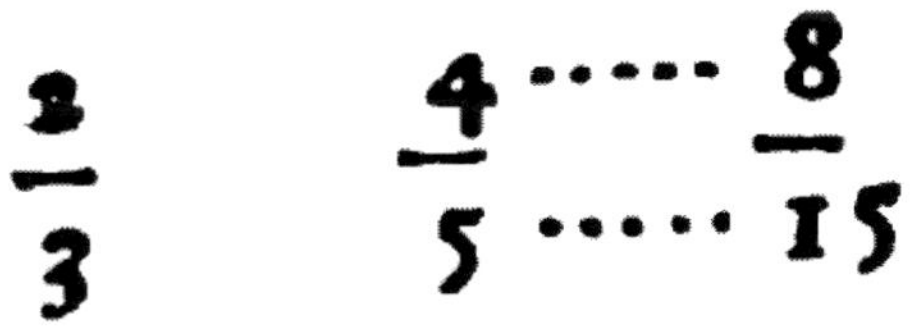

FIGURE 3.54: **Zuchetta's fractional multiplication. No. 101, 1600, p. 10**

the division of $\frac{20}{3}$ by $\frac{3}{4}$ is represented better thus [3.55]:

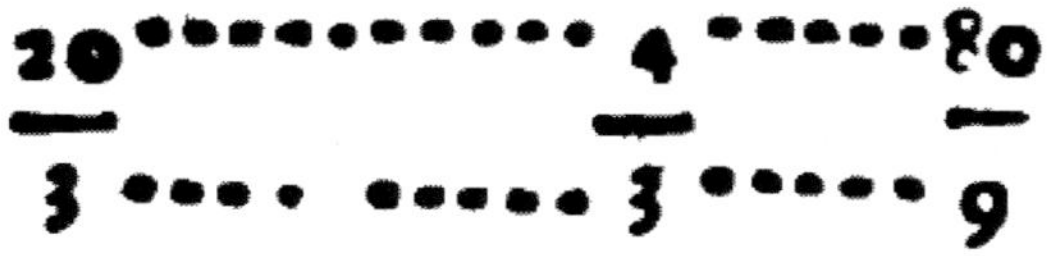

FIGURE 3.55: **Zuchetta's fractional division. No. 101, 1600, p. 13**

Bassi (No. 5, pp. 74, 75.) again uses horizontal lines as aids in multiplication and division of fractions; thus the division of $\frac{3}{4}$ by $\frac{2}{3}$ is written [3.56]

FIGURE 3.56: Bassi's fractional division. No. 5, 1645, p. 75

Cajori (p. 260.) refers to the use of × in division of fractions by Steinmetz (No. 86, f. 68, r.) in his 1568 edition. In 1575 Steinmetz uses an additional × when he divides $\frac{3}{4}$ by $\frac{2}{3}$ in this manner [3.57]:

FIGURE 3.57: Steinmetz's fractional division. No. 86, 1575, f. 68 r.

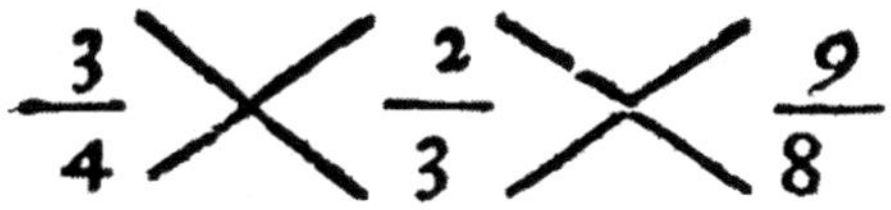

The scheme used by Manelli (No. 59, p. 33.) is different, for he writes the divisor first and then multiplies along the heavier bar of the capital X first; thus, in dividing $9\frac{1}{2}$ by $6\frac{2}{3}$ he writes [3.58]:

FIGURE 3.58: Manelli's fractional division. No. 59, 1659, p. 33

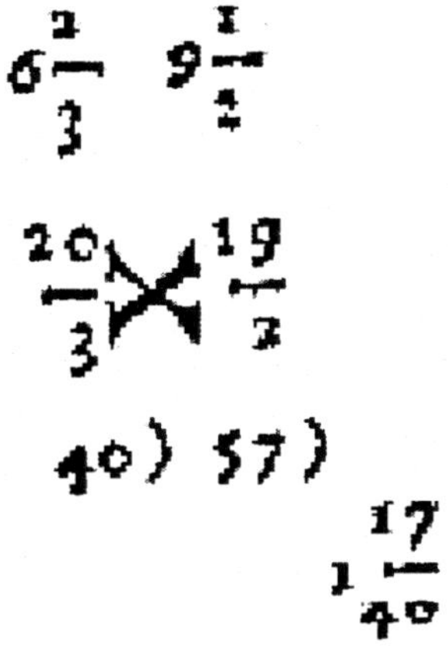

It is of interest to note the mnemonic signs for operations with fractions used by Graffenried (No. 42, p. 116.) [3.59],

FIGURE 3.59: Graffenried's fractional arithmetic. No. 42, 1619, p. 116

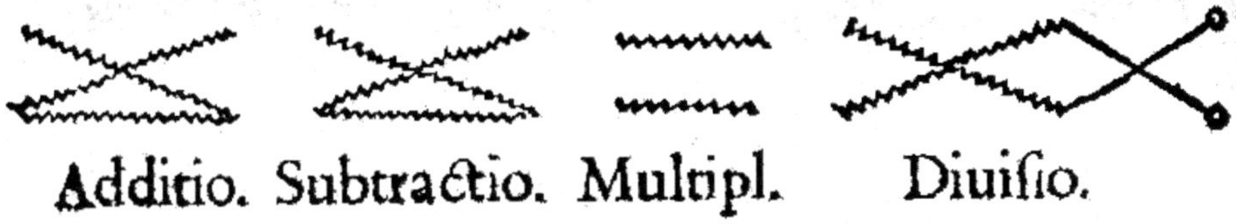

in which ⋈ denotes a scheme for addition and subtraction of fractions, = for multiplication, and ⋈⋈ for division. This same scheme was used by Apian in 1527 (See P.

Treutlein, "Das Rechnen im 16. Jahrhundert", *Abhandlungsn zur Geschichte der Mathematik,* v. 1, p. 80.).

3.3.5 Casting out the 9's

In many early arithmetics various settings of lines afforded a convenient grouping of the four results of an operation. The characteristic $\underline{\times}$ and X [3.60, 3.61]

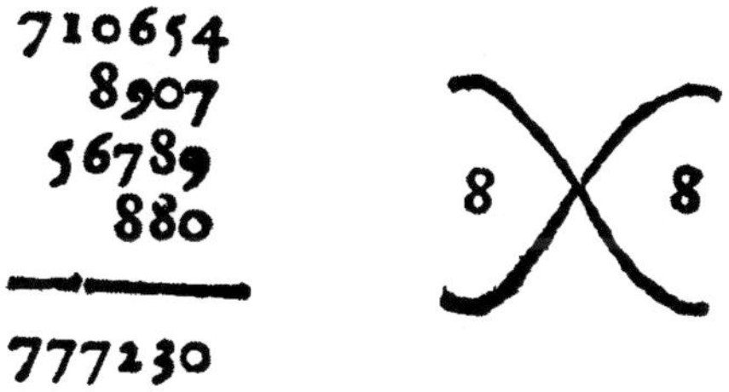

FIGURE 3.60: Clavius' early X. No. 24, 1585, p. 17

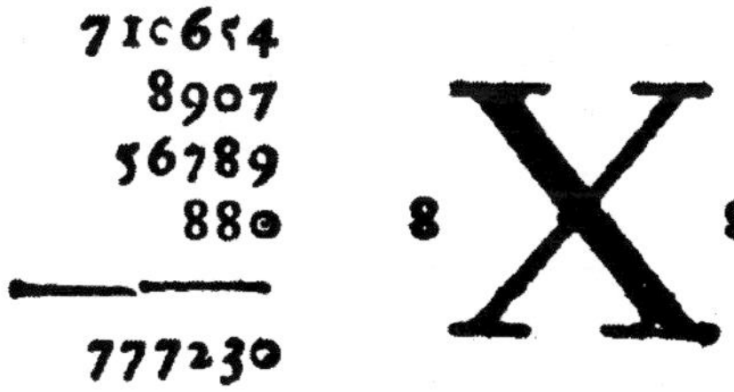

FIGURE 3.61: Clavius' later X. No. 25, 1618, p. 5

were again used by Clavius in casting out 9's, although Cajori (p. 262.) simply refers to his use of $\times$ for this purpose. Rudolff (No. 81, f. 12, r.) uses a small $\times$ in this manner [3.62]:

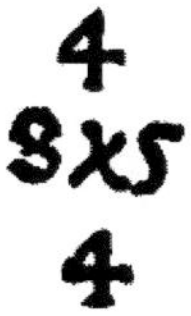

FIGURE 3.62: Rudolff's casting out 9s. No. 81, 1540, f. 12, r.

In Regius (No. 76, f. 56.) we find a small $\times$, but Micyllus (No. 60, p. 9.) uses the usual large X Suevus (No. 88, p. 52.) checks his multiplication only by casting out 9's and uses another X for this purpose [3.63].

FIGURE 3.63: Suevus' casting out 9s. No. 88, 1593, p. 52

We have referred[20] to Savonne's (No. 82, p. 52.) use of the Greek and Latin cross in casting out 9's in his 1571 and 1619 editions, respectively. It is strange, then, that he uses the usual ╳ in addition and subtraction, but no guiding lines of any kind for multiplication and division of fractions.

3.3.6 Multiplication of integers

In multiplication 'per crocetta', the product of two numbers is found mentally with the aid of guiding lines. Bassi (No. 5, p. 26.) uses this setting [3.64]:

Figure 3.64: Bassi's multiplication 'per crocetta'. No. 5, 1645, p. 26

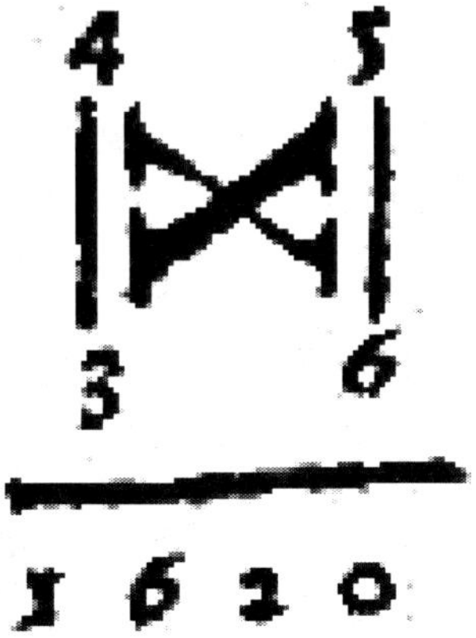

Figatelli (1678) and Manelli (No. 59, p. 14.) use a similar procedure.

3.3.7 To mark the place for 'thousands'

The use of ⋉ marked on the line on which a dot signified '1000' in computations upon lines is found in early arithmetics, as in those of Feme (No. 31.), Rudolff (No. 81, f. 30, v.), Regius, (No. 76, ff. 96, r., 61, v.) [3.65], and Sesen (No. 84, p. 5.).

Figure 3.65: Regius's notation for thousands.[21] No. 76, 1536, f. 96, r.

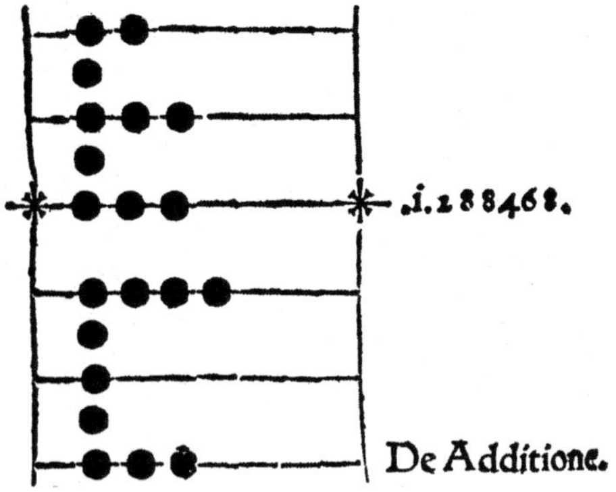

[20] Ed. See section 2.11 on page 84.

Reisch (No. 77, f. 94.) [3.66] also puts a cross at the seventh or million-line, when the latter occurs.[22] (See F. P. Barnard, *The Casting-Counter and the Counting Board,* Oxford, 1916, p. 271.).

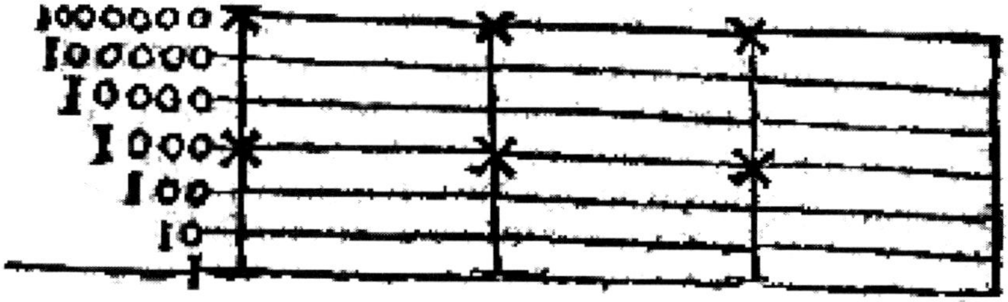

FIGURE 3.66: Reisch's notation for thousands. No. 77, 1503, f. 94, r.

3.3.8 In place of multiplication table above 5×5

Cajori (p. 264.) mentions the use of X in the so-called complementary multiplication. He refers to Regius (1543), fol. 56, whereas it should be fol. 61. The X is also used by Regius In a further complementary method of finding the product of two integers, say 8 and 7, thus [3.67]:

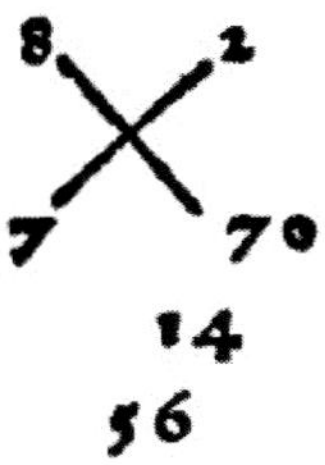

FIGURE 3.67: Regius' multiplication. No. 76, 1536, f. 53, v.

In this process he multiplies the complement of the larger number by the smaller number, and subtracts this result from the product of the smaller number and ten. Blondel (No. 9, p. 84.) illustrates the product of the same two numbers in the usual manner [3.68]:

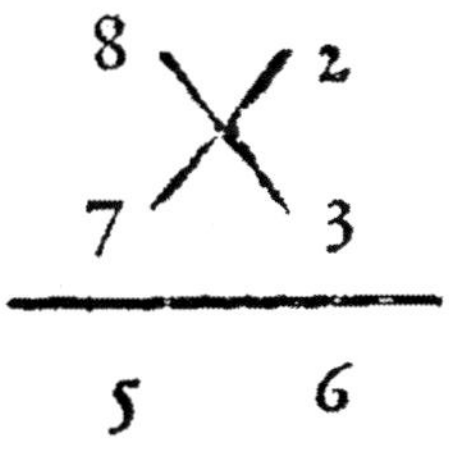

FIGURE 3.68: Blondel's multiplication. No. 9, 1683, v. 2, p. 84

[21] Ed. This figure (which represents the number 288,468) may not correspond to the figure that Sister Leontius referenced in the 1543 edition. The figure shown is taken from the first edition of the text (1536) and the folio number is not discernible.

[22] Ed. Regius (No. 76, f. 86, v.) also marks the 7th line with a cross to indicate the million-line.

Micyllus (No. 60, p. 17.) also gives an illustration of the product of numbers above 5, but Steinmetz (No. 86, f. 28, r.) and Monier (No. 62, p. 55.) both present the same problem in which one of the numbers is less than five, thus offering an added difficulty; for instance [3.69]:

FIGURE 3.69: Steinmetz's multiplication. No. 86, 1575, f. 28, r.

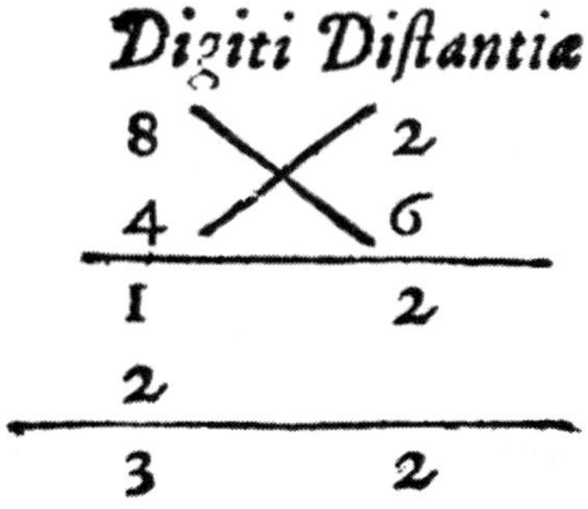

3.3.9 The St. Andrew's Cross as a symbol of multiplication

The St. Andrew's cross ╳ used for this purpose assumed various forms in its early history, after its first use by Oughtred (1631). We find large and small-sized symbols [3.70],

FIGURE 3.70: Schulte's Xs

a small elevated ˣ as in Moore (1660). In solving the proportion `HM:IK::BD:ED`, Ceva (No. 20, p. 12.) [3.71] uses ⋉ when he writes `HM⋉IK = BD⋉ED`.

FIGURE 3.71: Ceva's proportions. No. 20, 1682, p. 12

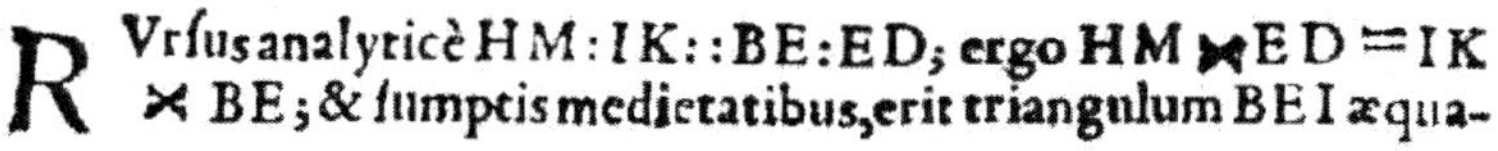
RVrſus analyticè HM : IK :: BE : ED; ergo HM ⋈ ED = IK ⋈ BE; & ſumptis medietatibus, erit triangulum BEI æqua-

However, we do not find × commonly used as a sign of multiplication in the texts of the seventeenth century.

3.4 Signs for Division

3.4.1 Early symbols

In the texts of the sixteenth and seventeenth centuries, division is most commonly expressed by separating the dividend from the divisor by such words as *per*, *por*, *a*, *by*, or *durch*. In illustrations, however, one or two curved bars, vertical and horizontal bars are used as symbolisms of division. The two signs of division, ÷ and :, accredited to Rahn (1659)[3.72] and Leibniz (1684) respectively, did not become common until the beginning of the eighteenth century.

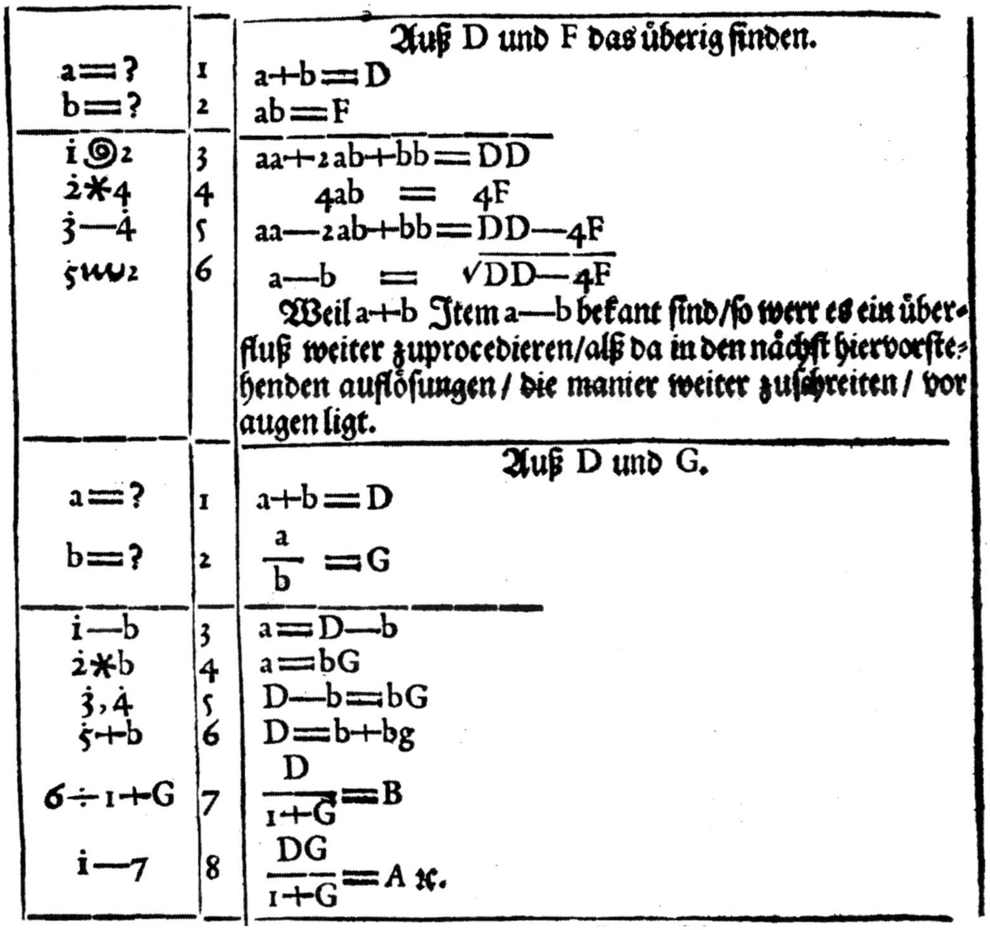

Auß D und F das überig finden.

a = ?	1	a + b = D
b = ?	2	ab = F
1 ⑨ 2	3	aa + 2ab + bb = DD
2 * 4	4	4ab = 4F
3 — 4	5	aa — 2ab + bb = DD — 4F
5 ѡ 2	6	a — b = √DD — 4F

Weil a + b Item a — b bekant sind/so werr es ein überfluß weiter zuprocedieren/alß da in den nächst hiervorstehenden auflösungen / die manier weiter zuschreiten / vor augen ligt.

Auß D und G.

a = ?	1	a + b = D
b = ?	2	$\frac{a}{b}$ = G
1 — b	3	a = D — b
2 * b	4	a = bG
3, 4	5	D — b = bG
5 + b	6	D = b + bg
6 ÷ 1 + G	7	$\frac{D}{1+G}$ = B
1 — 7	8	$\frac{DG}{1+G}$ = A ꝛc.

FIGURE 3.72: Rahn's *Teutsche Algebra,* with the first use of ÷ for division. No. 73, 1659, p. 73

3.4.2 Relative position of divisor and dividend

In these early books the divisor and quotient have been assigned various positions relative to the dividend. Thus below we cite eight authors (seven not in Cajori) whose works present eight different positions of divisor and quotient relative to the dividend. These variations are not noted by Cajori.

We begin with the arrangement most common for the 'scratch method' of division. In this case the divisor is placed below the dividend, and the quotient is usually placed to the right of the dividend. This, as noted above, was the case with Husswirt (1501),[23] Savonne (1571, 1619),[24] and Ramus (1599) [25]. Waser (No. 99, f. 26, r.), Bassi (No. 5, p. 58.) [3.73], Puig (No. 72, p. 75.), Prestet (No. 70, p. 30.), and Figatelli (No. 32, p. 30.) (who are somewhat representative of writers in Latin, Italian, Spanish, French, and

[23] Ed. See Figure 2.7 on page 65.
[24] Ed. See Figure 2.63 on page 86.
[25] Ed. See Figure 2.70 on page 88.

Italian respectively), follow a similar practice, placing a vertical bar, or curved bar, or some variation of it, to the right of the dividend.

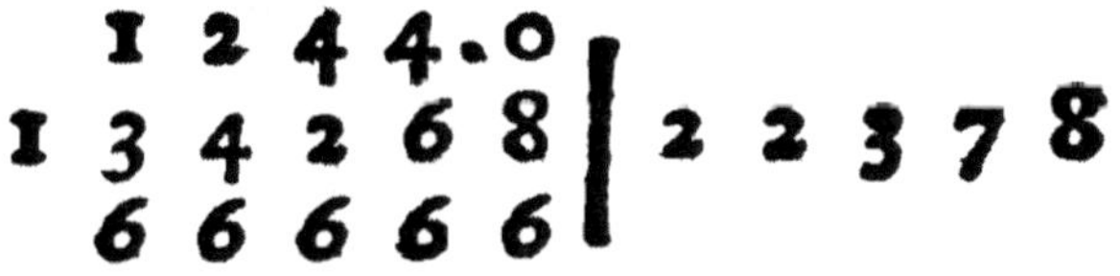

Figure 3.73: Bassi's division with quotient to right. No. 5, 1645, p. 59

Griminelli (No. 43, p. 67.) places the quotient to the left of the dividend in this method.

In the 'a Danda' method of division, as our method was called, we find various arrangements. The most common can be represented best as `a)b(c`, the mode followed by Bassi (No. 5. p. 55.) [3.74], Bonvicino (No. 12, pp. 11-14), and Figatelli (No. 32, p. 30.), who use vertical bars as symbolisms for division instead of curved bars.

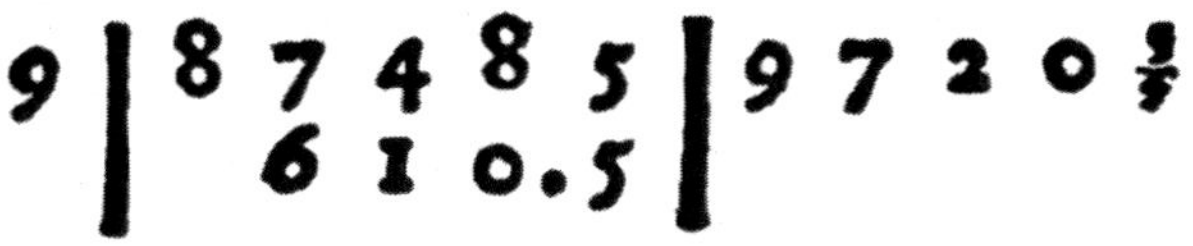

Figure 3.74: Bassi's division by a form of the 'a danda' method. No. 5, 1645, p. 55

Bonvicino (No. 12, p. 12.) presents an arrangement in short division in which the quotient is above the dividend, something like the following [3.75]:

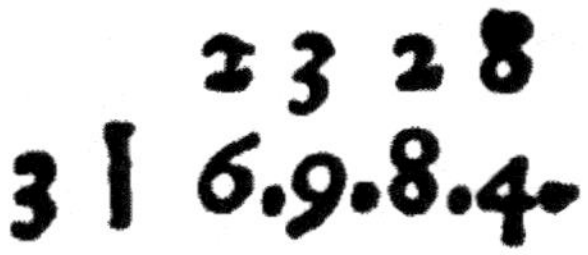

Figure 3.75: Bonvicino's division. No. 12, 1665, p. 12

Venturoli (No. 94, p. 7.) differs by having the quotient below the dividend as in [3.76]:

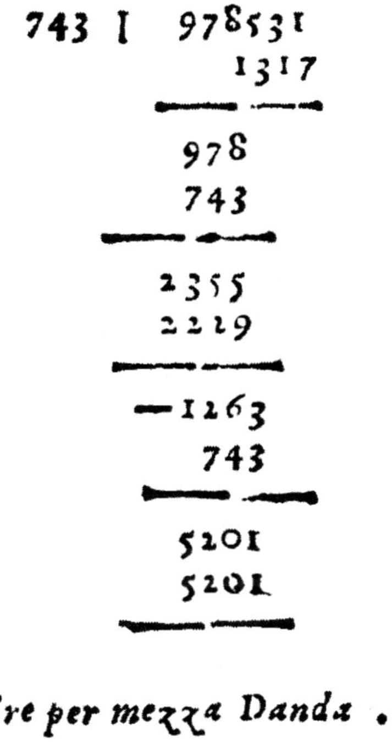

FIGURE 3.76: Venturoli's division. No. 94, 1681, p. 7

Still another variation is found in Griminelli (No. 43, p. 63), who places the quotient below the divisor in this manner [3.77]:

FIGURE 3.77: Griminelli's division. No. 43, 1656, p. 63

A more uncommon practice is also given by Bassi (No. 5, p. 61.), who places the divisor above the dividend [3.78].

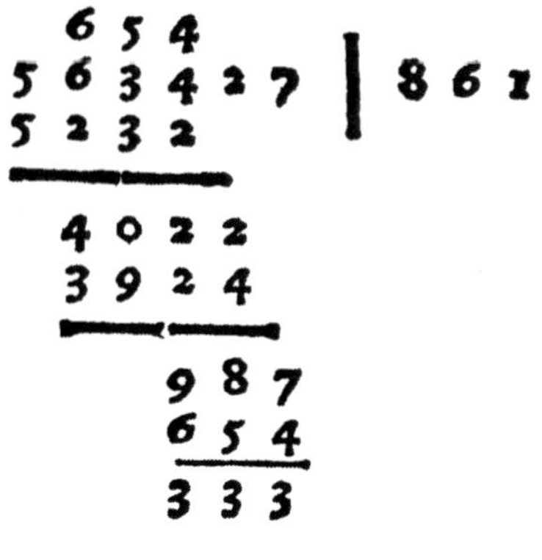

FIGURE 3.78: Bassi's division. No. 5, 1645, p. 61

This same method has been noted in Cataneo (1546), whereas Sfortunati (1561) placed the divisor either above or below the dividend. The very peculiar arrangement given by Finé of placing the quotient between the dividend and divisor, as given p. 15, is noteworthy. [3.79]

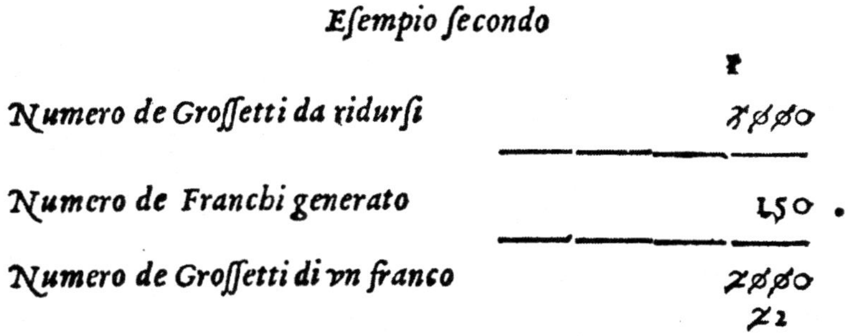

FIGURE 3.79: Finé's division. No. 34, 1587, f. 15, r.

Another instance of the so-called French practice of writing the divisor on the right is found in the Spanish Puig (No. 72, p. 82.), who writes [3.80]:

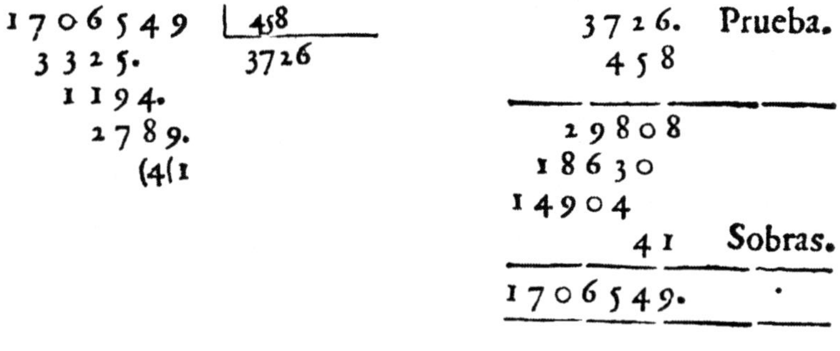

FIGURE 3.80: Puig's division. No. 72, 1672, p. 83

3.5 Signs of Proportion

3.5.1 Arithmetical and geometrical progression

Oughtred's notation for continued geometrical proportion was not commonly used until the latter part of the seventeenth century. We find only a few instances of its application in the texts under consideration. Cajori (p. 279.) has referred to the use of ∺ by Sir Jonas Moore in his 1688 *Arithmetick;* Moore's use of this symbol as early as 1660 has been cited previously [2.100, 2.101]. Isaac Barrow, according to Cajori, adopted Oughtred's symbol in his Latin (1655) and English (1660) editions of Euclid. We find this same symbol in Barrow's later work of 1675 (*Archimedis Opera Conicorum,* Libri III, London, p. 31.) [3.81].

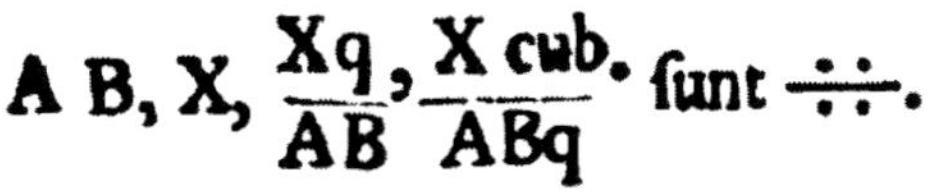

FIGURE 3.81: Barrow's continued geometric proportion. *Archimedis opera,* 1675, London, p. 31

In the same year, the French Prestet (No. 70, pp. 188, 170.) uses ∺ for geometrical progression, writing [3.82]

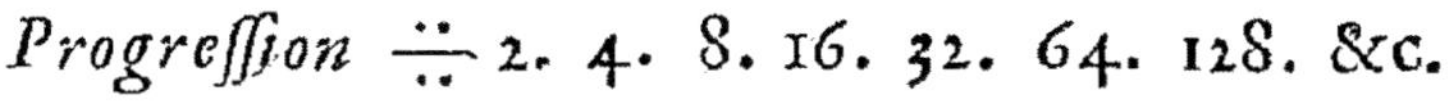

FIGURE 3.82: Prestet's geometrical progression. No. 70, 1675, p. 188

(This symbol is still sometimes seen in French textbooks. See D. E. Smith, *History of Mathematics,* v. 2, p. 413.) For arithmetical progression, however, he uses no symbol preceding the numbers, merely dots between them, as [3.83]

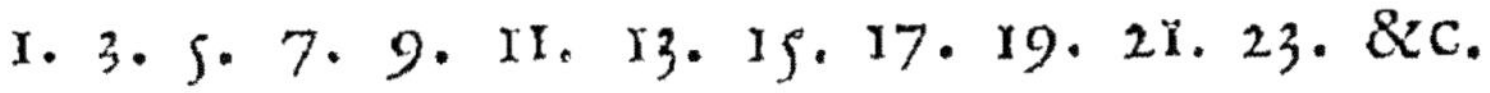

FIGURE 3.83: Prestet's arithmetical progression. No. 70, 1675, p. 170

Symbols for arithmetical progression are not in common use in the early books. The sign ÷ for arithmetical progression was not extensively applied until the eighteenth century. Ordinarily, for both kinds of progressions, we find a variety of symbols separating the numbers, as dots, commas, colons, or vertical bars, but no special symbol distinguishes arithmetical and geometrical progressions.

The use of dots simply as signs of separation has been mentioned in connection with Husswirt (1501)[26]. Followers of this practice are many: Noviomagus (No. 65, f. 43, v.), Vuelpius (No. 98, f. 38, v.), Micyllus (No. 60, p. 45.), Steinmetz (No. 86, f. 45.), Waser (No. 99, pp. 43, 45.), Clavius (No. 25, pp. 228, 243.), Alsted (No. 2, pp. 53, 62.), Bassi (No. 6, pp. 504, 505.), Puig (No. 72, pp. 267, 268.), Monier (No. 63, pp. 431, 436.), and Lucas (No. 58. pp. 180, 185.). Manelli (No. 59, p. 16.) writes [3.84, 3.85][27]

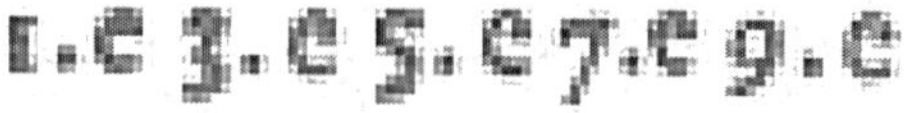

FIGURE 3.84: Manelli's arithmetic progression. No. 59, 1659, p. 16

and

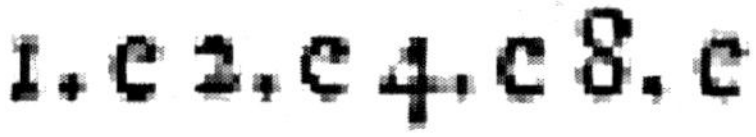

FIGURE 3.85: Manelli's geometric progression. No. 59, 1659, p. 16

Commas are used by Gemma (1549) and Peverone (No. 68, p. 20.). Blondel (No. 9, pp. 225, 229, 230.) uses commas or colons, and Unicorno (No. 93, f. 68.), vertical bars [3.86, 3.87].

FIGURE 3.86: Unicorno's arithmetic progression. No. 93, 1598, f. 68, v.

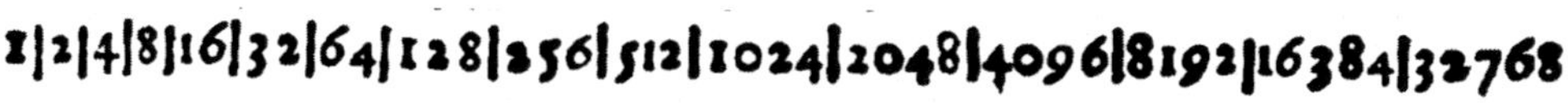

FIGURE 3.87: Unicorno's geometric progression. No. 93, 1598, f. 68, v.

3.5.2 Arithmetical proportion

Arithmetical proportion is not so commonly mentioned in the early texts, and a special notation for it is still less common. Just as in progressions, dots are used merely between the terms of the proportion. Alsted (No. 2, p. 53.) illustrates a disjunct arithmetical proportion by `10.8.4.2` [3.88].

[26] Ed. See section 2.1 on page 63.

[27] Ed. 'e', in Italian, translates to 'and' in English.

Proportio Arithmetica disjuncta est, quando duntaxat primus ad ſecundum ita ſe habet, ut tertius ad quartum, ita ut ſecundus ad tertium non habeat eundem etiam reſpectum. e.g. 10. 8. 4. 2. ſunt disjunctim proportionales, quia non continuatur differentia 2. inter 8. & 4.

FIGURE 3.88: Alsted's proportion. No. 2, 1641, p. 53

In Prestet (No. 70, p. 169.) there is a theorem on arithmetical proportion and the problem of finding the fourth term of the arithmetical proportion 15 . 10 . 18 as 13 [3.89].

COROLLAIRE.

En toute proportion arithmetique continuë, la ſomme des extrêmes eſt egale au moyen proportionel pris deux fois. Car ce terme tient la place de chacun des deux moyens. XXXVIII.

PREMIER PROBLEME.

Les trois premiers termes d'une proportion arithmetique eſtant connus, en trouver le quatriéme. XXXIX.

La ſomme des deux moyens moins le premier terme donnera ce terme. Soient par exemple 15, 10, & 18, les trois premiers termes d'une proportion arithmetique, la ſomme des moyens 10 & 18 eſt 28, dont le premier terme 15 eſtant retranché laiſſe pour reſte 13 qui eſt le quatriéme terme. Car par le Theoreme precedent ce terme 13 plus le premier qui eſt 15 donne une même ſomme que les deux moyens 10 & 18. 15. 10. 18. 13.

FIGURE 3.89: Prestet's proportion. No. 70, 1675, p. 169

3.5.3 Geometrical proportion

The 'Rule of Three' or of 'Proportion', as it is commonly called, was very popular in the early arithmetics, and we find various methods of presenting the numbers involved in a geometrical proportion. Short words were often used to write the proportion. Noviomagus (No. 65, f. 39, v.) writes [3.90]:

10 efficiunt 7. quot 17? 11. $\frac{2}{10}$.

FIGURE 3.90: Noviomagus' geometrical proportion. No. 65, 1544, f. 39, v.

In a similar way, Scheubel (No. 83, p. 136.), in the application of the rule of three, uses the form [3.91]:

Res	*aureis*		*Res*
21	18	*quanti*	29

FIGURE 3.91: Scheubel's geometrical proportion. No. 83, 1560, p. 136

and Clavius (No. 25, p. 118.) writes [3.92]:

FIGURE 3.92: Clavius' geometrical proportion. No. 25, 1618, p. 118

Sometimes horizontal lines are used to separate the three terms in the proportion, as by Alsted (No. 2, p. 55.), Lambeck (No. 55, p. 45.), Bassi (No. 6, p. 78.), Venturoli (No. 94, p. 81.), Behme (No. 7, p. 52.), and Blondel (No. 9, p. 177.). Rudolff (No. 81, f. 39, r.) separates them by dots and writes `3. 2. 12.` for `3:2::12:x`. Blondel also uses the colon in proportions as well as in progressions, thus: `2:4:5:10`, but Vuelpius (No. 98, f. 43, v.) simply separates the terms by spacing. As late as 1697, Lucas (No. 58, p. 84.) as Finé in the 1587 edition, designates the terms of a proportion by `A`, `B`, `C`, and `D`, respectively, thus [3.93]

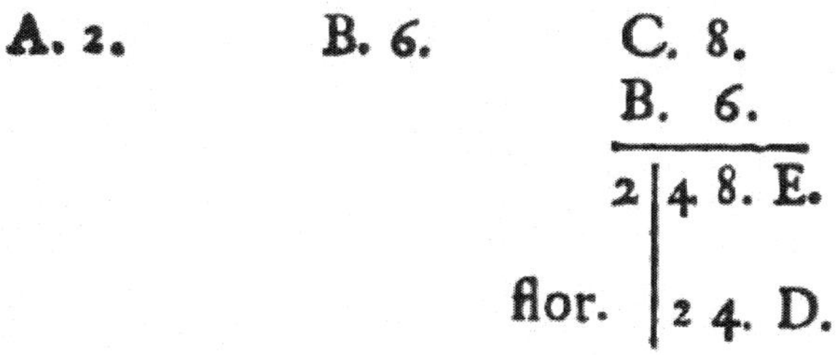

FIGURE 3.93: Lucas' geometrical proportion. No. 58, 1697, p. 84

In addition to the proportion notations given by Cajori, there is the use of the inclined bar by Finé in 1555, as well as Figatelli's clumsy notation for proportion, consisting of braces and words (illustrated above)[2.83]. Cajori refers (p. 289.) to a notation by Herigone (No. 49.) in which the letter π signifies ratio, but he does not mention that, in the explanation of the notes, Herigone writes the proportion `a:b::c:d` in these two ways [3.94, 3.95]:

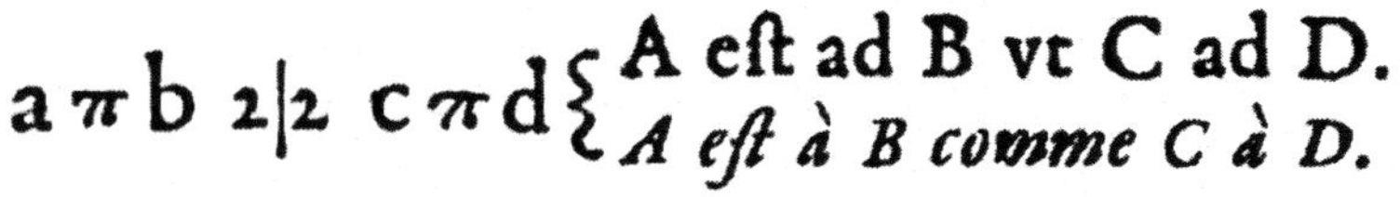

FIGURE 3.94: Herigone's symbol for proportion. No. 49, 1643, p. xx

or

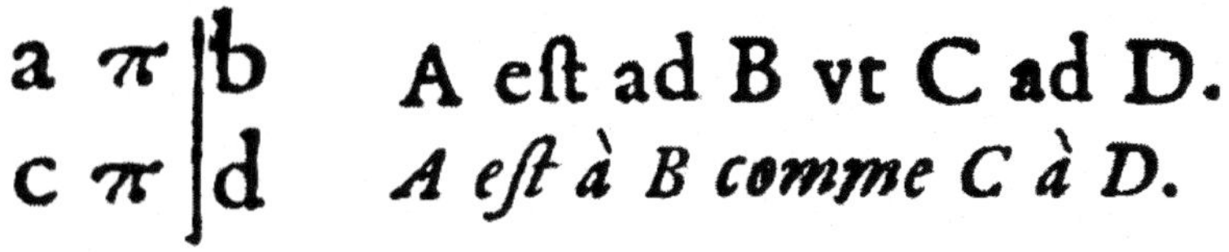

FIGURE 3.95: Herigone's symbol for proportion. No. 49, 1634, p. xx

(Cajori incorrectly says on page 289 that the first edition appeared in 1642, and on page 277 he refers to an 1834 edition, meaning 1634.)

3.5.4 Oughtred's and Wing's notations

The dot for geometrical ratio, and the four dots of proportion, . : : ., introduced by Oughtred (*Clavis Mathematicæ,* London, 1631.) [3.96],

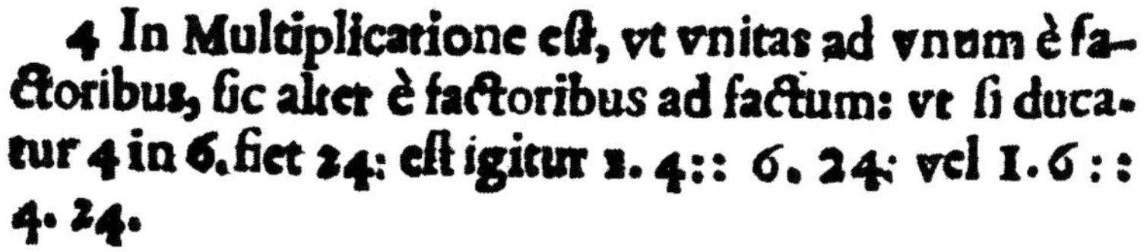
4 In Multiplicatione eſt, vt vnitas ad vnum è fa-
ctoribus, ſic alter è factoribus ad factum: vt ſi duca-
tur 4 in 6. fiet 24: eſt igitur 1. 4 :: 6. 24: vel 1. 6 ::
4. 24.

FIGURE 3.96: Oughtred's notation for proportion in *Clavis Mathematicae,* 1631, London, p. 5.

were used by Sir Jonas Moore (1660). This symbol, and a modification of it in the form $\mathtt{a} \mathbin{\overline{\cdot\cdot}} \mathtt{b} : \mathtt{b} \mathbin{\overline{\cdot\cdot}} \mathtt{X}$, were employed by Alexander (1693), noted by Cajori and above[28]. Prestet (No. 70, p. 187.) is again a follower of Oughtred in the use of the notation . : : . for geometrical proportion. It is of interest to add that, in using this notation, Prestet says, after writing $\frac{6}{3} = \frac{8}{4}$, that he is 'persuaded that this sort of expression, 6.3::8.4, has been chosen to avoid that of small characters which mark fractions, and in order not to use the sign of equality.'

The colon as the symbol for ratio, introduced by Vincent Wing (1651), was not generally adopted before 1700. Ceva (No. 20, p. 12.) [3.71] uses it, but instances of the use of either Oughtred's or Wing's notations in the late seventeenth-century arithmetics are very rare.

3.6 Signs of Equality

3.6.1 Equality expressed by words and abbreviations

For more than a century after the introduction of the symbol = by Recorde in 1557 [3.97], equality was still designated by words and abbreviations. In the German and Latin works of this early period, we find such expressions as *facit*, *faciunt*, *aequales*, and *aequantur*, among the authors Rudolff (1532), Cardan (1537), Regius (1543), Noviomagus (15440, Micyllus (1555), Scheubel (1560), Suevus (1593), Waser (1603), Vieta (1646), Behme (1695), and Lucas (1697). The Italian *eguale à*, is found in Bombelli (1579) *è* in Griminelli (1670), and *fanno* or *fà* in Unicorno (1598), Zuchetta (1600), Manelli (1659), and in the Italian edition of Clavius (1618). Aurel (1552) and Puig (1672) give the Spanish *ygual a*, *igual à* or *ig. à* respectively. In the French works we find the words *égal* and *font* as in Vieta (1630) and Blondel (1699).

[28] ED. See section 2.18 on page 102.

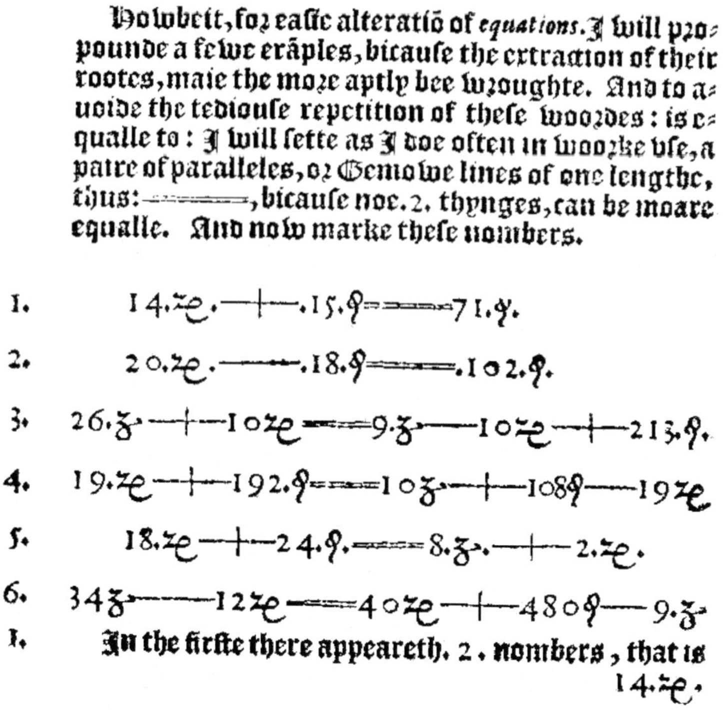
Howbeit, for easie alteratiō of *equations*. I will propounde a fewe exāples, bicause the extraction of their rootes, maie the more aptly bee wroughte. And to auoide the tediouse repetition of these woordes: is equalle to: I will sette as I doe often in woorke use, a paire of paralleles, or Gemowe lines of one lengthe, thus: ═══, bicause noe. 2. thynges, can be moare equalle. And now marke these nombers.

1. 14.ze.—+—.15.ϱ═══71.ϱ.
2. 20.ze.——.18.ϱ═══.102.ϱ.
3. 26.ʒ.—+—10ze═══9.ʒ.——10ze—+—213.ϱ.
4. 19.ze—+—192.ϱ═══10ʒ—+—108ϱ——19ze
5. 18.ze—+—24.ϱ.═══8.ʒ.—+—2.ze.
6. 34ʒ——12ze═══40ze—+—480ϱ——9.ʒ.

1. In the firste there appeareth. 2. nombers, that is 14.ze.

FIGURE 3.97: Recorde's *The Whetstone of Witte*, with the first use of = for equality, 1557, London.

3.6.2 Recorde's Sign of Equality

Cajori (p. 303.) cites the Swiss Johann Alexander as employing Descartes' sign of equality ꭗ; in reality, we find that although Alexander gives both = and ꭗ as equality symbols on page two of his 1693 edition [2.109], he uses Recorde's symbol throughout his work in problems. Harriot (1631) [3.98] and Prestet (1675) [3.99],

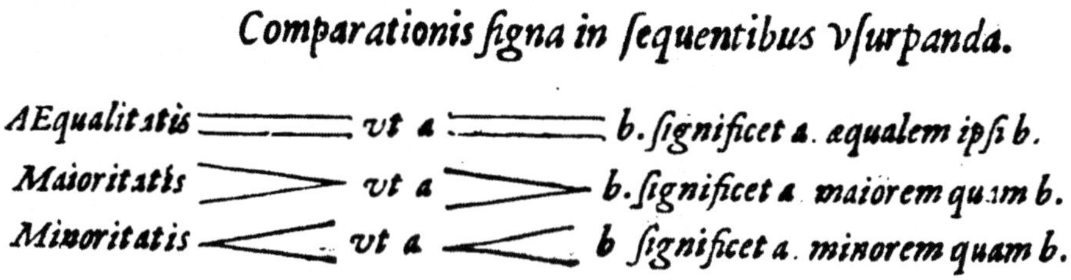
Comparationis signa in sequentibus vsurpanda.

AEqualitatis ═══ *vt a* ═══ *b. significet a. aequalem ipsi b.*
Maioritatis > *vt a* > *b. significet a maiorem quam b.*
Minoritatis < *vt a* < *b significet a. minorem quam b.*

FIGURE 3.98: Harriot's equality sign. No. 45, 1631, p. 10

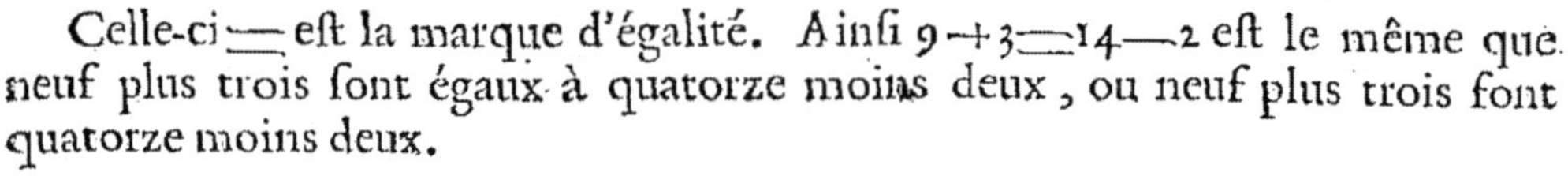
Celle-ci ═ est la marque d'égalité. Ainsi 9 + 3 ═ 14 — 2 est le même que neuf plus trois sont égaux à quatorze moins deux, ou neuf plus trois font quatorze moins deux.

FIGURE 3.99: Prestet's equality sign. No. 70, 1675, p. 2

by adopting Recorde's symbol, did much to give it a strong foothold in England and France.

In Italy, Recorde's symbol is used by Renaldini (No. 78, pp. 85, 86.) in 1665. Cajori does not include any mention of the early use of this symbol in Italy, although Renaldini is later listed by Cajori among the representative Italian writers. In Renaldini's work, the type used in printing the equality symbol is the letter `l`, placed horizontally, thus [3.100]

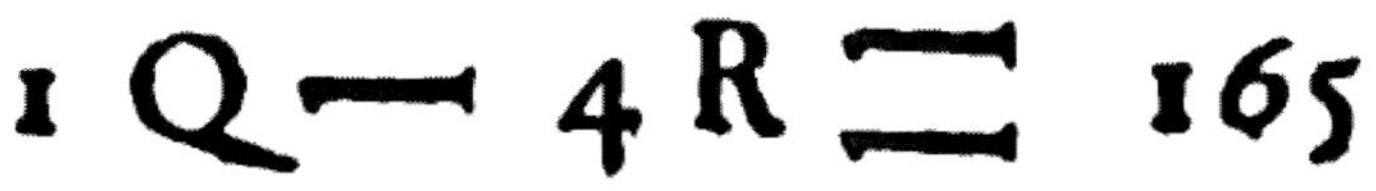

FIGURE 3.100: **Renaldini's long equality sign. No. 78, 1665, p. 85**

the `l`'s were sometimes shorter and closer together as [3.101]

FIGURE 3.101: **Renaldini's short equality sign. No. 78, 1665, p. 86**

This form of the symbol was very convenient, for one of the `l`'s generally served as his minus sign.

Our modern form of Recorde's sign of equality = is used also by the Spanish Ventura de Avila (No. 4, p. 2.), whose work is in dialog form much like Recorde's *Ground of Artes*. Ceva's symbol (No. 20, p. 9.) consists of very long lines close together.[29]

3.6.3 Symbols for equality before 1700

The majority of writers of the sixteenth century expressed equality by words, as noted above. Even in the seventeenth century, we find noted mathematicians using no regular symbol for equality. Recorde's sign, if any, was used by writers in the second half of the seventeenth century, but this sign was not practically established until the beginning of the eighteenth century.

3.7 Signs of Common Fractions

3.7.1 The fractional line

The fractional line was generally used by sixteenth and seventeenth century authors although there are several instances of its omission in the case of 'fractions of fractions.' Cajori's reference (pp. 152, 153.) to the omission of the fractional line after the first fraction by Clavius in his *Epitome arithmeticae* was to the 1583 edition. We also note this same feature in his 1585 and 1586 editions, usually in the case of the more common fractions. No such omission is made, however, in his later works of 1618 and 1686. Tunstall (1538)[30], Finé (1555)[31], and Ramus (1599)[32], as noted above, also have this

[29] Ed. This sentence is misleading. Contrasting Figures 3.39 and 3.71 of Ceva and Figure 3.98 of Harriot will verify this. These figures are on pages 126, 136, and 146 respectively.

[30] Ed. See section 2.8 on page 78.

[31] Ed. See section 2.10 on page 82.

[32] Ed. See section 2.12 on page 87.

distinct notation for 'fractions of fractions', and Savonne (1619) frequently omits the fractional lines altogether in the fundamental operations.[33] We find that Ghaligai (No. 38, pp. 15, 17.) sometimes writes his fractions without a separating line, but since this is not usual, it may possibly be the result of faintness in type.

Broken fractional lines, so very common in these early centuries, are an indication of the difficulty experienced in printing fractions with large numerators and denominators. We have remarked this concerning Husswirt (1501)[34] and Peurbach (1539)[35], and we may add Clavius in all his editions, Regius (1543), Noviomagus (1544), Tunstall (1544), Feliciano (1550, 1692), Gemma (1550), Micyllus (1555), Tartaglia (1556), Sfortunati (1561), Steinmetz (1575), Jacob (1612), Ramus (1613), Napier (1623), Chauvet (1627), Girard (1629), Alsted (1641), Moore (1660), Hemeling (1667), and Rahn (1668).

The length of the fractional line was not standardized, as may be seen in the case of Renaldini (No. 78, pp. 447, 296.) [3.102, 3.103, 3.104] who generally uses a very long line, but sometimes we find various lengths of lines on the same page. Renaldini leaves considerable space between the fractional line and the terms of a fraction. This feature is also noted in the early and rare German book of Felner (No. 30, f. 26.).

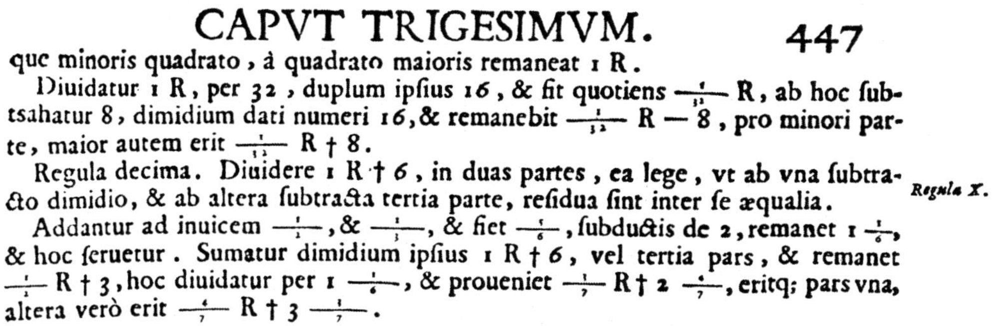
CAPVT TRIGESIMVM. 447

que minoris quadrato , à quadrato maioris remaneat 1 R.

Diuidatur 1 R, per 32 , duplum ipſius 16 , & fit quotiens $\frac{1}{32}$ R, ab hoc ſubtsahatur 8, dimidium dati numeri 16, & remanebit $\frac{1}{32}$ R — 8 , pro minori parte, maior autem erit $\frac{1}{32}$ R † 8.

Regula decima. Diuidere 1 R † 6 , in duas partes , ea lege , vt ab vna ſubtracto dimidio, & ab altera ſubtracta tertia parte, reſidua ſint inter ſe æqualia. Regula X.

Addantur ad inuicem $\frac{1}{2}$, & $\frac{1}{3}$, & fiet $\frac{5}{6}$, ſubductis de 2, remanet 1 $\frac{1}{6}$, & hoc ſeruetur. Sumatur dimidium ipſius 1 R † 6, vel tertia pars, & remanet $\frac{1}{2}$ R † 3, hoc diuidatur per 1 $\frac{1}{6}$, & proueniet $\frac{3}{7}$ R † 2 $\frac{4}{7}$, eritq; pars vna, altera verò erit $\frac{4}{7}$ R † 3 $\frac{3}{7}$.

FIGURE 3.102: Renaldini's fractional lines. No. 78, 1665, p. 447

[33] ED. See Figure 2.65 on page 86.
[34] ED. See section 2.1 on page 63.
[35] ED. See section 2.9 on page 80.

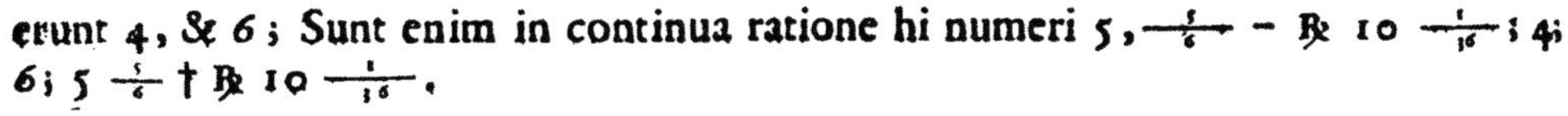
erunt 4, & 6; Sunt enim in continua ratione hi numeri 5, $\frac{1}{6}$ − ℞ 10 $\frac{1}{16}$; 4; 6; 5 $\frac{1}{6}$ † ℞ 10 $\frac{1}{16}$.

FIGURE 3.103: Renaldini's fractional lines. No. 78, 1665, p. 296

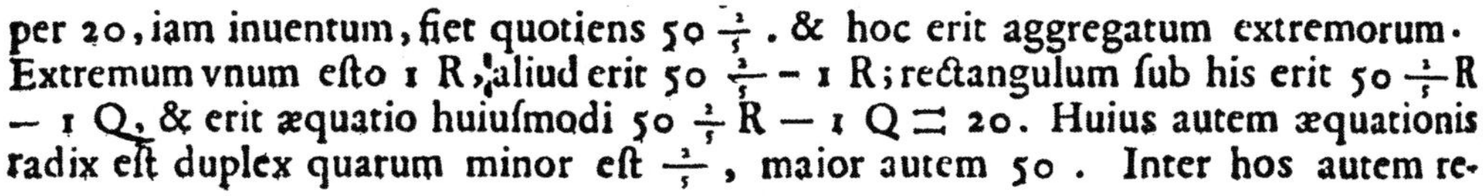
per 20, iam inuentum, fiet quotiens 50 $\frac{1}{5}$. & hoc erit aggregatum extremorum. Extremum vnum esto 1 R, aliud erit 50 $\frac{1}{5}$ − 1 R; rectangulum sub his erit 50 $\frac{1}{5}$ R − 1 Q, & erit æquatio huiusmodi 50 $\frac{1}{5}$ R − 1 Q ⊐ 20. Huius autem æquationis radix est duplex quarum minor est $\frac{1}{5}$, maior autem 50. Inter hos autem re-

FIGURE 3.104: Renaldini's fractional lines. No. 78, 1665, p. 296

An extraordinary fractional notation is brought out as late as 1675 in the work of Ciacchi (No. 22, pp. 48–50.). In his arithmetic he generally employs fractions whose terms consist of only one digit, and these numbers are placed sidewise in the running text, thus: $\frac{1}{2}$ or $\frac{1}{2}$ and $\frac{2}{3}$ or $\frac{2}{3}$ [3.105].

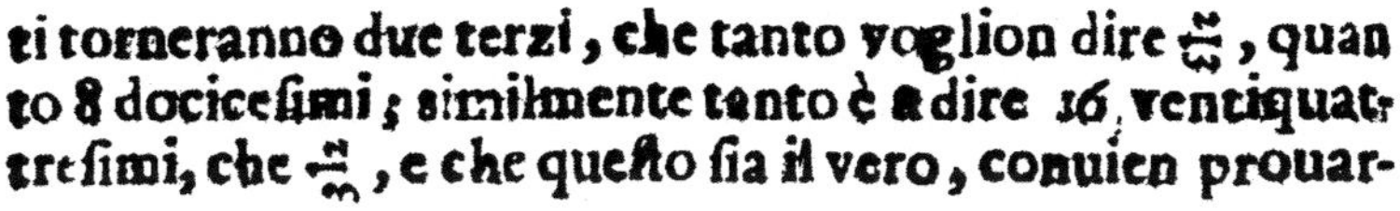
ti torneranno due terzi, che tanto voglion dire $\frac{2}{3}$, quan to 8 docicesimi; similmente tanto è a dire 16. ventiquat- tresimi, che $\frac{2}{3}$, e che questo sia il vero, conuien prouar-

FIGURE 3.105: Ciacchi's fractions. No. 22, 1675, p. 48

Somehow he prefers $\frac{1}{2}$ and $\frac{2}{3}$ to the alternative forms, probably because these can be written and read with greater facility and naturalness. In illustrations, however, he sometimes uses a heavy horizontal line, and writes the numerator and denominator very far apart (*Ibid.*, p. 58.) [3.106].

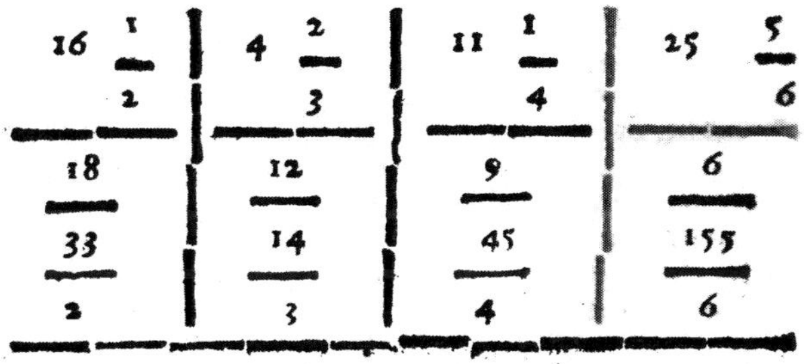

16 $\frac{1}{2}$	4 $\frac{2}{3}$	11 $\frac{1}{4}$	25 $\frac{5}{6}$
18	12	9	6
33	14	45	155
2	3	4	6

FIGURE 3.106: Ciacchi's fractions. No. 22, 1675, p. 58

In case of fractional terms of more than one digit he uses the ordinary form, as in $\frac{232}{299}$ (*Ibid.*, p. 182.) [3.107].

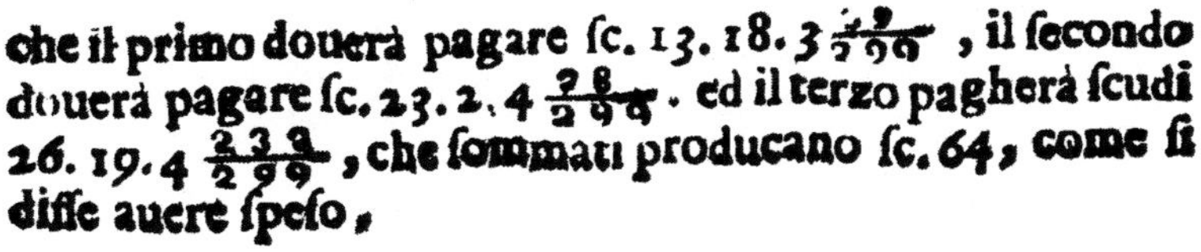
che il primo douerà pagare sc. 13. 18. 3 $\frac{39}{299}$, il secondo douerà pagare sc. 23. 2. 4 $\frac{28}{299}$. ed il terzo pagherà scudi 26. 19. 4 $\frac{232}{299}$, che sommati producano sc. 64, come si disse auere speso.

FIGURE 3.107: Ciacchi's fractions. No. 22, 1675, p. 182

3.8 Signs of Decimal Fractions

3.8.1 Early notations

Many of the early writers, in the division of numbers by 10, used a vertical bar to separate the quotient from the remainder. We might interpret this symbol as a decimal separatrix, but these writers themselves did not have the full concept of decimal fractions. (Rudolff's *Exempelbüchlein,* Augsburg, 1530, is the only known case of the use of the decimal bar at that date. See D. E. Smith's article in *Teachers College Bulletin,* s. 1, v. 5, 1910–11, p. 21, in which Rudolff's discovery was first announced. Eneström, however, thinks that Rudolff did not realize the full significance of his work. See *Scripta Mathematica,* Dec., 1932, pp. 168, 169). In addition to the writers mentioned by Cajori we have noted above the use of the vertical stroke by Tagliente (1520)[36], Wolphius (1534)[37], Cataneo(1546)[38], and Sfortunati (1561)[39], all of which appeared before the first printed explanation of decimal fractions by Stevin in 1585 (See D. E. Smith's *Source Book in Mathematics,* New York, 1929, for a selection from "Stevin on Decimal Fractions", pp. 20–34). Ramus (1599) also made this close approach to decimal fractions.

We quote from Rudolff's *Künstliche rechnung* (No. 81. f. 10, v.) [3.108]:

> Zu mercken das 1 weder multiplicirt noch dividirt. Demnach wañ ein zal geteilt sol werden in 10/ schneid ir ab rait einer virgel die erst ziffer/ die figuren gegen der lincken hand sein der quocient / uñ die erst abgeschnitten figuren das rest ...

Geteilt in 10	Geteilt in 100	Geteilt in 1000
678|5	67|85	6|785

FIGURE 3.108: Rudolff's decimal bar. No. 81, 1540, f. 10, v.

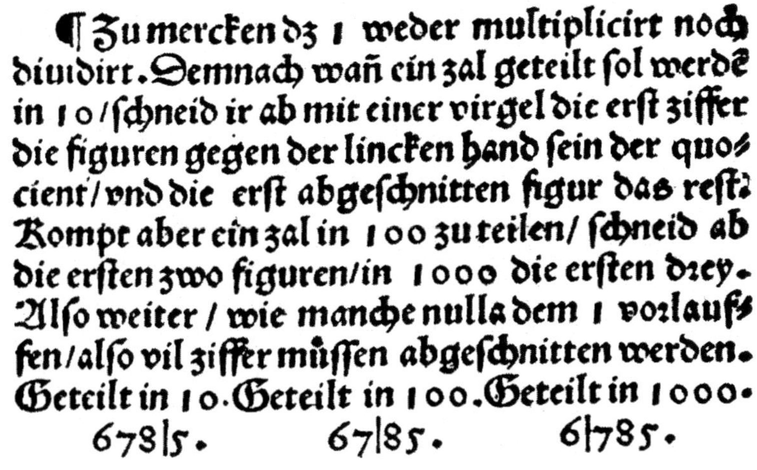
¶ Zu mercken dz 1 weder multiplicirt noch
diuidirt. Demnach wañ ein zal geteilt sol werdē
in 10/ schneid ir ab mit einer virgel die erst ziffer
die figuren gegen der lincken hand sein der quo-
cient/ vnd die erst abgeschnitten figur das rest.
Kompt aber ein zal in 100 zu teilen/ schneid ab
die ersten zwo figuren/ in 1000 die ersten drey.
Also weiter/ wie manche nulla dem 1 vorlauf-
fen/ also vil ziffer müssen abgeschnitten werden.
Geteilt in 10. Geteilt in 100. Geteilt in 1000.
678|5. 67|85. 6|785.

[36] Ed. See section 2.4 on page 69.
[37] Ed. See section 2.2 on page 66.
[38] Ed. See section 2.5 on page 72.
[39] Ed. See section 2.7 on page 77.

Similarly, Felner (No. 30, f. 15, v.) divides 67845 by 100 thus:

```
678|45 facit|678
1  |00
```

das Rest 45. macht am kleinsten $\frac{9}{20}$

The figure 45 looks like $\frac{45}{100}$, but Felner regards it simply as a remainder. Tartaglia (No. 90, f. 30, v.), Scheubel (No. 83, p. 52.), Steinmetz (No. 86, f. 43, v.), and Unicorno (No. 93, p. 38.) also use the vertical bar in illustrating division by 10 and 100. Peverone's (No. 68, p. 18.) separatrix takes the form of an inclined bar [3.109]

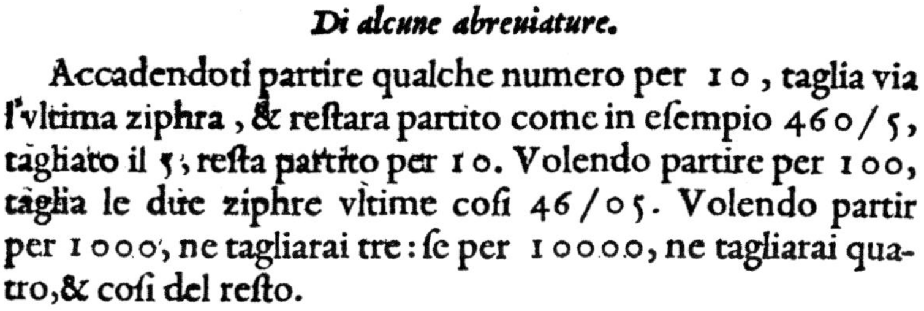
Di alcune abreuiature.

Accadendoti partire qualche numero per 10, taglia via l'vltima ziphra, & restara partito come in esempio 460/5, tagliato il 5, resta partito per 10. Volendo partire per 100, taglia le due ziphre vltime cosi 46/05. Volendo partir per 1000, ne tagliarai tre: se per 10000, ne tagliarai quatro,& cosi del resto.

FIGURE 3.109: Peverone's decimal fractions. No. 68, 1558, p. 18

but serves the same purpose. Even as late as 1659 we find Manelli (No. 59, p. 17.) employing the notation 87|0 [3.110],

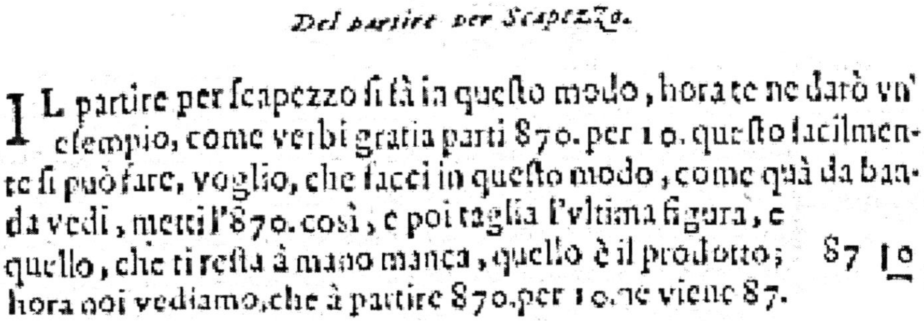
Del partire per Scapezzo.

IL partire per scapezzo si fà in questo modo, hora te ne darò vn' essempio, come verbi gratia parti 870. per 10. questo facilmente si può fare, voglio, che facci in questo modo, come quà da banda vedi, metti l'870. così, e poi taglia l'vltima figura, e quello, che ti resta à mano manca, quello è il prodotto; 87|0 hora noi vediamo, che à partire 870. per 10. ne viene 87.

FIGURE 3.110: Manelli's decimal fractions. No. 59, 1659, p. 17

in the division of 870 by 10, a close approach to the rectangular line symbol ∟ [3.111] used by Oughtred (1631) in the writing of decimals.

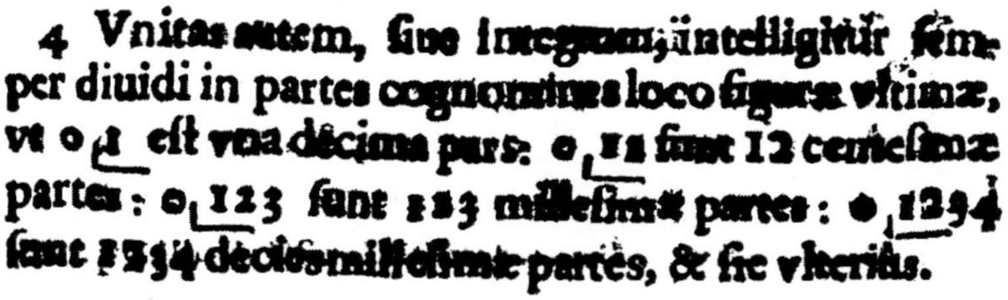
4 Vnitas autem, siue Integrum, intelligitur semper diuidi in partes cognominiales loco figuræ vltimæ, vt 0|1 est vera decima pars: 0|12 sunt 12 centesimæ partes: 0|123 sunt 123 millesimæ partes: 0|1234 sunt 1234 decies millesimæ partes, & sic vlterius.

FIGURE 3.111: Oughtred's sepratrix in *Clavis Mathematicæ*. 1631, London, p. 1

3.8.2 Seventeenth-century notations

An illustration of use of the decimal comma and point as separators of units and tenths by Napier in his *Rabdologia* of 1617 is given by Cajori (p. 323.). In it the quotient of 861094 and 432 is written as 1993, 273. In the 1623 edition of the *Rabdologia,* which is an Italian translation of the 1617 edition, we find this same quotient written as (No. 64, p. 56.) [3.112]:

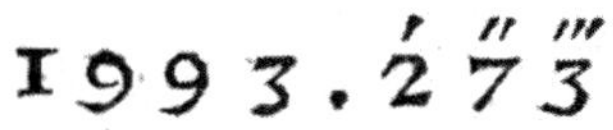

FIGURE 3.112: Napier's quotient. No. 64, 1623, p. 56

Again, Cajori (p. 324.) states that Napier writes $1\,9\,9\,4\,,\,\overset{\prime}{9}\,\overset{\prime\prime}{1}\,\overset{\prime\prime\prime}{6}\,\overset{\prime\prime\prime\prime}{0}$ for 1994.9160, but in the 1623 edition he merely uses the comma, as 1994,9160 (*Ibid.,* p. 94.) [3.113].

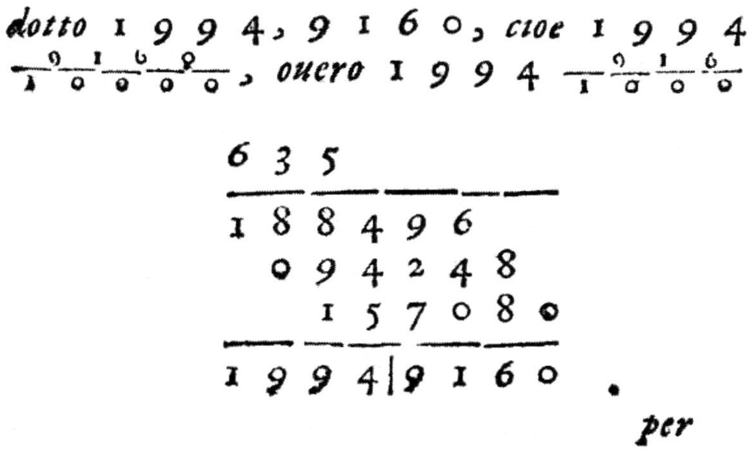

FIGURE 3.113: Napier's comma. No. 64, 1623, p. 94

Another use of the comma for the same purpose occurs when Napier writes 821,25 for $821\frac{1}{4}$. (*Ibid.,* p. 88.). These three examples indicate his vacillation in the use of period, comma, and sexagesimal exponents as decimal symbols.

Reference has been made to Sir Jonas Moore's five different decimal notations[40], whereas Cajori notes for this author only one symbol, the inverted comma. Voigtel (No. 97, p. 2.) [3.114, 3.115] devotes the first chapter of his *Marckscheide-Kunst* to 'Decimal-Rechnung', and in it he has a preference for the sexagesimal exponents for decimals; he usually writes $3\;6\;\overset{\circ}{5}\;\overset{\prime}{7}\;\overset{\prime\prime}{3}\;\overset{\prime\prime\prime}{5}$, [3.114] occasionally $\overset{\circ}{5}.\overset{\prime}{8}.\overset{\prime\prime}{9}.\overset{\prime\prime\prime}{2}$, [3.115]

[40] ED. See section 2.17 on page 95.

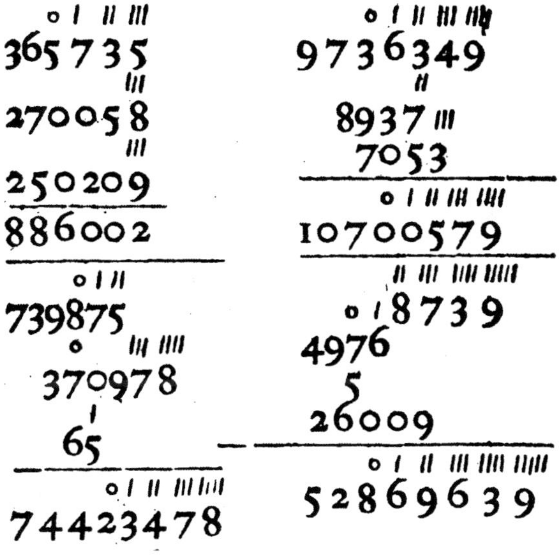

FIGURE 3.114: Voigtel's decimal fractions. No. 97, 1686, p. 2

• ı ıı ııı
mit zweyen etc. bezeichnet; als zum Exempel: dieſe Zahlen 5.8.9.2.
werden alſo ausgeſprochen: 5. Lachter/8. Erſtens/9. Zweytens / und
2. Drittens.

FIGURE 3.115: Voigtel's decimal fractions. No. 97, 1686, p. 2

In Lucas (No. 58, p. 225.) there is a chapter of eight pages on 'Arithmetica geometrica decimalis'; in it he uses the tautological notation $11^{o}.4^{\prime}.5^{\prime\prime}.6^{\prime\prime\prime}$. [3.116]

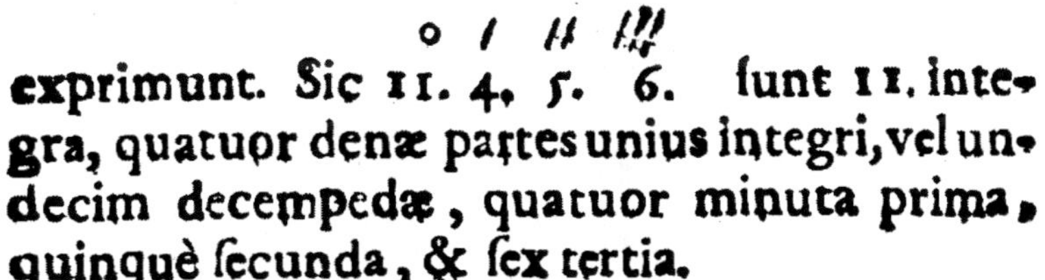

o ı ıı ııı
exprimunt. Sic 11. 4. 5. 6. ſunt 11. inte-
gra, quatuor denæ partes unius integri, vel un-
decim decempedæ, quatuor minuta prima,
quinquè ſecunda, & ſex tertia.

FIGURE 3.116: Lucas's decimal fractions. No. 58, 1697, p. 225

Much earlier, Graffenried (No. 42, p. 648.) discusses decimals on seven pages and writes them in the form $3|\ 9\ \dot{5}\ \ddot{8}\ \dddot{7}$ [3.117]

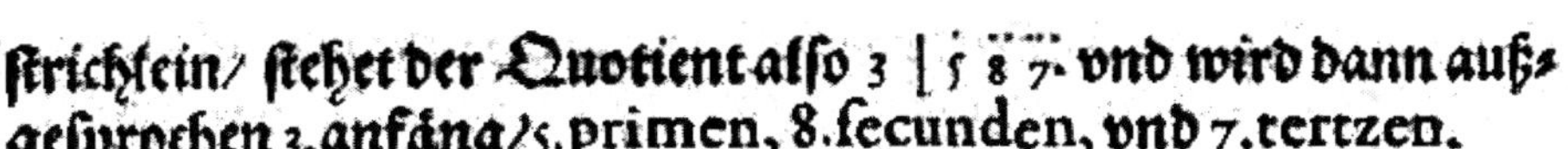

ſtrichlein/ ſtehet der Quotient alſo 3 | 5 8 7. vnd wird dann auß-
geſprochen 3. anfäng/5. primen, 8. ſecunden, vnd 7. tertzen.

FIGURE 3.117: Graffenried's decimal fractions. No. 42, 1619, p. 648

An entire text devoted to the doctrine of decimals, including tables of interest and rebate, is that of Hodder (No. 50, pp. 12, 22.). The comma is given as a symbol once, but the period is otherwise used throughout. The decimal 0.35 is read as 3 Primes and

5 Seconds according to custom. This text is of particular interest as Hodder's was the first separate treatise on arithmetic published in English in America (1719).

3.9 Signs of Powers

3.9.1 Early symbolisms

In the early Latin books, in connection with the extraction of roots, one often finds the expressions *numerus quadratus* and *numerus cubicus* to represent the square or cube of a given number, as in Vuelpius (No. 98, f. 57, r.). Micyllus (No. 60, p. 194.) labels the 4 and 8 of the geometrical progression 1,2,4,8,. . . as quad., Cub. respectively [3.118].

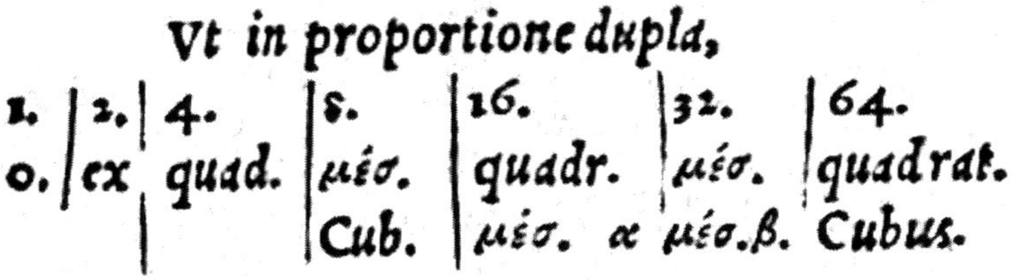

Vt in proportione dupla,

1.	2.	4.	8.	16.	32.	64.
0.	ex	quad.	μέσ.	quadr.	μέσ.	quadrat.
			Cub.	μέσ. α	μέσ.β.	Cubus.

FIGURE 3.118: Micyllus' symbols. No. 60, 1555, p. 194

Very frequently a table of the squares and cubes of numbers is given with the column headings R., Q., C.; we find this in Bonvicino (No. 12, p. 33.) [3.119]

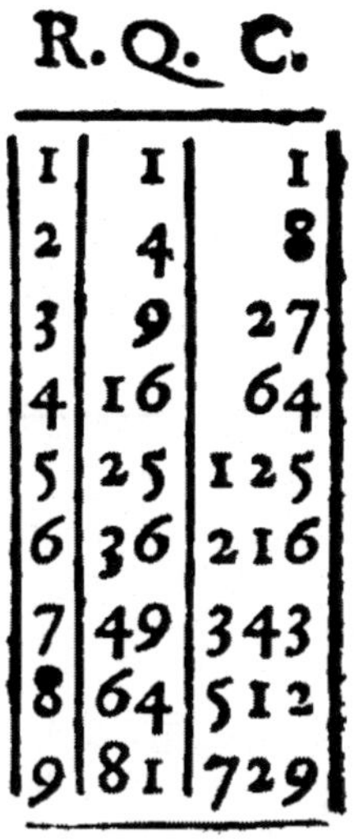

R.	Q.	C.
1	1	1
2	4	8
3	9	27
4	16	64
5	25	125
6	36	216
7	49	343
8	64	512
9	81	729

FIGURE 3.119: Bonvicino's symbols. No. 12, 1665, p. 33

as late as 1665. The powers of known numbers did not need any symbolism until the latter came to be represented by letters; there was need, however, of a symbolism for powers of the unknown.

3.9.2 Additive principle

The additive principle of the Greeks in combining powers was not so popular as the multiplicative principle of the Hindus. Cajori (p. 342.) has referred to the use of the additive principle by Vieta in several of his publications, but not to his important French work, *L'Algebre,* of 1630. In the introduction Vieta (No. 95, pp. 7, 8.) lists the names of the powers according to the additive principle as follows [3.120, 3.121]:

La premiere des grandeurs ſcalaires eſt,
Le coſté ou la racine.
2. Le quarré.
3. Le cube.
4. Le quarré quarré.

FIGURE 3.120: Viète's additive principle. No. 95, 1630, p. 7

5. Le quarré cube.
6. Le cube cube.
7. Le quarré quarré cube.
8. Le quarré cube cube.
9. Le cube cube cube.

FIGURE 3.121: Viète's additive principle. No. 95, 1630, p. 8

In the second book of 'Zetetiques' (*Ibid.,* pp. 80, 81.) he uses the symbols N, Q, and C [3.122] for the unknown number and its powers;

Soit D pl. 104. B pl. 96. le plus grand coſté 1 N. 1 Q eſt egal à 100.
Soit le plus petit coſté 1 N. 1 Q eſt egal à 4.

FIGURE 3.122: Viète's unknown numbers to the first and second power. No. 95, 1630, p. 80

in the third book, the abbreviations are *Aq.*, *Ac.* [3.123]

de A c. — 3 B pl. par A, egal à B pl. par Z. C'eſt à dire

FIGURE 3.123: Viète's unknown numbers to the first and third power. No. 95, 1630, p. 114

Another French writer, Blondel (No. 9, p. 18.), uses the additive principle, but he has a preference for capital letters thus: ℞, Q, C, QQ, CC, etc. [3.124, 3.125]

FIGURE 3.124: Blondel's additive principle. No. 9, 1683, v. 2, p. 18

me le nombre 4 eſt √ ou Racine quarrée de 16 ; √C ou √ 3 ou Racine cubique de 64 ; √√ ou √QQ ou √. 4 c'eſt à dire Racine de racine ou

FIGURE 3.125: Blondel's additive principle. No. 9, 1683, v. 2, p. 18

SPECULATIVE. 19

racine quarréquarrée de 256 ; √ QC, ou √ 5 c'eſt à dire Racine quarrécube de 1024 ; √CC ou √ 6 c'eſt à dire Racine cubecube de 4096 ; √QQC ou √ 7 c'eſt à dire racine quarréquarrécube de 16384. & ainſi des autres.

3.9.3 Multiplicative principle

Pacioli has many followers in the use of his power symbols, *co.*, *ce.*, *cu.*, *ce.ce.*, $p^o.r^o.$, *ce.cu.*, etc. Feliciano's use of these Italian symbols in the 1550 edition has been noted above[41], but in 1669 the editor of his work substituted *A*, *AA*, etc., for them [3.126].

FIGURE 3.126: Feliciano's variables to first and second power. No. 29, 1669, p. 367

Suppoſto, che la Dote di Claudia ſij ſcudi 1 A. Dunque il peculio d'Orlando è 1 A — 300. e moltiplicando in ſe ſteſſa la Dote 1 A, ne viene 1 AA, e moltiplicando in ſe il peculio 1 A — 300. ne viene 1 AA — 600 A ✣ 90000. come vedete.

		1 A — 300	
		1 A — 300	
1 A		— 300 A	✣ 90000
1 A	1 AA	— 300 A	
1 AA	1 AA	— 600 A	✣ 90000

[41] Ed. See section 2.6 on page 73.

Unicorno (No. 93, f. 136, r.) uses *cosa*, *censo*, *cubo*, *ce.ce.*, *l.rel.*, *ce.cu.*, etc., merely a slight modification of Pacioli [3.127].

Ordine

coſa	2---1
cenſo	4---2
cubo	8---3
ce.ce.	16---4
1. rel.	32---5
ce.cu.	64---6
2. rel.	128---7
ce.ce.c.	256---8
cu.cu.	512---9
ce.rel.	1024-10
3. rel.	2048-11
c.ce.ce.	4096-12
	8192-13
	16384-14
	32768-15

r.q.c.s.p.r.e.q.2

$\frac{2}{8}$

FIGURE 3.127: Unicorno's powers. No. 93, 1598, f. 136, r.

In Bassi (No. 6, p. 545.) we find *numero*, *cos.*, *quad.*, *cub.*, *quad.qu.*, *prim.rel.*, *quad.cub.* [3.128],

FIGURE 3.128: Bassi's powers. No. 6, 1666, p. 545

Titoli dell' Algebra.

0	Numero.	16	Quadrato di quad. di quad. di quad.
1	Cosa, cioè Radice.	17	Quinto Relato.
2	Quadrato.	18	Quadrato cubo di cub.
3	Cubo.	19	Sesto Relato.
4	Quadrato di quadrato.	20	Quadrato di quad. primo Relato.
5	Primo Relato.	21	Cubo secondo Relato.
6	Quadrato cubo.	22	Quadrato terzo Relato.
7	Secondo Relato.	23	Settimo Relato.
8	Quadrato di quad. di quad.	24	Cubo quad. di quad. di quad.
9	Cubo di cubo.	25	Ottavo Relato.
10	Quadrato primo Relato.	26	Quadrato quarto Relato.
11	Terzo Relato.	27	Cubo di cub. di cub.
12	Cubo quadrato di quad.	28	Quadrato di quad. secondo Rel.
13	Quarto Relato.	29	Nono Relato.
14	Quadrato secondo Relato.		
15	Cubo primo Relato.		

the only variation of note being *qu.* for *ce.* These same forms with more intense abbreviations were noted in Figatelli (1678). The Spanish Puig (No. 72, p. 369.) capitalizes the symbols of Pacioli, thus: *N.*, *Co.*, *Ce.*, *Cu.*, *Cce.*, *1R.*, *CeCu* [3.129].

FIGURE 3.129: Puig's powers. No. 72, 1672, p. 369

Progression Arithmetica.

0	1.	2.	3.	4.	5.	6.	7.	8.	9.	10.	11.
N.	Co.	Ce.	Cu.	Cce.	1 R.	CeCu.	2 R.	Ccce.	Ccu.	Ce 1 R.	3 R.
1.	2.	4.	8.	16.	32.	64.	128.	256.	512.	1024.	2048

Progression Geometrica.

The German symbols, φ (*dragma* or *numerus*), ϰ (*radix*), ʒ (*zensus*), ϲꝛ (*cubus*), ʒʒ (*zensdezens*), β (*sursolidum*), ʒϲꝛ (*zensicubus*), as noted above, may be found in Jacob (1612) when he writes *rad.*, *zens.*, *cub.*, *zensdezens*, but uses no further abbreviations.[42] The same symbols are used by the Dutch writers, Sesen (No. 84, p. 654.), Coutereels (No. 26, p. 334.) and Hansz (No. 44.), and by many others. Lambeck (No. 55. p. 299.) uses in the

[42] ED. This sentence is confusing. Sister Leontius begins by mentioning 'The German symbols' but then, correctly, states that Jacob writes the abbreviated names of the symbols. As section 2.15 on page 93 states, and consulting Jacob (No. 54, 1612) confirms, Jacob does not use the German symbols, rather he uses their names or abbreviations.

Appendix these symbols of Rudolff with slight differences for *dragma*, *radix*, and *cubus*, thus: Dr., rr., c3, cs., 33., ß., 3cs [3.130].

		Dupla.	Tripla.	Quadr.
0.	Dr. Dragma.	1.	1.	1.
1.	rr. Radix.	2.	3.	4.
2.	z. Zensus.	4.	9.	16.
3.	cs. Cubus.	8.	27.	64.
4.	zz. Zens de zens.	16.	81.	256.
5.	ß. Sursolidus.	32.	243.	1024.
6.	zcs. Zensicubus.	64.	729.	4096.
7.	Bß. Bsursolidus.	128.	2187.	16384.
8.	zzz. Zens zens de zens.	256.	6561.	65536.
9.	cs. cs. Cubus de cubo.	512.	19683.	262144.

Philosophische Rechnung.

1. Ich habe eine Zahl, wenn ich dieselbe plus, 2 R + 1 mit 3 R ÷ 6 multipl. des Products Quadrat ferner in 4 R + 7 augir, so kommen diese Algebraische Quantitäten: 144 ß + 396 zz + 2340 R + 6300 ÷ 1152 cs. ÷ 3177 z?

FIGURE 3.130: Lambeck's powers. No. 55, n.d., p. 299

We find these symbols in Alsted (No. 2, p. 69.) with some alternate terms: *cubus*, *biquadratus* and *zensizensus*, *solidus*, *quad-raticubus*, *zensicubas*. In a chapter on 'Algebra vulgaris', Lucas (No. 58. p. 285.) uses this curious combination of symbols, N, R, Q, C, bq, Ss, cq [3.131].

Num. Progr. Geom.	Num. Progr. Arithm Expon.	Charactes res.	Denominationes Characterum
1	0	N.	Abſolutus.
2	1	R.	Radix, vel latus.
4	2	Q.	Quadratus.
8	3	C.	Cubus.
1 6	4	bq.	Biquadratus.
3 2	5	Ss.	Surdiſolidus.
6 4	6	cq.	Cubiquadratus
1 2 8	7	bss.	Surdiſolidus ſecundus
2 5 6	8	triq.	Triquadratus.
5 1 2	9	CC.	Cubicubus

FIGURE 3.131: Lucas' powers. No. 58, 1697, p. 285

Thus we see that even a half century after Descartes published his *Geometrie* (1637), authors were still adhering to the old notations of the 'Abbreviate Plan', using either the additive principle, or the multiplicative principle with its German or Italian symbolism. The so-called 'Index Plan' was a step towards our modern notation, and we find various authors developing and perfecting this plan. Cajori (p. 243.) gives attention to a new symbolism of crossed out numerals as exponents used by Pietro Cataldi in an algebra of 1610. The present writer also finds limited examples of the 'crossed out' numeral exponents, as 1.$\not 1$ and 1.$\not 3$ for X and X^3 respectively, in the 1619 work (No. 17, p. 2.) [3.132],[43] for Cataldi reverts to the Italian symbols after page 7.

FIGURE 3.132: Cataldi's powers. No. 17, 1619, p. 2

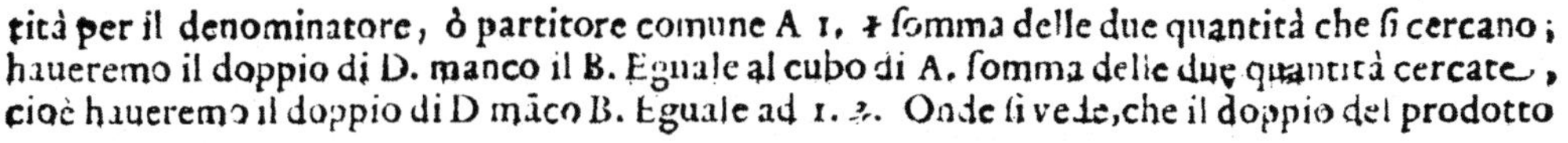
tità per il denominatore, ò partitore comune A 1, + somma delle due quantità che si cercano; haueremo il doppio di D. manco il B. Eguale al cubo di A. somma delle due quantità cercate, cioè haueremo il doppio di D mãco B. Eguale ad 1.$\not 3$. Onde si vede, che il doppio del prodotto

3.9.4 New exponential notations

The Cartesian notation became generally adopted in England and in Continental Europe during the second half of the seventeenth century. We are able to give additional references for this notation to the English and French works given by Cajori, and also to German, Swiss, and Spanish works.

The exponential symbolisms employed by Sir Jonas Moore as early as 1660 are passed over in silence by Cajori. Previous reference[44] has been made to his three different notations, Ac, aaa, and a^3, according to Oughtred, Harriot, and Descartes, respectively. In the last few pages of Lambeck's *Rechenbuch* (No. 55, p. 308.), in which Johann Wohlgemuth adds solutions of six algebraic problems, our modern exponents occur, except that he has no preference for xx or x^2 [3.133][45].

FIGURE 3.133: Lambeck's powers. No. 55, n.d., p. 308

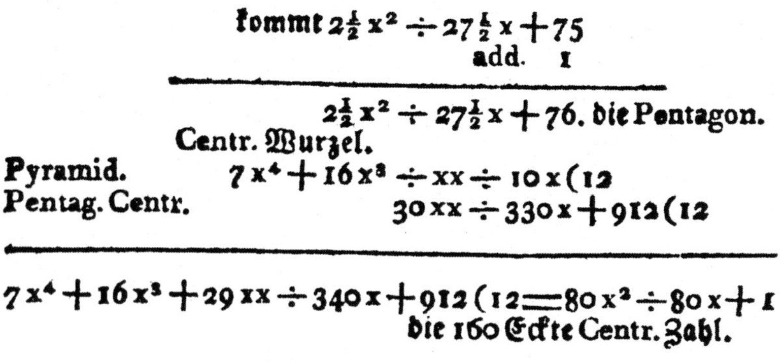
kommt $2\frac{1}{2}x^2 \div 27\frac{1}{2}x + 75$
add. 1

$2\frac{1}{2}x^2 \div 27\frac{1}{2}x + 76$. die Pentagon.
Centr. Wurzel.
Pyramid. $7x^4 + 16x^3 \div xx \div 10x$ (12
Pentag. Centr. $30xx \div 330x + 912$ (12

$7x^4 + 16x^3 + 29xx \div 340x + 912$ (12 $= 80x^2 \div 80x + 1$
die 160 Eckte Centr. Zahl.

An early trace of the Cartesian notation in France is found in Jean Prestet (No. 70, pp. 65, 66.), who follows Descartes also in the writing of aa for a^2 [3.134].

[43] ED. The 'cross out' on the 3 is at a lower angle and difficult to discern. See section 3.11 on page 166 for other examples of this notation including 'crossed out' 2s for x^2s.

[44] See page 100.

[45] ED. Recall Lambeck uses $\div$ for subtraction [3.43].

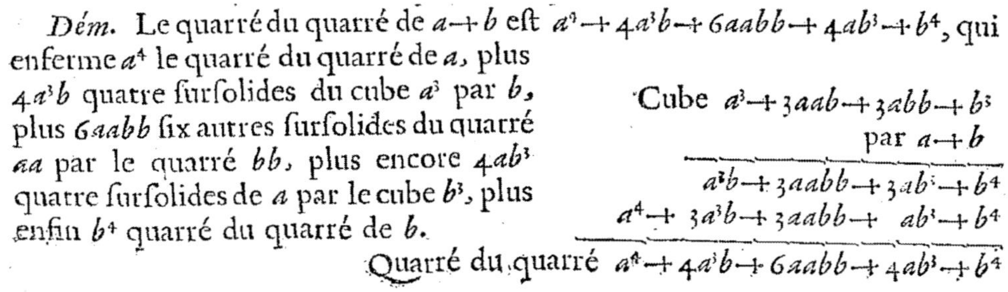

Dém. Le quarré du quarré de $a+b$ eſt $a^4+4a^3b+6aabb+4ab^3+b^4$, qui enferme a^4 le quarré du quarré de a, plus $4a^3b$ quatre ſurſolides du cube a^3 par b, plus $6aabb$ ſix autres ſurſolides du quarré aa par le quarré bb, plus encore $4ab^3$ quatre ſurſolides de a par le cube b^3, plus enfin b^4 quarré du quarré de b.

Cube $a^3+3aab+3abb+b^3$
par $a+b$

$a^3b+3aabb+3ab^3+b^4$
$a^4+3a^3b+3aabb+ab^3$

Quarré du quarré $a^4+4a^3b+6aabb+4ab^3+b^4$

FIGURE 3.134: Prestet's powers. No. 70, 1675, p. 66

The algebra of the Swiss John Alexander in 1693, as noted above[46], designates involution by means of exponents rather than with the Archimedean spiral used by Rahn in 1659.

At Barcelona, the Spanish Ventura de Avila (No. 4, pp. 69, 95, 103.) brought out an algebra in which the equations contain the unknowns `m`, `n`, and `t` with their integral exponents in the modern form, with only an occasional nn for n^2. The Italians, outside of Renaldini (1665), do not seem to adopt the new exponential notation so early, probably due mostly to the very great influence of Pacioli and his followers, and to some extent, to the novel notations of Ghaligai, Bombelli, and Cataldi.

3.10 Signs for Roots

3.10.1 Early expressions for roots

In the extraction of square and cube roots, common in the early arithmetics, such expressions as *radix quadrata, radix cubica,* and their equivalents, were sufficient and needed no symbolism. These terms may be found in the works of Vuelpius (1544), Micyllus (1555), Steinmetz (1575), Waser (1603), and Lambeck (c. 1661). Late into the seventeenth century, Griminelli (1670), Ciacchi (1674), and Monier (1693) were satisfied with these expressions. Even such prominent writers as Scheubel (1560), Clavius (1586), and Napier (1603) used no further notation for radical expressions.

3.10.2 The basic symbol R

As noted above,[47] Jacob (1612) simply expressed the second, third, and fourth roots by *radix quadrata, radix cubica,* and *radix zensdezens,* respectively. Unicorno (No. 93, ff. 126 v. 127.) again follows Pacioli in his manner of designating the order of the roots, but uses the small letter *r* for *radix*, thus: `r.7`, `r.cu.7`, `r.r.5`, `r.rel 5`, `r.cu.q.9` [3.135].

[46]Ed. See section 2.18 on page 102.
[47]Ed. See Section 2.15 on page 93.

FIGURE 3.135:
Unicorno's roots.
No. 93, 1598,[48]

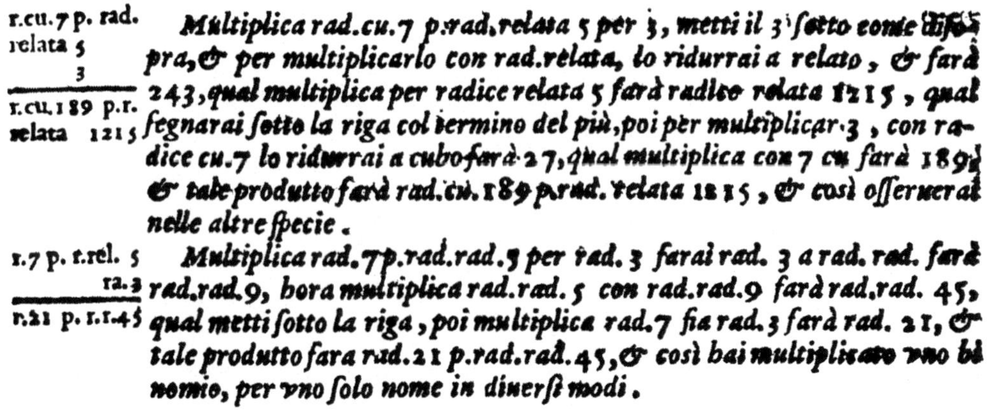
r.cu.7 p. rad. relata 5
3
r.cu.189 p.r. relata 1215

Multiplica rad.cu.7 p.rad.relata 5 per 3, metti il 3 sotto come di sopra, & per multiplicarlo con rad.relata, lo ridurrai a relato, & farà 243, qual multiplica per radice relata 5 farà radice relata 1215, qual segnarai sotto la riga col termino del più, poi per multiplicar 3, con radice cu.7 lo ridurrai a cubo farà 27, qual multiplica con 7 cu farà 189, & tale produtto farà rad.cu.189 p.rad.relata 1215, & così osseruerai nelle altre specie.

r.7 p. r.rel. 5
ra.3
r.21 p. r.r.45

Multiplica rad.7 p.rad.rad.5 per rad. 3 farai rad. 3 a rad.rad. farà rad.rad.9, hora multiplica rad.rad. 5 con rad.rad.9 farà rad.rad. 45, qual metti sotto la riga, poi multiplica rad.7 sia rad.3 farà rad. 21, & tale produtto fara rad.21 p.rad.rad.45, & così hai multiplicato vno binomio, per vno solo nome in diuersi modi.

A variety of notations is contained in Bassi (No. 6, pp. 530, 541, 542.) [3.136, 3.137, 3.138].

FIGURE 3.136:
Bassi's roots.
No. 6, 1666,
p. 530

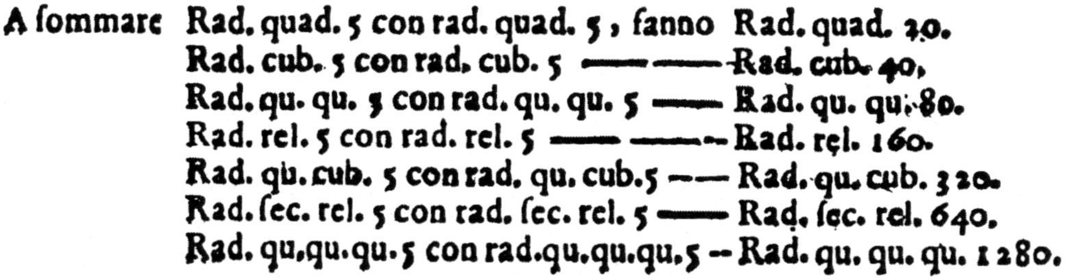
A sommare Rad. quad. 5 con rad. quad. 5, fanno Rad. quad. 20.
Rad. cub. 5 con rad. cub. 5 —— Rad. cub. 40.
Rad. qu. qu. 5 con rad. qu. qu. 5 —— Rad. qu. qu. 80.
Rad. rel. 5 con rad. rel. 5 —— Rad. rel. 160.
Rad. qu. cub. 5 con rad. qu. cub. 5 —— Rad. qu. cub. 320.
Rad. sec. rel. 5 con rad. sec. rel. 5 —— Rad. sec. rel. 640.
Rad. qu. qu. qu. 5 con rad. qu. qu. qu. 5 -- Rad. qu. qu. qu. 1280.

FIGURE 3.137:
Bassi's roots.
No. 6, 1666,
p. 541[49]

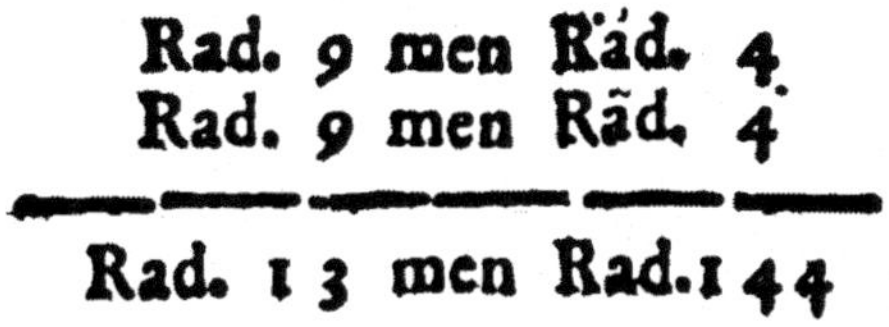
Rad. 9 men Rad. 4
Rad. 9 men Rad. 4
Rad. 13 men Rad. 144

FIGURE 3.138:
Bassi's roots.
No. 6, 1666,
p. 542

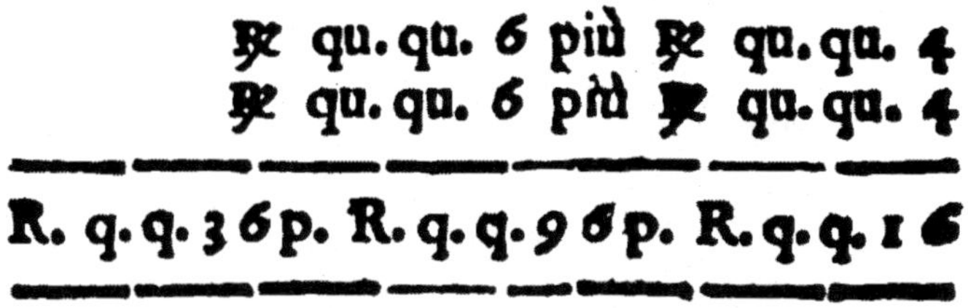
℞ qu. qu. 6 più ℞ qu. qu. 4
℞ qu. qu. 6 più ℞ qu. qu. 4
R. q. q. 36 p. R. q. q. 96 p. R. q. q. 16

[48] ED. Note the typographical error in the lower of the two examples printed in the margin of the text. As can be discerned from the text, to the right of the example, the top line of the marginal work should be: r.7 p. r.r.5.

[49] ED. Note the typographical error in the bottom line. The 'Rad.' prior to the 13 is incorrect. The bottom line should read: 13 men Rad.144 (which is equivalent to 1).

He uses *rad. quad.*, *rad. cub.*, *rad. qu. qu.*, *rad. rel.*, *rad.qu.cub.*, later ℞, *R*, or *Rad.* In Puig (No. 72, p. 484.) we find *rq.*, *rc.*, and *rrq* [3.139].

rq. quiere dezir raiz quadrada. Eſta rc. quiere dezir raiz cu-bica. Eſta rrq. quiere dezir raiz quadrada de raiz quadrada. Eſ.

FIGURE 3.139: Puig's roots. No. 72, 1672, p. 484

As late as 1697, Lucas (No. 58, p. 319.) gives this unusual notation: *radix Q*, *radix C.*, *radix bq.*, *radix ss.*, *radix C.*, *radix ssb.*, *radix trig.*, and *radix CC.*[50]

The ℞, used by Pacioli as a fundamental symbol for indicating roots, reached its zenith in Italy during the sixteenth century. It was used by Feliciano in 1550, and still survived in Figatelli in 1678. In 1619, Cataldi (No. 17, p. 5.) follows the general symbolism ℞, ℞*c*, ℞*q.c.*, for square, cube, and sixth roots respectively [3.140].

℞ c.3 $\frac{3}{4}$.p̄ ℞ q.c.c. $\frac{2}{4}$ $\frac{5}{8}$.
via ℞ 2.m̄ ℞ 3.

Prodotto. ℞ c.30.p̄ ℞ q.c.33 $\frac{1}{3}$.
m̄ ℞ q.c.379. $\frac{1}{1}$ $\frac{1}{6}$.m̄ ℞.c.3 $\frac{3}{4}$.

FIGURE 3.140: Cataldi's roots. No. 17, 1619, p. 5

We have found an instance of the use of the crossed capital ℞ in Vieta (No. 95, pp. 183, 184.) [3.141],

prendra F plus grande que S + ℞ S q. — G pl. Au-rebours, R par 2 E — E q. ſera plus grande que G pl. ſi bien que F ſera priſe plus grande que R + ℞ R q. — G plan.

Soit G pl. 60. B 5. D 8. A. 1 N. 1 N ſera moindre que ℞ 76 + 4, & plus grand que ℞ $\frac{265}{4}$ + $\frac{5}{2}$ mais 12 eſt moindre que ℞ 76 + 4. car la valeur du quarré de 12 eſt ℞ 64 + 4. & 11 eſt plus grand que ℞ $\frac{40}{8}$ + $\frac{5}{2}$. Soit

FIGURE 3.141: Viète's roots. No. 95, 1630, p. 184

who used 𝓁 (latus), or the word *radix,* for indicating roots. We must mention Renaldini (No. 78, p. 21.), who represented square root by ℞*Q* or simply *R*, cube root by ℞*C*.

[50] ED. On page 319 Lucas uses the word 'radice', not 'radix', prior to the the various letters which denote the index of the root (e.g. radice bq.). As shown above [3.104] Lucas also uses these same letters (e.g. Q., C., bq., etc...) to represent the associated powers of a variable.

3.10.3 The basic symbol $\surd$

Sir Jonas Moore (1660), as noted above[51], marked the order of the radical as did Oughtred, $\surd$, $\surd$c, $\surd$qq. Just as many authors used the sign □ as an operator to indicate squaring, Lambeck (No. 55, App. 308, 309.) used it in the extraction of square root; he writes $\surd$□ and $\surd$cub [3.142].

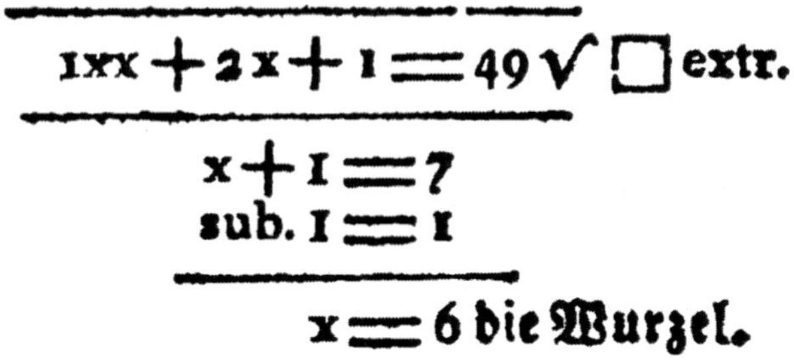

1xx + 2x + 1 = 49 √ □ extr.

x + 1 = 7

sub. 1 = 1

x = 6 die Wurzel.

FIGURE 3.142: Lambeck's square root symbol. No. 55, n.d., p. 309

In France, Prestet and Blondel used $\surd$ extensively, differing only in the mode of designating the order of the root. Prestet (No. 70, p. 108.) expresses square roots simply by $\surd$, but he has two forms for cube root, $\surd$C and $\surd 3^e$; however, he gives preference to the former. Likewise fifth and seventh roots are $\surd 5^e$, and $\surd 7^e$, and in sixth and tenth roots he shows the commutative order of the radicals when he writes $\surd$C.$\surd$ or $\surd\surd$C., and $\surd 5^e \surd$ or $\surd\surd 5^e$, respectively. Fourth, eighth, and ninth roots similarly follow the multiplicative principle [3.143].

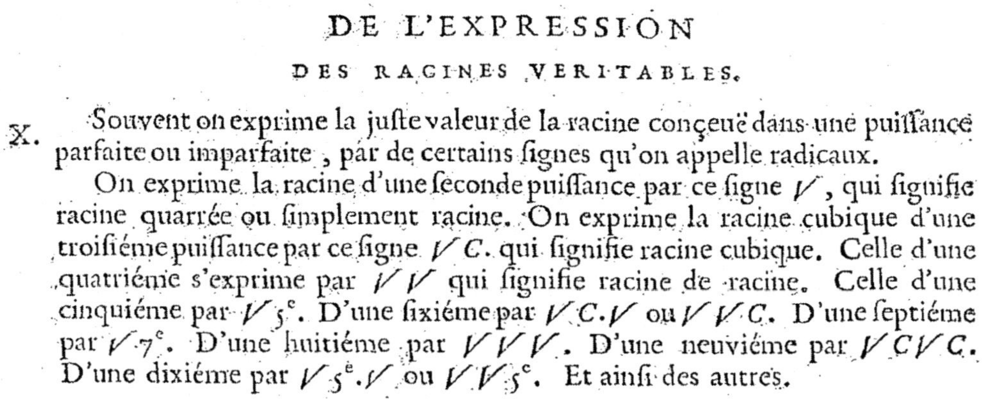

DE L'EXPRESSION

DES RACINES VERITABLES.

X. Souvent on exprime la juſte valeur de la racine conçeuë dans une puiſſance parfaite ou imparfaite, par de certains ſignes qu'on appelle radicaux.

On exprime la racine d'une ſeconde puiſſance par ce ſigne √, qui ſignifie racine quarrée ou ſimplement racine. On exprime la racine cubique d'une troiſiéme puiſſance par ce ſigne √C. qui ſignifie racine cubique. Celle d'une quatriéme s'exprime par √√ qui ſignifie racine de racine. Celle d'une cinquiéme par √5e. D'une ſixiéme par √C.√ ou √√C. D'une ſeptiéme par √7e. D'une huitiéme par √√√. D'une neuviéme par √C√C. D'une dixiéme par √5e.√ ou √√5e. Et ainſi des autres.

FIGURE 3.143: Prestet's square root symbol. No. 70, 1675, p. 108

Blondel (No. 9, pp. 18-20.), however, is consistent when he uses the additive principle in his notation for roots as well as for powers, for he says in French that 'the roots derive their name from the relation to the power to which each of them is compared.' For expressing roots, he uses his notation for corresponding powers preceded by the radical sign, thus: $\surd$, $\surd$C, $\surd$QQ, $\surd$QC, $\surd$CC, $\surd$QQC. It is noteworthy that he gives as alternate forms the symbolism used by John Wallis in 1655, in which the root indices in numerals

[51] Ed. See section 2.17 on page 95.

are elevated and placed after the sign $\surd$ in this manner, $\surd^3$, $\surd^4$, $\surd^5$, etc. [3.144, 3.145, 3.146][52]

	1	2	3	4	5	6	7	8	9
	R	Q	C	QQ	QC	CC	QQC	QCC	CCC
1.	2.	4.	8.	16.	32.	64.	128.	256.	512.
1.	3.	9.	27.	81.	243.	729.	2187.	6561.	19683.
1.	4.	16.	64.	256.	1024.	4096.	16384.	65536.	262144.
1.	5.	25.	125.	625.	3125.	6625.	33125.	165725.	828625.

FIGURE 3.144: **Blondel's notation for radicals. No. 9, 1683, v. 2, p. 18**

SPECULATIVE. 19

racine quarréquarrée de 256 ; √ QC , ou √ 5 c'eſt à dire Racine quarrécube de 1024 ; √CC ou √ 6 c'eſt à dire Racine cubecube de 4096 ; √QQC ou √ 7 c'eſt à dire racine quarréquarrécube de 16384. & ainſi des autres.

FIGURE 3.145: **Blondel's notation for radicals. No. 9, 1683, v. 2, p. 19**

res √, comme √ 2, qui ſignifie racine 2, √ 5 ou racine quarrée de 5 : Ainſi √ C 6, ſignifie racine cubique de 6 : √√ 8 veut dire racine de racine ou racine quarréquarrée de 8, &c. Or

FIGURE 3.146: **Blondel's notation for radicals. No. 9, 1683, v. 2, p. 20**

The Dutch writers, Sesen (No. 84, p. 667.), Coutereels (No. 26, p. 429.), and Hansz (No. 44.), use instead the German symbols to mark the order of the root.

The Swiss Alexander (1693), as noted before[53], represents roots as did Moore, but he follows the letters by dots. In Spain, Ventura de Avila, (No. 4, pp. 99, 102, 103.) uses the modern sign for square root on a few pages. It is interesting to note in passing that in solving the problem $nn + 13 = 6n$, he gets the roots $n = 3 + \sqrt{-4}$ and $n = 3 - \sqrt{-4}$. He proceeds to say that since these roots are impossible, the problem is impossible.

[52] ED. Blondel depicts another form form of the fourth root that Sister Leontius didn't mention. Namely, the two adjacent "check-type" radical symbols.

[53] ED. See section 2.18 on page 102.

3.11 Signs for Unknown Numbers

3.11.1 Early forms

In our survey of powers of the unknown we met with the term *cosa* as the Italian word for the first power of the unknown x. *Cosa*, or some abbreviated form of it, is used by Feliciano (1550), Unicorno (1598), Bassi (1666), Puig (1672), and Figatelli (1678).

Radix, or some equivalent symbol, is found in Jacob (1612), Graffenried (1618), Bonvicino (1665), Lambeck (c. 1661), Hemeling (1667), Lucas (1697), and Blondel (1699).

Cajori (p. 380.) has referred to the use of N (numerus) for x by Vieta in 1615; Vieta further uses also IN [3.147].

FIGURE 3.147: Viète's unknowns. No. 95, 1630, p. 70

Soit B plan 20. R 1. S 5. A 1 N. 1 Q. eſt egal à 100.
Ou ſoit E 1 N. 1 Q eſt egal à 4.

in the 1630 edition (French). As noted above[54], Feliciano changed from the Italian symbolism to that of Vieta (1591) when, in 1669, he wrote A for the unknown. Likewise Moore (1660) uses the large or small characters, A or a, for x. Cataldi, as noted [3.132], exhibits the notation of the crossed 'one' for x, but Cajori (p. 344.) says he made only very limited use of this notation. It has been pointed out previously that in 1619, after only a few pages of the use of these 'crossed out' numeral exponents, Cataldi resumes the Italian symbol co. for unknown.

A still different notation appears in the algebra of the Spanish Ventura de Avila (No. 4, p. 64.). He speaks there of the 'tres incognitas, n, m, t,' and uses them throughout his work. He makes no explanation for his use of m and t as the second and third unknowns.

3.11.2 Descartes' z, y, x

A prominent early follower of Descartes' way of signifying known quantities by the first letters of the alphabet, and unknowns by the last, z, y, x, is the French writer Prestet (No. 70, p. 150.). He says [3.148]:

FIGURE 3.148: Prestet's unknowns. No. 70, 1675, p. 150

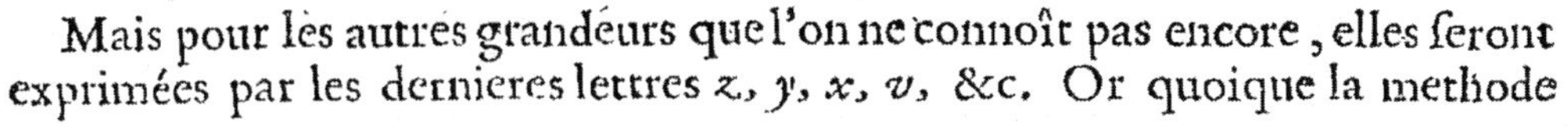

Mais pour lès autres grandéurs que l'on ne connoît pas encore, elles ſeront exprimées par les dernieres lettres *z, y, x, v,* &c. Or quoique la methode

In Brancker's translation of Rahn's *Algebra* (No. 73.), the x, y, and z are also used freely. Thus the Cartesian notation for the unknown was being adopted in France in the late seventeenth century in purely algebraic works (also in Ozanam's *Nouveaux Elemens d'Algebre,* Amsterdam, 1702; see Cajori, *Notations,* p. 370.) as well as in editions of Descartes' geometry.

---

[54] ED. See Figure 2.39 on page 76.

3.12 Signs of Aggregation

3.12.1 Aggregation expressed by letters

Although Cajori makes the statement (p. 385.) that the use of letters for aggregation practically disappeared in the seventeenth century, we have on hand several noteworthy instances of their use in Italian and Spanish arithmetics. Unicorno (No. 93, f. 142, v.) uses the abbreviations *rad. v.* and *rad. legata* in marking the *radix universale* (aggregate) and *radix legata* (single term), for he writes `rad. v. 5, p. rad. v. 3, p. rad. legata 49, p. rad. 16, rad. 4` [3.149]

FIGURE 3.149: Unicorno's aggregation. No. 93, 1598, f. 142, v.

for our[55]

$$\sqrt{5+\sqrt{3+\sqrt{49}+\sqrt{16}+\sqrt{4}}}$$

Following Bombelli, the inverted capital letter `L` was employed by Cataldi (No. 17, p. 13.) to express aggregation. Thus he writes [3.150]:

FIGURE 3.150: Cataldi's aggregation. No. 17, 1619, p. 13

for

$$\sqrt[3]{7\tfrac{1}{2}-\sqrt{52\tfrac{1}{12}}}$$

Cajori (p. 135.), however, mentions Cataldi's use of the dot to express aggregation.

The Spanish Puig (No. 72, p. 484.) explains the symbols `rv.`, `rcv.`, and `rrv.` as indicating 'raiz quadrada universal, 'raiz cubica universal,' and 'raiz quadrada de raiz quadrada universal.' It is important to note Figatelli's use of ℞V to signify universal root, the root of a polynomial, which he curiously follows by the left part of a parenthesis, thus [3.151]:

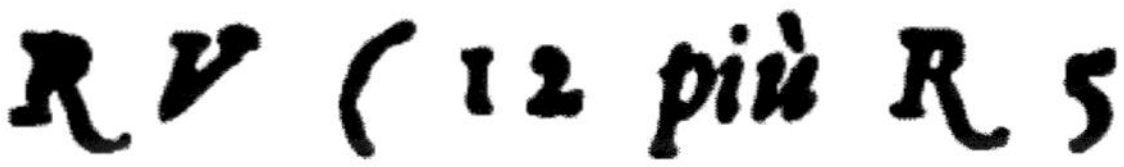

FIGURE 3.151: Figatelli's aggregation. No. 32, 1678, p. 195

for $\sqrt{12+\sqrt{5}}$ (No. 32, p. 195.).

[55] ED. In Sister Leontius' work this expression was mistyped. It has been corrected here.

3.12.2 Aggregation expressed by the vinculum or bar above

As does Cajori, we also find the vinculum adopted by J. Prestet (No. 70, pp. 108, 136, 121.), but it is important to add that in connection with the vinculum and radical sign he sometimes uses a dot in cases like the following [3.152]:

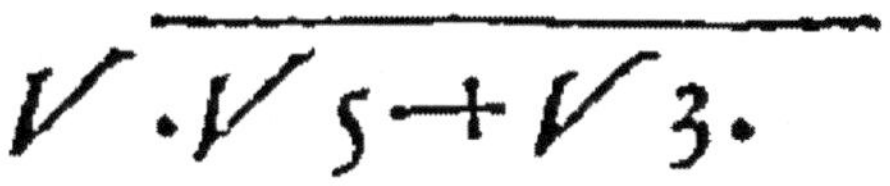

FIGURE 3.152: **Prestet's aggregation. No. 70, 1675, p. 108**

and [3.153]

FIGURE 3.153: **Prestet's aggregation. No. 70, 1675, p. 136**

for

$$\sqrt{\sqrt{5}+\sqrt{3}}$$

and

$$\sqrt{\sqrt{1\tfrac{1}{4}}+\sqrt{\tfrac{1}{2}}}$$

respectively. However, he does not frequently have occasion to use this combination of symbols, but he is very liberal in using the vinculum to indicate aggregation in the treatment of radicals as well as in its general use in such expressions as [3.154]

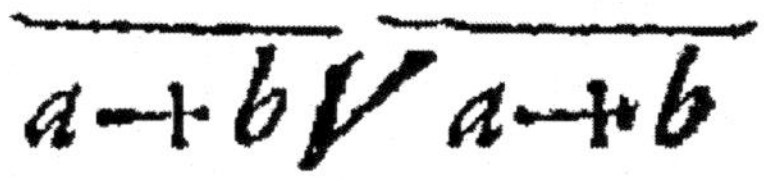

FIGURE 3.154: **Prestet's aggregation. No. 70, 1675, p. 121**

Mention has been made of the use of the horizontal bar in connection with the radical sign by John Alexander (1693).[56] Ventura de Avila (No. 4, pp. 60, 106.) is sparing in his general use of the vinculum; occasionally such expressions

$$3 \times \overline{8m - 6} = m$$

and

$$\frac{\overline{7v + 60}}{3} \times \frac{\overline{v - 3}}{3}$$

occur.

[56] ED. See Figures 2.112 on page 104, 2.113 on page 104, and 2.114 on page 104.

3.12.3 Aggregation expressed by parentheses

Renaldini (No. 78, 79.) is listed by Cajori (p. 392.) among the Italian writers using parentheses [3.156, 3.157]. In the *Ars analytica mathematum* (No. 78, pp. 21, 317.), to which all of Cajori's references to Renaldini are made, this statement is given:

Signum autem hoc ℞ (), ſignificat radicem vniuerſalem, ſiuè ligatam ; itaq; ℞ Q (), denotabit radicem quadratam vniueſalem

FIGURE 3.155: Renaldini's aggregation. No. 78, 1665, p. 21

$$\frac{\dfrac{\dfrac{1}{2}p \dagger \text{℞}\left(\dfrac{1}{4}p^2 - q\right)}{\dfrac{1}{2}p \dagger \text{℞}\left(\dfrac{1}{4}p^2 - q\right)}}{\dfrac{\text{℞}\left(\dfrac{1}{16}p^4 - \dfrac{1}{4}p^2 q\right) \dagger \dfrac{1}{4}p^2 - q}{\dfrac{1}{4}p^2 \dagger \text{℞}\left(\dfrac{1}{16}p^4 - \dfrac{1}{4}p^2 q\right)}}$$

$$\frac{1}{4}p^2 \dagger \text{℞}\left(\frac{1}{4}p^4 - p^2 q\right) \dagger \frac{1}{4}p^2 - q$$

$$\frac{1}{2}p^2 \dagger \text{℞}\left(\frac{1}{4}p^4 - p^2 q\right) - q.$$

FIGURE 3.156: Renaldini's aggregation. No. 78, 1665, p. 317

Propoſita igitur æquatio ſi fuerit $a^3 \dagger b^2 a = b^2 d$, ita vt tranſmutata induat hanc formam $a^3 = * - p\,a \dagger q$, eius radix dicebatur ℞ $C\,(\dagger \frac{1}{2}q \dagger$ ℞ $[\frac{1}{4}q^2 \dagger \frac{1}{27}p^3]$ $-$ ℞ $C\,[-\frac{1}{2}q \dagger$ ℞ $[\frac{1}{4}q^2 \dagger \frac{1}{27}p^3]$.

FIGURE 3.157: Renaldini's aggregation. No. 78, 1665, p. 317

However, in his *Artis analyticae mathematum* (No. 79, p. 96.) is the explanation [3.158]:

Quod ſi in animo , ſit extrahere radicem ex binomio , vt ex $a^2 \dagger b^2$, præfigendum eſt character radicalis , priùs clauſo binomio intra parent veſin , vt hinc , qui ſequitur in modum ℞$(a^2 \dagger b^2)$, vel loco parenteſis adhiberi etiam ſolet lineola deſuper vt ℞ $\overline{a^2 + b^2}$.

FIGURE 3.158: Renaldini's aggregation. No. 79, 1669, p. 96

Thus in the second volume Renaldini employs parentheses and vinculums interchangeably. Occasionally also, in the first part, brackets are employed as aggregation symbols, sometimes brackets and parentheses on the same page. These same symbols are also used to enclose parenthetical statements.

It is worthy of note that the second volume of Renaldini is an additional source for the use of the usual star to mark the absence of terms, as in this expression:

$$a^3 \dagger ba^{2*} - z^3$$

Chapter 4

Completion of Certain Topics in Tropfke

4.1 Numbers

4.1.1 Numeral words

In numeration the word for one thousand is commonly *mille* (Latin), *mil* (French), *mille* (Italian), *tausend* (German), and *thousand* (English) in the texts of the various countries during the sixteenth and seventeenth centuries. Numbers higher than one thousand are named largely from their formation by the multiplicative principle; thus ten thousand may be found expressed as *decem millia,* and one hundred thousand as *centena millia.*

In some early Latin books 10^6 (million) is occasionally called *decies centena millia,* as by Regius (No. 76, f. 51, v.) and Steinmetz (No. 86, f. 12, v.); but a form corresponding to *mille millia* is used by Noviomagus (No.65, f. 23, r.) and Vuelpius (No. 98, f. 8, r.) and *millena millia* by Waser (No. 99, f. 10, v.) and Alsted (No. 2, p. 25.). *Tausend mal tausend* was general in Germany until the word million appeared in the arithmetic of Christoff Rudolff (1526). (See Tropfke, v. 3, p. 7.). We also find *tausend mal tausend oder million* mentioned in Rudolff's 1532 arithmetic (No. 81, f. 3, r.), but in reading of a large number he uses *tausendmal tausend.* About the same time the rare arithmetic of Felner (No. 30, f. 4, r.) gives only *tausent mal tausent.* Suevus (1593), in devoting nineteen pages to numeration, with his customary prolixity, does not once use the word *million,* nor does Behme as late as 1695. (According to L. L. Jackson, *Sixteenth Century Arithmetic,* New York, 1906, p. 39, the sixteenth century records the struggle of the word *million* for its place in numeration.)

In Lambeck's (No. 55, p. 17.) arithmetic (c. 1661) we read:

> Nach der neuen und bequemen Art zu numeriren, da man tausend mal tausend, eine Million; tausend mal tausend Millionen, eine Billion; tausend mal tausend Billionen, ein Trillion, nennet etc.; stehet die vorgegebene Zahl also eingetheilet 41,234567,894.

In Italy *millioni* or *millione,* a variation in spelling, occurs in Ghaligai (1552), Peverone (1558), Unicorno (1598), and in the seventeenth century in Griminelli, Venturoli, Bonvicino, Bassi, and Ciacchi. The French writers Chauvet, Prestet, Monier, and Blondel give only the form *Millions,* while the works of Aurel and Puig give the Spanish *cuentos.* (Cantor, *Vorlesungen über Geschichte der Mathematlk,* v. 2, p. 387, says that Ciruelo (1505) called 10^6 *cuentos,* and 10^{12} millon, according to the Spanish custom.)

For 10^9 Chauvet (No. 21, f. 2, r.) uses the multiplicative principle when he writes *mil millions;* Prestet (No. 70, p. 7.) has the special form *milliars,* and Lucas (No. 58, p. 3.) uses *milleni.* We have in Blondel (No. 9, p. 52.) an instance of the use of *bimillion* for 10^9, and *trimillion* for 10^{12}, which is still customary in France today. (See Tropfke, v. 1, 3d. ed., p. 9.) It has been noted that[1] Lambeck (No. 55, p. 17.) mentions 10^{12} read as *Billion* according to the newer and easier method; Lucas (No. 38, p. 3.) also uses *biliones* in preference to the older forms.

In the translation of the numeral word for *zero* from the Arabic *as-sifr* into the Latin, we have *zephirum* (Leonard of Pisa), *cifra* (Nemorarius), and their many forms. Although the word *chiffre* gradually assumed a new meaning in France in the fourteenth century, we still find an instance of its use as a numeral word for *zero* in 1627. In Chauvet (No. 21, ff. 1, v., 3, r.) we read: "le dixiesme s'appelle vulgairement chiffre, les uns s'appellent zero, comme les Arabes." Chauvet actually uses *chiffre* for zero.

The use of *siphram* by Tunstall (1538), *cyphram* by Gemma (154.9) and *cypher* by Moore has been noted above.[2] Noviomagus (No. 65, f. 22, r.) writes *sipheram* or *zyphram*, which latter form is found also in Regius (No. 76, f. 49, r.). In Reisch (No. 77.) and in Micyllus (No. 60, p. 2.) the *cifra* of Nemorarius is mentioned alone. Napier (No. 65, p. 5.) gives cifra or zero, whereas Wolphius (1534), Peverone (No. 68, p. 6.), and Forestani (No. 35.) give *cifra* (*zifra*) or *nulla* respectively. *Nulla* was used exclusively by Rudolff (No. 81, f. 2, v.), Felner (No. 30, f. 3, v.), Suevus (No. 88, p. 3.), and, as noted before,[3] by Jacob (1612). Among those who give *nulla* and *zero*, as did Feliciano (1550), are Bonvicino (No. 12, p. 1.), and Bassi (No. 6, p. 4.). Together with these words we find Bonvicino using the peculiar term *Nota*. Graffenried (No. 42.) and Curtius (No. 27.) give no alternate forms for *nulla*. However, Lambeck (No. 55, p. 16.) (c. 1661) uses *Null* for the tenth 'Ziffer',[4] and underneath it writes *Nichts,*[5] thus [4.1]:

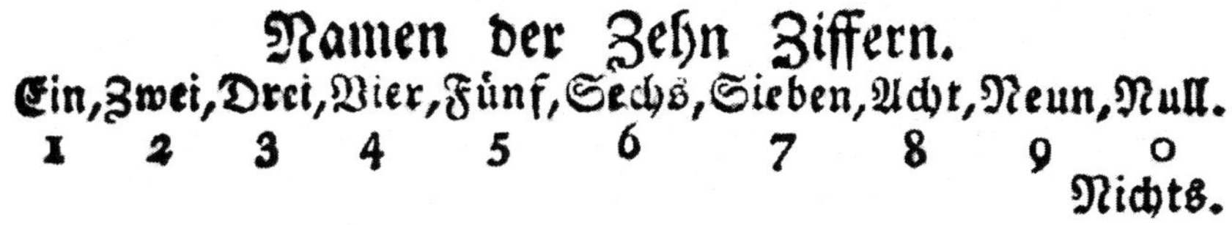
Namen der Zehn Ziffern.
Ein,Zwei,Drei,Vier,Fünf,Sechs,Sieben,Acht,Neun,Null.
1 2 3 4 5 6 7 8 9 0
Nichts.

FIGURE 4.1: Lambeck's names for the numeral 0. No. 55, n.d. (c. 1661), p. 16

in the century before the writers mentioned by Tropfke as using the term *null* or *nulla*. (See Tropfke, v. 1, 3d ed., p. 12.) An equivalent term for *Nichts* is that of *nihil* listed in Cardan (No. 16, f. 5, r.), but spoken of as *nullus* in the running text. Steinmetz (No. 86, ff. 8, v., 6, r.) gives *cyphram* or *nihili* but uses the former. Ramus in both the 1599 and

[1] ED. See Lambeck's quote on page 171.

[2] ED. See section 2.8 on page 79, Figure 2.14 on page 67 and Figure 2.88 on page 95.

[3] ED. See section 2.15 on page 93.

[4] ED. 'Ziffer' is German for 'number.'

[5] ED. 'Nichts' is German for 'nothing.'

1615 editions says *circulus* for 0, which form was common in the early treatises before the introduction of printing. (See Susan R. Benedict, *A Comparative Study of the Early Treatises Introducing into Europe the Hindu Art of Reckoning,* p. 52.).

We have mentioned[6] the change from *nulle* or *tziphra* to zero in the 1587 and 1619 editions of Finé and Savonne respectively. Clavius (No. 24, p. 8; No. 25, p. 11.) writes *cifra* in 1585, but in the following year and in his seventeenth-century editions he says *cifra o zero.*

Zero was used exclusively, as noted above,[7] by Cataneo (1546), and also by the Spanish writers Aurel (No. 5, f. 1, v.) and Puig (No. 72, p. 60.), by the Italians Manelli (No. 59, p. 2.) and Venturoli (No. 94, p. 5.) and the French Prestet (No. 70, p. 6.), Monier (No. 62, p. 5.), and Blondel (No. 9, p. 50.). The latter writes:

> ...il ne sert qu' à remplir les sieges ou places entre les chiffres, qui demeurant vides pourraient causer de la confusion.

The German Waser (No. 99, f. 10, r.) in his Roman notations, gives five word forms for zero, the Hebrew *en,* Arabic *Siphra,* German *Ziffer,* Greek $\gamma\delta\omega$, the Latin *Nulla* or *Nullum*; but he himself resorts to a sixth word, *circulus.* Alstea (No. 2, p. 24.) lists *nulla, ciphra,* and *circulus,* but uses *ciphra.* Blasius (No. 8.) calls 0 *theta, circulus,* and *cifra,* as does Husswirt. Both Unicorno (No. 95, p. 5.) and Griminelli (No. 45, p. 2.) give *nulla, cifra* (*ziffra*) and *zero,* but the former uses commonly *nulla,* and the latter, *zero.*

4.2 Fundamental Operations

4.2.1 Fundamental operations in general

In his translation of Rahn's Algebra in 1668 Thomas Brancker (No. 73, p. 2.) states:

> There are in Arithmetick six distinct single operations, all of which are performed in Algebra, viz., Addition, Subtraction, Multiplication, Division, Involution, Evolution;...

The German Felner (No. 30, ff. 5, r.-9, r.) speaks of 'die Species' as 'Addiren, Subtrahiren, Duplirn, Medirn, Multiplicirn, Dividirn.' Scheubel (No. 83, pp. 71-78.), in 1560, like Brancker, lists six species, but he substitutes progressions for involution.

Followers of Sacrobosco (See Susan R. Benedict's thesis on the early treatises) in teaching the following nine fundamental operations are Husswirt (1501), Micyllus No. 60, p. 2.), and Steinmetz (No. 86, f. 7, v.): numeration, addition, subtraction, duplation, mediation, multiplication, division, progressions, roots. There are others in the sixteenth century who give seven, omitting duplation and mediation, as Cardan (No. 16, f. 6, r., who thus early speaks of them as operations rather than species), Regius (No. 76, f. 50, v.), Noviomagus (No. 65, f. 8, r.), and Unicorno (No. 93, ff. 4, v., 5, r.). (See D. E. Smith, *Source Book in Mathematics,* p. 18, for Recorde's account of the seven operations.). Still others, like Tagliente as early as 1520, and Vuelpius (No. 98, f. 5, r.), teach only five, excluding progressions and roots. Aurel (No. 3, f. r.) speaks of the same

[6] Ed. See Figure 2.58 on page 84 and Figure 2.60 on page 85.

[7] Ed. See Figure 2.26 on page 72.

five *Species,* but he calls addition, subtraction, multiplication, and division the 'quarto reglas principales.' By the end of the seventeenth century the number of fundamental operations had quite generally been reduced to at most these five: numeration, addition, subtraction, multiplication and division, as in Puig (No. 72, p. 578.) and Lucas (No. 58, p. 1.). Tartaglia (No. 90, f. 3.) discusses the question of the number of fundamental operations at some length:

> Le specie, ... sono noue, ... Numeratione, ouer Representatione, ... Additione, (cioe Sommare, ouer Raccogliere) ... Sottratione (cioe Sottrare, ouer abbatere, ouer Cavare, ouer Restare) ... Dupplatione, (cioe doppiare) ... Multiplicatione, (cioe Multiplicare) ... Mediatione (cioe dimezzare, ...) ...divisione, (cioe Dividere, ouer Partire) ... Progression, ...é detta estratione de Radice, Et queste noue specie alcuni le chiamano, Atti, altri gli dicono Passioni del Numero, & perche lo indoppiare non si distingue dal Multiplicare, ne lo Dimezzare dal partire, molti hanno detto, & determinato, Le sopradette noue specie Atti, ouer Passioni esser solamente sette, cioe Numerare, Sommare, Sottrare, Moltiplicare, Partire, Progressioni, & estration, ouer cauar de Radice, Ma perche le Progressioni, & l'estration de Radice, non sono necessarie in alcun caso, realmente accadente in tutte l'arte Negotiaria, della quale solamente quiui trattar intendo ...

In the second half of the sixteenth century there are those who reduce the number of operations to four, omitting numeration. These include Cataneo (1546), as noted above[8], Feliciano (1550), Finé (1555), Sfortunati (1561), and Savonne (1571). As early as 1515 Ortega treats the four *species* with integers and fractions. (However, D. E. Smith in his *History of Mathematics,* v. II, p. 33, gives Gemma Frisius the credit of being one of the first of the writers of any note to reduce the number to four.). There are these representative works during the seventeenth century which give four operations: Waser (1603), Chauvet (1627), Lambeck (c. 1661), Griminelli (1670), Prestet (1675), Monier (1693), and Blondel (1699). The first three speak of them as 'species'; Griminelli (No. 43, p. 1.) says that arithmetic is divided 'in quattro parti, o specie'; Prestet (No. 70, pp. 10, 19, 25.) uses the expressions: 'De l'Addition simples', 'De la Soustraction simples', 'De l'Addition composée ou Multiplication', 'De la soustraction composée, ou Division.' On the other hand, by the terms 'Addition and Soustraction composée' Monier (1693) refers to addition and subtraction of denominate numbers. Blondel (No. 9, pp. 60, 61) says;

> il ya [sic] quatre Operations de l'Arithmetique pratique que l'on appelle *les quatre premieres regies* ... Ces quatre regies sont le fondement non seulement de tout ce que se peut pratiquer en Arithmetique, mais mêm des principales operations du reste des Mathematiques; ...

Duplation and mediation are among the fundamental operations, as noted above,[9] in the works of Husswirt (1501), Wolphius (1534), and Peurbach (1539). Husswirt has

[8] Ed. A previous reference at which Sister Leontius mentions that Cataneo omits "numeration" as an operation could not be found. This would reduce Cataneo's number of operations to four.

[9] Ed. See section 2.1 on page 65.

duplation and mediation follow multiplication and division respectively, whereas Wolphius has them precede. Felner (1535), Peurbach (1539), Micyllus (1555), and Steinmetz (1575) place both duplation and mediation before multiplication and division, and Husswirt does the same in counter reckoning. Rudolff (No. 8, f. 35.) lists doubling and halving only in counter reckoning, while Scheubel (No. 83, p. 112.) gives them only for fractions; Tunstall (1538), Suevus (No. 88, p. 2.), and Alsted (No. 2, p. 27.) merely mention these operations as included by some among the fundamental ones. By the end of the sixteenth century these two operations had largely fallen into disuse; an exception noted above,[10] as late as 1660, is Moore, who still speaks of 'bypartition and duplication.' (D. E. Smith, *History of Mathematics,* v. II, p. 35, says that these processes are rarely found in printed arithmetics of Italy, Spain, France, or England.).

The order of the four fundamental operations advocated by Köbel (1515), as addition, subtraction, multiplication, and division, was generally adopted by the writers of the sixteenth and seventeenth centuries. Calandri (1518), however, and Tagliente (1520), as noted above,[11] followed the order of Borghi (1484); namely, multiplication, division, addition, and subtraction.

The use of symbols for these fundamental operations during the two centuries in question has been discussed at some length above,[12] and it has been shown[13] that the symbols + and − were in abundant use in the seventeenth century. The references to such popular and influential works as those of Vieta (1630), Descartes (1649), Moore (1660), Renaldini (1665), Feliciano (1669, 1692), Prestet (1675), Hemeling (1678), and Blondel (1699) are quite essential in the discussion of this topic. These writers are not mentioned by Tropfke (v. I, 3d. ed., p. 81.), who ascribes incorrectly the spread of these symbols among the common people to the German writers of the eighteenth century.

In regard to checks, Rudolff (1526) uses 6, 7, and 8; however, in 1532 he checks largely by 9, mentioning also the checks by 7 and 11. He adds (No. 8, f. 12, v.):

> Die gewissest prob / so man gehaben mag / ist / wañ ein Species die ander probirt / wie nachvolgt,

indicating that the contrary operation is the sure test. This last method of checking is used exclusively by Felner (1535) and Waser (1603). The majority of the writers of these centuries check by 9 alone, as do Aurel (1552), Micyllus (1555), Suevus (1595), in addition to Tagliente (1520), Gemma (1549), and Moore (1660), mentioned above.[14] Many check by the more common way, by 9 and 7, as Cardan (1539), Regius (1543), Tartaglia (1556), Bassi (1666), Griminelli (1670), Husswirt (1501), Cataneo (1540), Feliciano (1550), Sfortunati (156I), and Lapazzaia (1601). Wolphius (1534) has been noted[15] as giving the checks with 7 only, while Jacob (1612) mentions the check by 9, 3, 7, 8, 11, etc. A further peculiarity is introduced by Steinmetz (No. 86, f. 18, r.), who uses the check by 9 but gives a table for checks by 7, 11, and 13. Tartaglia (No. 90, ff. 8-12.) gives a lengthy discussion of checks, especially by 9 and 7.

[10]Ed. See Figure 2.97 and section 2.17 on page 97.
[11]Ed. See section 2.4 on page 69.
[12]Ed. See sections 3.2–3.43.4 on pages 123–136.
[13]Ed. See sections 3.2.3–3.2.4 on pages 124–126.
[14]Ed. See Figure 2.24 on page 71, section 2.3 on page 68 and section 2.17 on page 96.
[15]Ed. See section 2.2 on page 66.

4.3 Fundamental Operations in Particular

4.3.1 Addition

In the treatment of tables, Tropfke (v. I, 3d. ed., pp. 140–148) omits reference to the addition tables found not only in Tartaglia (1556), as noted above,[16] but also in Tunstall (1538), Peverone (No. 68, p. 7.), and in Hemeling as late as 1678. (L.L. Jackson, *Sixteenth Century Arithmetic,* p. 41, says that the writers who used addition tables were the exception, and he cites Tartaglia as the exception. F. Unger, *Die Methodik der Praktischen Arithmetik,* Leipzig, 1888, p. 73, refers to the use of addition tables up to 90 + 90 by Leonard of Pisa.) Noviomagus (No. 65, f. 23, v.), Chauvet (No. 21, f. 3, v.), Puig (No. 72, p. 62.), Prestet (No. 70, p. 13.), Hemeling (1678), and Monier (No. 62, p. 10.) teach addition downwards. Lambeck (No. 55, p. 18.), however, instructs the reckoner:

> Fange unten bei der rechten Hand an und gebrauche das Worts Und,

writing the word *und* between each of the corresponding digits of two summands. A survival of the Hindu method of adding from left to right is seen in Ciacchi (No. 22, p. 61.), who uses this procedure as one of his checks for addition. Monier (No. 62, p. 10.) writes the numbers to be carried below the digits of the sum and scratches them out after adding them, thus [4.2]:

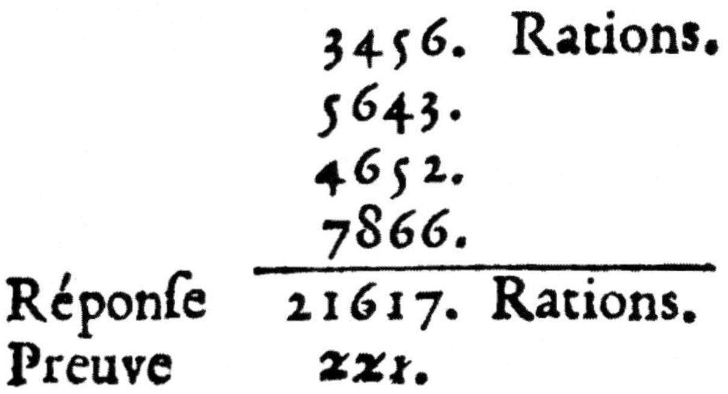

FIGURE 4.2: Monier's carry digits below the sum.[17]
No. 62, 1693, p. 10

Regius (No. 76, f. 54) illustrates addition of only two numbers, and scratches out the digits as added in this manner [4.3]:

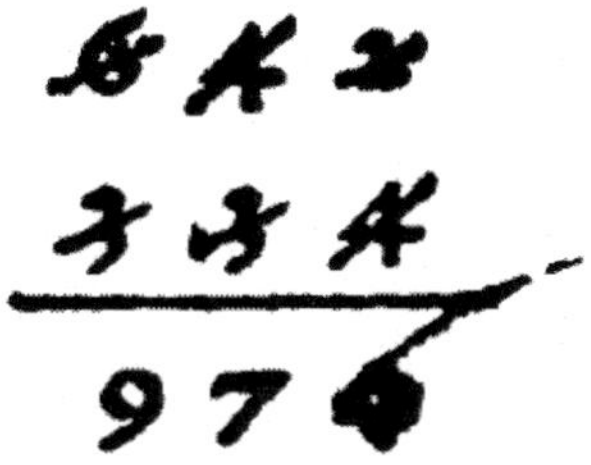

FIGURE 4.3: Regius' scratching out of digits for addition.
No. 76, 1536, f. 46, v.

[16] ED. This previous reference to Tartaglia's addition tables could not be found.

[17] ED. Figure 4.2 is from a 1693 edition published in Paris. Sister Leontius used a 1693 edition published in Amsterdam.

This process is connected with the old scratch method taught before the introduction of printing. Save for such variations, the addition problems of these early texts appear in the same form as they do today.

4.3.2 Subtraction

Three methods of subtraction may be found simultaneously in the arithmetics of these two centuries; namely, the two borrowing methods and the so-called complementary method.

Tunstall (1538) illustrates the first borrowing method very clearly in the following manner [4.4]:

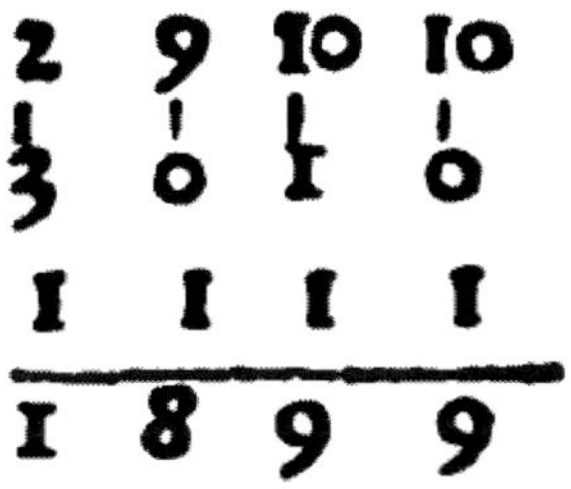

FIGURE 4.4: **Tunstall's first borrowing method of subtraction. No. 91, 1538, p. 30**

In addition to the eight writers of the sixteenth century mentioned by Tropfke (v. I, 3d. ed., p. 90.) as using this first method, we wish to mention eight others of importance: Noviomagus (No. 65, f. 25, r.), Ghaligai (No. 38, f. 4, v.), Peverone (No. 68, p. 11.) Fine (1587), Clavius (No. 25, p. 19.), Bonvicino (No. 12, p. 6.), Bassi (No. 6, p. 14.), and Prestet (No. 70, p. 15.). Bassi (No. 6, p. 14.) says that this method was used 'dalli nostri Antichi, benche alcuni ancora al presente se ne servono.'

Hemeling (1667, 1678) performed subtraction by a second method, that is, the addition of one to the next highest number in the subtrahend, introduced into Europe by Leonard of Pisa and one of the most common methods of subtraction in the sixteenth century. (See L. L. Jackson, *The Educational Significance of Sixteenth Century Arithmetic,* New York, 1906, p. 51.). Followers of this method in the seventeenth century are Manelli (No. 59, p. 10,), Bonvicino (No. 12, p. 6.), Bassi (No. 6, p. 14.), Griminelli (No. 43, p. 29.), Prestet (No. 70, p. 16.), and Monier (No. 62, p. 30.).

These two methods are compared by Bonvicino in the subtraction problem

$$\begin{array}{r} 84065 \\ 68478 \\ \hline 15587 \end{array}$$

when he says:

> Altri in vece di aggiungere la decina al numero susseguente la levano realmente dal numero di sopra, come sarebbe levato il primo 8. di 15. poiche la decina fù levata dal 6. nel secondo numero, in vece di dire 7. dal 16. dicono 7. dal 15. mà è una cosa istessa, e perche genera qualche confusione, ritengo la prima via per più facile.

The additive method of subtraction is also suggested by him in the following words:

> Altri in vece di dire 8. di 15. riman 7. dicono 8. ad andare al 15. vene vogliono 7. poi operano come è detto, mà tutto è una cosa istessa. Può adoprarsi qual modo à cadauno riesce più facile, che nul la rileva.

Complementary subtraction, in which the difference of the digit in the subtrahend from ten is added to the respective minuend digit, was very popular during these centuries. It appeared, as noted above,[18] in the works of Gemma Frisius (1549), Fine (1555, 1587), Savonne (1571), Lapazzaia (1601), and Jacob (1612). Tropfke (v. I, 3d. ed., p. 94.) has mentioned the earlier works of Finé (1535) and Jacob (1565). Besides the nine authors of the sixteenth century listed by Tropfke we find the complementary method used by Reisch (No. 77.), Felner (No. 30, f. 10, v.), Steinmetz (No. 86, f. 20, v.), Unicorno (No. 93, p. 12.), and in the seventeenth century by Clavius (No. 25, p. 19.), Alsted (No. 2, p. 31.), Bassi (No. 6, p. 14.), Puig (No. 72, p. 65.), Lucas (No. 58, p. 8.), and Blondel (No. 9, p. 72.). Clavius (No. 25, 1618, p. 20.) calls this method 'Piú facil regola di sottrarre', whereas he states that the first borrowing method 'é usata da molti Aritmetici.' Likewise Tartaglia (No. 90, f. 15, v.), to whom Tropfke has referred (v. I, 3d. ed., p. 94.), makes this statement with reference to this method:

> ...il terzo dapoi che si ha inteso é alquãto piu leggiadro & accommodo da maneggiare da cadauno de gli altri duoi ...;

and Bassi says:

> ...e questo terzo modo quasi da tutti viene usato, per essere grandemente facile.

Tropfke has noted (*Ibid.,* p. 92.) that Ramus in his arithmetic of 1592 begins his subtraction on the left, according to the Hindu custom. We find this true also in the 1599 edition. Concerning this order of subtraction we quote from Prestet (No. 70, p. 17.):

> Cette methode me semble plus d'usage, & je m'en sers plus volontiers que des precedentes, parceque je trouve qu'elle charge beaucoup moins la memoire que celle du probleme, & qu'outre cela il ne faut effacer ni ajoûter aucuns chiffres, ce qui est fort commode pour les supputations longues & frequentes, où le meilleur est toujours d'embarasser ses nombres & ses calculs le moins qu'on le peut.

Alsted (No. 2, p. 31) gives directions that the remainder may be written above or below. Chauvet (No. 21, f. 13, r.) makes his subtraction clear in this manner:

```
 799
 8000   (8 struck out)
 3459
 ----
 4541
```

[18] Ed. A previous reference could not be found for Gemma and Moore. For Finé, Savonne and Lapazzaia see section 2.10 on page 82, section 2.11 on page 84 and section 2.13 on page 89.

Various symbolic aids in the subtraction of a larger number from a smaller number are introduced in the early books. We find a dot placed above or to the right of the numbers in the minuend to indicate a borrowing in the complementary method, as by Blondel (No. 9, p. 72.) [4.5] and Savonne (1571) [4.6] respectively.

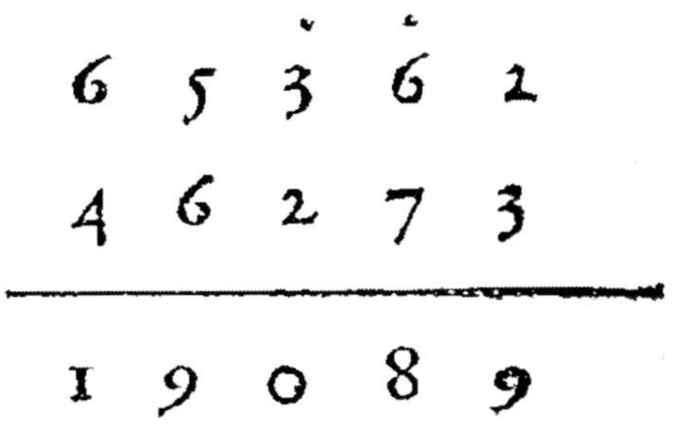

FIGURE 4.5: **Blondel's dots for 'borrowing'. No. 9, 1683, vol. 2, p. 72**

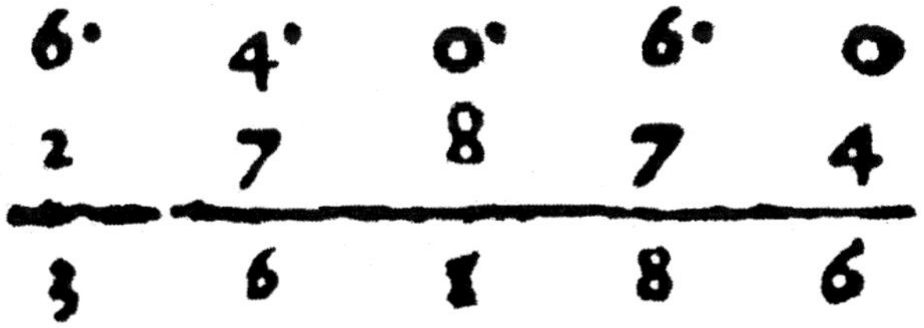

FIGURE 4.6: **Savonne's subtraction.[19] No. 82, 1571, f. 6, r.**

Alsted (No. 2, p. 31.), Felner (No. 30, f. 10, v.), Graffenried (No. 42.), and Lucas (No. 58, p. 8.) place the dot above, to the right, or below the numbers in the subtrahend respectively, to denote the addition of one in this same method. In other respects the subtraction problems resemble those of today.

4.3.3 Multiplication

Italian writers of these centuries, following Pacioli, include commonly various methods of multiplication in their texts. This was the case with Tagliente (1520), Feliciano (1550), and as late as 1678 with Figatelli. Unicorno (1598), Manelli (1659), and Bassi (1666) display many methods, among them our modern method called 'per Baricocolo', the so-called lightning method 'per Crocetta', the chess-board method 'per Quadrato', and 'per modo allo Indietro' or 'alia Roverscia.' Besides these we find in texts of this time the methods 'per Ripiego', 'per Scapezzo', complementary multiplication, 'per Triangolo', 'per Piramide', and others.

Our method, 'per Baricocolo', is very popular and may be found among other methods in practically all of the works of this period, with the important exception of Ghaligai, (1552) who seems to have a preference for multiplication 'allo Indietro.' Felner (No.

[19]Ed. From the Plimpton Collection and used with the permission of the Rare Book & Manuscript Library, Columbia University.

30, f. 13, r.), Aurel (No. 3, f. 6, r.), Girard (No. 39, f. 6, r.), and Prestet (No. 70, p. 21.) follow Cardan (1539) in teaching only our method.

Multiplication 'per Ripiego', in which the multiplicand is multiplied by one factor of the multiplier and this result by the other factor, is included by Steinmetz (No. 86, 33, v.), Unicorno (No. 93, ff. 19, v., 21, v.), Puig (No. 72, p. 71.), and Venturoli (No. 94, p. 4.). The 'assai ammaestreuole et ingenioso' (No. 93, f. 19, v.) method 'per Crocetta' is given by Cantone (No. 15, p. 38.), Steinmetz (No. 86, f. 35, r.), and Venturoli (No. 94, p. 5.). A way of multiplication 'molto bella & facile', multiplication 'per Quadrata', is displayed by Peverone (No. 68, p. 14.); Regius (No. 76, f. 64, r.) calls this 'modus elegans.'

In multiplication 'allo Indietro', the partial products are obtained by beginning with the highest number of the multiplier. Ghaligai (No. 38, f. 6, r.) multiplies 978 by 876 thus [4.7]:

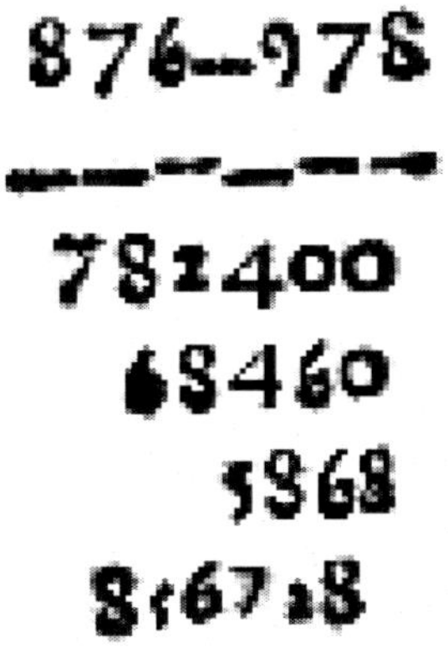

FIGURE 4.7: **Ghaligai's 'allo Indietro' method of multiplication. No. 38, 1552, f. 6, r.**

This method may be explained in this manner:

$$978 \times 876 = 978(800 + 70 + 6) = 782400 + 68460 + 5868 = 856728.$$

This problem and this form are almost exactly given by Juan Diez Freyle in Mexico in 1556 in the first arithmetical work printed in the New World. Napier (No. 64, p. 39.) and Ciacchi (No. 22, p. 17.) use this method also, but, like Hemeling, omit the zeros. The method 'per Scapezzo' is similar in procedure, in that the multiplicand is multiplied in turn by the digits whose sum is equal to the multiplier, and the results thus obtained are added.

We find this example, 34 × 16, in Unicorno [4.8]:

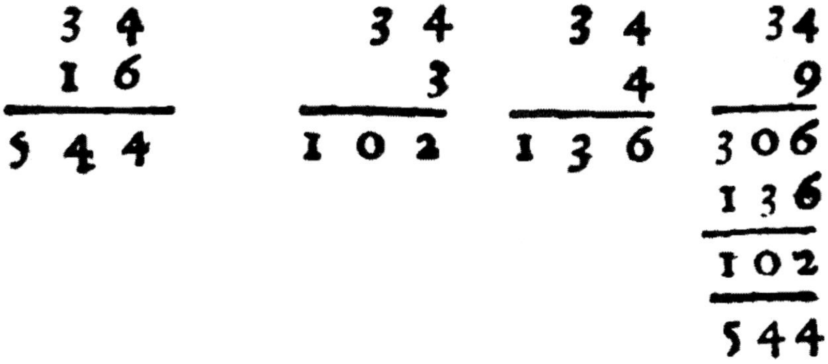

FIGURE 4.8: Unicorno's 'per Scapezzo' method of nultiplication. No. 93, 1598, f. 21, v.

This same process is used by Alsted (No. 2, p. 34.) in multiplying 8 by 7 in this fashion [4.9]:

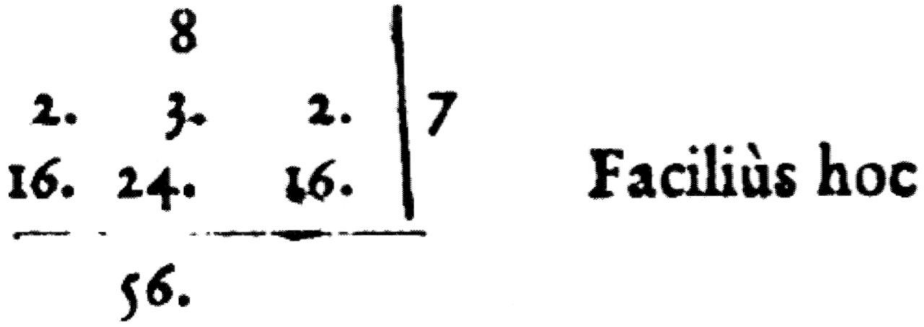

FIGURE 4.9: Alsted's method of multiplication. No. 2, 1641, p. 34

In the texts examined, the complementary method is based upon these formulas, both of which may be found in treatises of the thirteenth century (See Susan R. Benedict, *A Comparative Study of the Early Treatises Introducing into Europe the Hindu Art of Reckoning,* pp. 82, 88, 89, 99, 100, 103):

1 $a \cdot b = 10a - a(10 - b)$

2 $a \cdot b = 10[a - (10 - b)] + (10 - a)(10 - b)$.

Not only Regius (1543), as noted above,[20] uses the first formula, but also Scheubel (No. 83, p. 37.). Both formulas are used by Rudolff (No. 81, ff. 5, r., 6, r.) in his work of 1532. (Tropfke, v. 1, 3d. ed., p. 101, refers to the 1526 edition.). Formula two is more frequently employed in the majority of the works of this time, as in Husswirt (1501), Gemma (1549), Finé (1555), Micyllus (1555), Steinmetz (1575), and Blondel (1699), mentioned previously,[21] also in Vuelpius (No. 98, f. 15, v.), Noviomagus (No. 65, f. 25, v.), Clavius (No. 25, p. 28.), Graffenried (No. 42, pp. 21, 22.), and Monier (No. 62, p.

[20] ED. See Figure 3.67 on page 135.

[21] ED. See section 2.1 on page 63, section 2.3 on page 67, section 2.10 on page 82, section 3.3.8 on page 135, Figure 3.68 on page 135, and Figure 3.69 on page 136.

55). A similar principle is used by Monier (ibid., p. 56.) in the 'methode par les doigts', and before him by Moore (1660).

A sort of 'filling-in' process, as used by Steinmetz (No. 86.) and Behme (No. 7, p. 38.), is illustrated by the following example (No. 86, f. 31, r.) [4.10]:

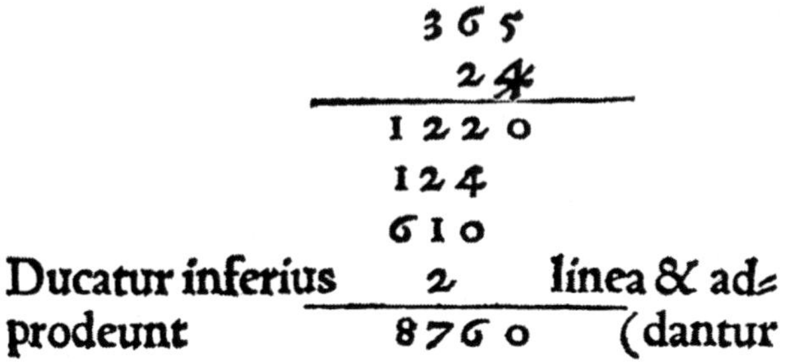
365
24
1220
124
610
Ducatur inferius 2 linea & ad-
prodeunt 8760 (dantur

FIGURE 4.10: Steinmetz's 'filling-in' method of multiplication. No. 86, 1575, f. 31, r.

Lucas has a chapter entitled 'Arithmetica Rabdologica', which teaches the use of Napier's rods for multiplication. The use of 'Nepairs bones', as Napier's rods are vulgarly called, was also taught and recommended in the arithmetics of Jonas Moore (1660) and William Leybourn (1660, p. 265.).

An instance of actual multiplication with zero is found in Vuelpius (No. 98, f. 16, r.) who gives this note [4.11, 4.12]:

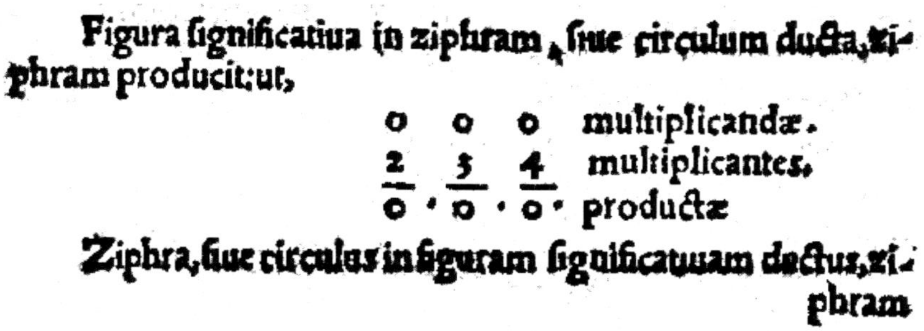
Figura significatiua in ziphram, siue circulum ducta, ziphram producit:ut,
0 0 0 multiplicandæ.
2 3 4 multiplicantes.
0·0·0· productæ
Ziphra, siue circulus in figuram significatiuam ductus, ziphram

FIGURE 4.11: Vuelpius' multiplication with zero. No. 98, 1544, f. 16, r.

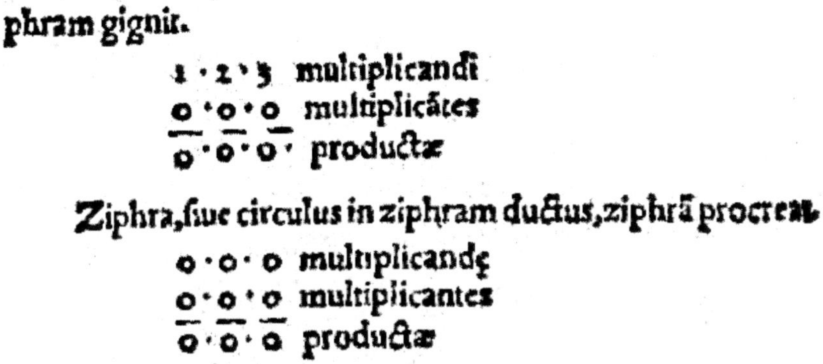
phram gignit.
1·2·3 multiplicandi
0·0·0 multiplicãtes
0·0·0· productæ
Ziphra, siue circulus in ziphram ductus, ziphrã procreat.
0·0·0 multiplicandę
0·0·0 multiplicantes
0·0·0 productæ

FIGURE 4.12: Vuelpius' multiplication with zero. No. 98, 1544, f. 16, v.

It may be of interest to add here that zero combinations are recently being emphasized in instruction in arithmetic.

4.3.4 Division

It may be of interest here to quote Rahn's definition of division (No. 73, p. 2.):

> Division is to take the same quantity from another as oft as may be, as

$$\begin{array}{cccl} 6 & 4 & 2 & \\ \underline{2} & \underline{2} & \underline{2} & \underline{\vert 3} \\ 4 & 2 & 0 & \end{array}$$

Remnants of a Hindu method of division, consisting of decomposition of the divisor into factors, may be found in Tartaglia (No. 90, f. 35, v.), Unicorno (No. 93, f. 26, v.), Manelli (No. 59, p. 18.), and Venturoli (No. 94, P- 7.). Tartaglia's explanation of this division 'per Repiego' may be seen in the problem,[22]

$$5867 \div 48 = 122\frac{11}{48}$$

where the chosen factors are 8 and 6:

$$\begin{array}{l} 8\,\underline{\vert\ 5\ 8\ 6\ 7} \\ \quad 6\ \underline{\vert\ 7\ 3\ 3}\ {-3} \\ \qquad\ \ 1\ 2\ 2\ {-1} \end{array}$$

To obtain the final remainder add the product of the first divisor and the last remainder to the first remainder, in this case, $8 \cdot 1 + 3 = 11$.

Among other methods of division, division 'per Galera' is rarely missing in the arithmetics of these centuries. In the sixteenth century, Rudolff (No. 31, f. 10, r.), Felner (No. 30, f. 15, r.), Regius (No.. 76, f. 68, r.), Noviomagus (No. 65, f. 30, r.), Aurel (No. 3, f. 8, v.), Micyllus (No. 60, p. 29.), Peverone (No. 68, p. 17.), Scheubel (No. 83, p. 49.), and Steinmetz (No. 86, f. 41, r.) teach only this method. In the seventeenth century, Gautruche (No. 36, p. 10.) (a work used in Harvard College), and Blondel (No. 9, p. 110.) still cling to it. On the other hand, a few of the later writers like Bonvicino (1665), Venturoli (1681), and Lucas (1697) do not advocate it. In fact, Bonvicino remarks (No. 12, p. 10.):

> Lasciarò da parte il partire per Galea, perche è oggidi poco usato, come quello, che per la moltiplicità de numeri, e per il molto scassamento, genera confusione, e darò il modo di dividere, per la danda, prima lunga, poi corta.

[22]Ed. The following equation and, subsequent, organized rows of numerals is Sister Leontius's construct and is not a reproduction of Tartaglia's work. Tartaglia explained this process verbally, in written out sentences. Tartaglia's explanation is in (No. 90, 1556, ff. 35 v., 36 r.). Sister Leontius had a typographical error stating the work was on f. 25, v.

Division 'per Batello', according to Manelli (No. 59, p. 22.), is similar to division 'per Galera', except that the numbers in the process are written below the dividend instead of above. Manelli illustrates as follows [4.13]:

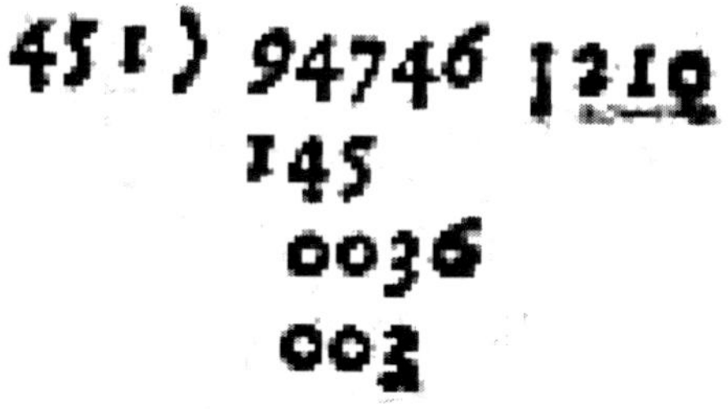

FIGURE 4.13: Manelli's 'per Batello' method of division. No. 59, 1659, p. 22

We must not confuse the latter method with 'per Danda corta', or 'mezza Danda', taught in many Italian works of the seventeenth century, as in Clavius (No. 25, p. 59.), Napier (No. 64, p. 54.), Manelli (No. 59, p. 20.), Bonvicino (No. 12, p. 15.), Bassi (No. 6, p. 44.), Griminelli (No. 43, p. 64.), and Venturoli (No. 94, p. 7.). In both, the work is carried out beneath the dividend and the partial products are not written; but in division 'per Danda corta', the multiplications and subtractions are performed from right to left, and we may see this from the following problem in Manelli [4.14]:

FIGURE 4.14: Manelli's 'per Danda corta' method of division. No. 59, 1659, p. 21

(For the methods used by Clavius, see L. G. Simons, "Historical Notes on Arithmetical Division", *Scripta Mathematica,* June, 1933, pp. 362, 363.).

In 'per Danda longa', our present-day method, the partial products are recorded. It is used particularly in the works of these Italian and Spanish writers who offer so many divers methods of operation: Unicorno (No. 93, f. 34, r.), Forestani (No. 35, ff. 16-25.), Manelli (No. 59, p. 21.), Bassi (No. 6, p. 49.), Griminelli (No. 43, p. 68.), Puig (No-81-), and Venturoli (No. 94, p. 7.). Manelli, in comparing the two most used methods, 'per Danda', and 'per Galera', says (No. 59, p. 19.):

> ...e questo li pare più facile il partire à Danda, per li principàti è più sicuro, perche si vede il moltiplicare è il sottrare; mà il partire à galera è più corto, ma fà di bisogno star bene avvertito nel tagliar le figure, perche errando, non si può si facilmente correggere, come la persona non è ben prattica.

Prestet (No. 70, p. 31.) also calls our method 'un peu plus longue à la verité, mais plus facile a observer.'

In Prestet (*Ibid.*), also, we find an instance of the repetition of the divisor below each remainder, similar to the practice of Karsten (1767). (See Tropfke, v. I, 3d. ed. p. 109.). There are still a few, like Coutereels (No. 26, p. 30.), Griminelli (No. 43, p. 68.), and Behme (No. 7, p. 37.), who follow Finé as noted above,[23] in writing down in advance the partial products of the divisor. It should be noted further that Ghaligai does not treat division of integers, but speaks of division 'per Ripiego', 'per Danda', and 'per Galera', only in connection with denominate numbers.

4.4 Fractions

4.4.1 Common fractions

It is usual to find a fraction defined as 'pars integri.' Cardan (No. 16, f. 7, r.), however, says:

> Unde nihil aliud est dicere `3/5` quam `.3.` divisum per `.5.`

Prestet (No. 70, p. 42.), too, regards it as a division or ratio. A little later, Monier (No. 62, p. 142.) remarks:

> Les fractions ne sont que l'expression de la raison que est entre l'entier & sa partie.

Girard (No. 39, p. 8.) goes further and defines it as 'une division imparfaite.'

Fractions in general are still classified as vulgar or common, and astronomical or sexagesimal by Finé (1555), Regius (No. 76, f. 83, v.) and Bassi (No. 6, p. 70.). Ordinarily writers of these centuries subdivide common fractions into simple fractions, and fractions of fractions, as do Regius (No. 76, f. 83, v.) and Aurel (No. 3, f. 10, v.). Lambeck (No. 55, p. 72.), however, says that fractions of fractions belong to multiplication. Rudolff (No. 81, f. 18, v.) and Felner (No. 30, f. 27, v.) call them 'schlechte bruch' and 'bruch von bruchen', whereas Vuelpius (No. 98, f. 29, v.) and Lucas (No. 58, p. 55.) use the terms 'primaria' and 'secundaria.' Girard (No. 39, p. 8.) speaks of proper and improper fractions, and in the latter he includes such fractions as `5/4`, `6/2`, and even `6/1`. Lambeck (No. 55, p. 72.) says that in trade there occur 'Gemengte Brüche' as [4.15]

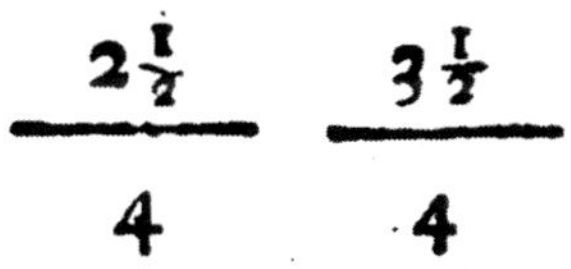

FIGURE 4.15: **Lambeck's complex fractions. No. 55, n.d. (c. 1661), p. 72**

which he reads as 'Drittehalb Viertheil' and 'Viertehalb Viertheil.'

The majority of writers mention the value of a fraction according as the numerator is greater than, equal to, or less than the denominator. Some, like Cardan (No. 16, f. 21, v.), Alsted (No. 2, p. 45.), Bonvicino (No. 6, pp. 18, 19.), Bassi (No. 6, p. 55.), and

[23] Ed. See Figure 2.53 on page 82.

Lucas (No. 58, p. 56.), observe that a fraction decreases in value if, keeping the same numerator, the denominator is increased, and on the contrary, increases if, keeping the same denominator, the numerator is increased.

In listing the operations for fractions, most authors treat numeration, reduction, and the four fundamental operations. Some, like Felner (No. 30, f. 27, v.) and Scheubel (No. 83, p. 91.), include doubling and halving, and a few, like Cardan (No. 16, 21, v.) and Scheubel (No. 83, p. 91.), add roots and progression. The notation used in writing fractions has been discussed previously.[24] Scheubel (*Ibid.*) and Alsted (No. 2, p. 44.) give very definite directions for the reading of a fraction in saying that the numerator is read as a cardinal number, the denominator as an ordinal numeral.

Reduction of fractions to higher or lower denominations depends upon the property that multiplying or dividing both terms of a fraction by the same number does not alter the value of the fraction. This is clearly enunciated by Lucas (No. 58, p. 57.). According to Bassi (No. 6, p. 56.), Gautruche (No. 36, p. 13.), and Blondel (No. 9, p. 150.), those fractions are defined as equivalent whose numerators have the same ratio to their denominators. In order to find the greater of two fractions recourse is usually had to reduction to a common denominator, but Steinmetz (No.86, f. 65, v.), in determining which is the greater, 3/4 or 1/2, does what is equivalent to reduction to decimals, thus [4.16]:

FIGURE 4.16: Steinmetz's comparison of fractions. No. 86, 1575, f. 65, v.

Girard calls fractions reduced to their lowest terms, primitive, and the fractions themselves, derivative, as 1/2 and 5/10 respectively.

The process of obtaining the primitive fraction by finding the greatest common divisor, called by Lucas (No.58, p. 57.) 'aureum numerum', is usually performed by the method of Euclid, Book 7, Prop. 2. This may be illustrated briefly from the explanation given by Clavius (No. 25, p. 79.) in the reduction of 60/96[25]:

96 ÷ 60, rem.	36
60 ÷ 36, rem.	24
36 ÷ 24, rem.	12
24 ÷ 12, rem.	0
G.C.D.=	12

There are others who find the common divisor by trial, and a few, like Jacob (1612) mentioned previously,[26] present tests for divisibility by 2, 3, 4, etc. Rudolff (No. 81, f. 22, r.) gives the familiar tests for divisibility by 2 and 5; and his test for 3 is that 0, 3, or 6 remains after the check by 9. Lucas (No. 58, pp. 57, 58.) gives the usual tests for 2,

[24] ED. See section 3.7.1 on page 147.

[25] ED. The following lines of work, to find the G.C.D., are Sister Leontius's construct and are not a reproduction of Clavius' work. Clavius explained this process verbally, in written out sentences, on page 79 of the 1618 edition.

[26] ED. See section 2.15 on page 93.

5, and 10, to which Blondel (No. 9, p. 153.) adds these rules, some of which are rarely used:

> Tout nombre dont les caracteres ajoutés ensemble font 5 ou un multiple de 3, peut être divisé par 6, si le dernier chiffre est nombre pair.
> Et par 15, si le dernier caractere est 5.
> Et par 30, si le dernier caractere est 0.
> Tout nombre dont tous les chiffres ajoutés ensemble font 9 ou un multiple de 9, peut être divisé par 18 si le dernier caractere est nombre pair,
> Et par 45, si le dernier chiffre est 5.
> Et par 90, si c'est un 0.

There is nothing extraordinary in the treatment of addition and subtraction of fractions in the textbooks of these centuries. Rudolff (No. 81, f. 20, v.) treats multiplication before the other three operations and he explains it simultaneously with 'Teil suchen.' We may add the writers Bonvicino (No. 12. p. 23.) and Blondel (No. 9, pp. 171, 172.) to those given by Tropfke as enunciating the fact that the product of two fractions is less than either. As regards division of fractions we quote from Bonvicino (No. 12, p. 24.):

> Perciò ogni qual volta la minuzia, che divide è minore di quella, che deve esser divisa, il quoziente farà maggiore, perche tante più volte v'entra: come all'incontro, quando il divisore è più della frazione, che deve esser divisa, il quoziente è minore, e quanto è maggiore, tanto minore farà il quoziente, perche tante meno volte v'entra.

Multiplication is usually performed with the aid of parallel lines, and division by means of crossed guiding lines. In dividing 3/4 by 2/3, many writers, like Cardan (No. 16, f. 21, v.), Peverone (No. 68, p. 28.), Unicorno (No. 93, f. 46, v.), Manelli (No. 59, p. 33.), and Venturoli (No. 94, p. 16.), place the divisor on the left in an arrangement similar to this [4.17]:

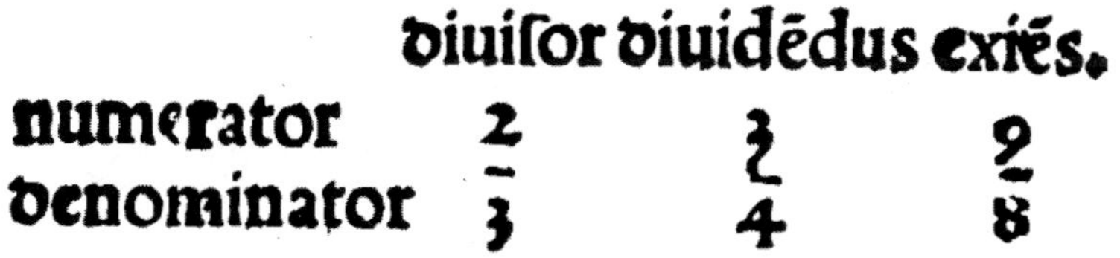

	diuisor	diuidēdus	exiēs.
numerator	2	3	9
denominator	3	4	8

FIGURE 4.17: Cardan's division of fractions. No. 16, 1539, f. 21, v.

Some place the divisor on the right, thus:

$$\frac{3}{4} \times \frac{2}{3} \quad \frac{9}{8}$$

We may mention many other writers besides Clavius who make use of the principle of inversion: Waser (No. 99, f. 40, v.), Graffenried (No. 42, p. 112.), Alsted (No. 2, p. 49.), Lambeck (No. 55, p. 86.), Bonvicino (No. 12, p. 24.), Bassi (No. 6, p. 62.), Venturoli (No. 94, pp. 15, 16.), and Lucas (No. 58, p. 78.). Alsted (No. 2, p. 49.) remarks:

> Facilius hoc praestiteris, si divisore inverso procedas ut in multiplicationes.

Again Bonvicino (No. 12, p. 23.) says:

> Conviene scrivere il partitore al riverscio di quello si scrivono li rotti.

Lucas (No. 58, p. 78.) directs briefly as follows: "In praxi invertitur dividens." The method of inversion is rarely used in the sixteenth century (See L. L. Jackson, *Sixteenth Century Arithmetic,* p. 107.).

4.4.2 Decimal fractions

The history of decimal fractions in the sixteenth century has been completely treated by Tropfke. In this connection, Oughtred's *Clavis Mathematicæ* is cited by him (v. I, 3d. ed., p. 181.). Another important reference for this subject is to the arithmetic of Sir Jonas Moore (No. 63.), who was a close follower of Oughtred. In his development of decimal fractions, Moore refers several times to 'The Learned Mr. Oughtred in his Clavis Limata.' (*Ibid.,* p. 14 in preface.).

In speaking of fractions, Moore states (*Ibid.,* p. 6.)

> ...we use a more briefe expression of some kindes of them by omitting the Denominators, when the parts are commonly known, ...
>
> As because the twentieth parts of a pound are famously known by the name of shillings, we use to say and write 2s. 3s ... briefly with the Numerators onely, and not fully with the Denominators, thus, 2/20, 3/20, ...of a pound.
>
> The same we do also in the twelve parts of a shilling saying, 1d. 2d. ...briefly with the Numerators onely, and not fully thus, 1/12, 2/12,..., of a shilling with the Denominator.

Later he sums up by saying (Ibid., p. 10.):

> From this Abbreviation of Fractions ariseth that late most useful invention of Decimall Arithmetique, where the Denominators are always 1 with Cyphers, 10, 100, 1000, 10000, &c. and therefore the Denominators are quite ommitted (as alwayes certainly knowne being still more by one place than the figures of the Numerator) and the Numerators onely set down.

We need not stop here to give the various forms of the decimal separatrix, since that has been treated adequately above.[27] Suffice it to say that the vertical stroke played an important part as forerunner of a decimal separatrix in the writings of many prominent sixteenth-century authors already mentioned. Voigtel (No. 97, pp. 1-16.) who begins his *Markscheide-Kunst* with an explanation of 'so genannten Decimal-Rechnung', and Lucas (No. 53, pp. 224-231.) who devotes a chapter to 'Arithmetica Geometrica Decimalis', both still adhere to the sexagesimal exponents[28] as their decimal notation, a slight deviation from the symbolism used by Stevin.

Like Stevin, Moore in his introduction to decimal arithmetic, writes a plea for the introduction of decimals into the systems of money, weights, and measures (No. 63, pp. 14–15.):

[27] Ed. See section 3.8 on page 150.

[28] Ed. See Figures 3.114, 3.115, and 3.116 on page 153.

In practicall Arithmeticke, if the first institution had been in Decimalls, we had never been troubled with so many fractions; it were yet worthy the name of Reformation to cause the fractions of mony and weight to be altered: And as concerning the ease in measure, Surveyors and Land-meaters, who use the decimall chaine, and those who use a decimall foot, yard or scale can best certifie upon experience: If it had not been the Stationer's fault, I had long since published some Tables, after the most easie and methodicall way hitherto thought on, towards the ready obtaining the true place of any of the planets in this decimall way, which would have been a sufficient testimony of the ease thereof in Astronomicall calculations. Thus much for an Introduction into Decimall Arithmetick.

(F. Unger, *Die Methodik der Praktischen Arithmetik,* Leipzig, 1388, pp. 104–106, likewise quotes Christian Schessler, 1692, on the advantages of a decimal system.).

As regards the fundamental operations with decimals, we find that they differ little from our modern methods. Directions are usually given to proceed as with integers, and rules are set forth for determining the number of decimal parts in the result. Moore performs all the fundamental operations with decimals, and besides has a section on the reduction of fractions into decimals and vice versa, and gives reduction tables for money, weight, time, and measure. He applies decimals to proportions, compound interest, and logarithms. In fact, his entire work gives quite the modern treatment of decimals, 'for the great facility it brings with its practice in all parts of Arithmeticke, Astronomy, and the rest of the Mathematicks' (*Ibid.,* p. 14.).

4.4.3 Proportions

As a general rule, proportions arise chiefly because of their use in the 'rule of three', so very popular in these early centuries. Some writers distinguish between a continued and discontinued proportion, as do Aurel (No. 3, f. 18, r.), Ghaligai (No. 33, f. 22, v.), Unicorno (No. 93, f. 156, v.), Alsted (No. 2, p. 53.), and Puig (No. 72, p. 31.). A continued proportion, a proportion whose two means are equal, usually termed a progression, does not concern us here so much as a discontinued or disjunct proportion. Very few authors treat the three kinds of proportion taught by Pythagoras, namely, arithmetic, geometric, and harmonic. The work of Bradwardin (No. 13.), however, is devoted chiefly to proportion in general.

Harmonic proportion, defined by a:c::(a - b):(b - c), is rarely given attention in these arithmetic texts. Puig (No. 72, p. 33.) mentions it. Aurel (No. 2, ff. 15. v., 16, r.) gives the very interesting story of the invention of harmonic ratio by Pythagoras. Harmonic proportion as a combination of geometric and arithmetic proportions is illustrated by Aurel (No. 3, f. 18, r.) in the example 6.3.2., so that, according to the rule expressed above, 6:2::(6 - 3):(3 - 2). He is one of the few writers who treat the harmonic mean and give instructions how to find it. An explanation in ten lines of harmonic proportion may be found in Bassi (No. 6, p. 509), something given seldom (Noted also by Tropfke, v. III, 2nd. ed., p. 6.).

Although arithmetic proportion is mentioned by these same writers, and as well by Cardan (No. 16, Caput 45.), Alsted (No. 2, p. 53.), and Prestet (No. 70, p. 168.), yet little use is made of it. Of all these men, Prestet treats it at the greatest length. He speaks of it as 'l'egalité de deux differences', which we may express by the equation a - b = c - d. In the first theorem, he proves that in every arithmetic proportion the sum of the extremes is equal to the sum of the means, thus: a+d=b+c; and in the corollary (*Ibid.,* p. 169.), that in every continued arithmetic proportion the sum of the extremes is equal to the mean proportional taken twice.

Geometric proportion is an integral part of practically every early arithmetical text, and is given special treatment under the 'rule of three.' Micyllus (No. 60, p. 80.) refers to 'Regula Proportionum' and says it is called 'Regula De Tre' by the Italians. Others, like Suevus (No. 88, p. 245.), Clavius (No. 25, p. 117.), Griminelli (No. 43, p. 131.), Ciacchi (No. 22, p. 121.), and Prestet (No. 70, p. 194-) also call it the 'golden rule', as the latter says 'à cause du grand usage que l'on en fait.' T Hodder (No. 51, p. 87.) remarks this:

> As Gold transcends all other Metals, so doth this Rule all other in Arithmetick.

The term 'golden rule' is also used by Recorde, Baker, Moore, and other early writers in English. Waser (No. 99, f. 48, v.) and Bohme (No. 7, p. 49.) add still another name 'Regula Mercatorum', for the latter says:

> Weil die Kauffleute von wegen ihres grossen Handels dieselben nicht können ohne sein.

The great majority of the authors of those centuries speak of the 'rule of three' *simple,* and *composée* or *composta.* The latter is only an application of the 'rule of three' two or three times, and consists of five terms, of which the sixth may be found by dividing the product of the last three terms by that of the first two. The 'rule of three' may be direct and indirect (inverse or reciprocal), as illustrated by 2:3::4:6 and 2:6::4:3 respectively. In the former, the product of the means equals the product of the extremes according to Euclid, Prop. 19, Book 7, a fact enunciated again by Rudolff (No. 81, f. 37, v.), Clavius (No. 25, p. 119.), Alsted (No. 2, p. 58.), and Prestet (No. 70, p. 191.). In the latter, the first term is to the third as the fourth is to the second, so that in this case the product of the first two terms is equal to the product of the last two, according to Clavius (No. 25, p. 129.), Alsted (No. 2, p. 59.), Bassi (No. 6, p. 113.), and Prestet (No. 70, p. 193.).

The four common forms of writing a proportion, by inversion (*conversa*), alternation (*permutata*), composition (*componendo, conjunta*). and division (*dividendo, disjunta*), may be found complete in Cardan (No. 16, Caput 45.), Unicorno (No. 93, f. 156, v.), Puig (No. 72, p. 32.), and Prestet (No. 70, p. 192.). Clavius (No. 25, pp. 122, 123.) and Alsted (No. 2, p. 59.) mention mainly the first two, while Rudolff (No. 81, f. 37, v.), Regius (No. 76, f. 100, r.), and Vuelpius (No. 98, f. 43, v.) give only proportion by alternation. To the four forms mentioned above Unicorno (No. 93, f. 157, r.) and Puig (No. 72, pp. 32, 33.) add proportion *eversa* expressed by a:a-b::c:c-d, proportion *egual,* a:c::d:f obtained from a:b::d:e and b:c::e:f. Puig (Ibid., p. 33.) extends the list still further and illustrates as follows:

Ordinata, b:a-b::d:c-d, as 12.9.8.6,

Inordinata, $b:a-b::\frac{bc}{a-b}:c$, 12.9.24.6,

Extensa, $a:b::c:d::b:e::d:f$, 12.6.8.4.3.2,

Perturbata, $a:b::c:d::b:e::f:c$, 12.6.8.4.3.16.

Thirty-nine other properties of proportions follow in Puig (*Ibid.,* pp. 43–52.). Among these is included the Euclidean statement (*Ibid.,* p. 44.) that if

$$a:b::c:d::e:f,$$

then

$$(a+c+e):(b+d+f)::a:b,$$

also given by Chauvet (No. 21, f. 44, v.) and Prestet (No. 70, p. 188.).

Rudolff (No. 81, ff. 38, v., 39, r.) gives express directions for setting down the terms of a proportion and of finding any one missing term of a proportion. He follows this by giving rules for reduction of a proportion to simpler terms, which he accomplishes by dividing the first and the third, or the first and second terms by a common divisor, as illustrated below (*Ibid.,* f. 40, r.) in the proportion $32:36::18:x$: [4.18, 4.19]

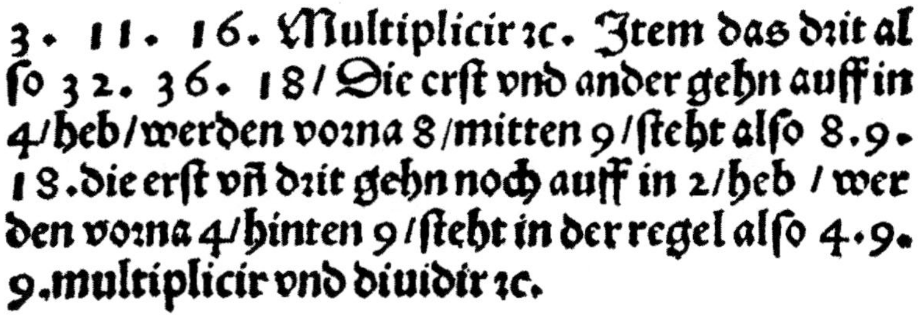
3. 11. 16. Multiplicir ꝛc. Item das drit al so 32. 36. 18/ Die erst vnd ander gehn auff in 4/heb/werden vorna 8/mitten 9/steht also 8.9. 18. die erst vñ drit gehn noch auff in 2/heb / wer den vorna 4/hinten 9/steht in der regel also 4.9. 9.multiplicir vnd diuidir ꝛc.

FIGURE 4.18: **Rudolff's explanation of reducing a proportion. No. 81, 1540, f. 46, r.**

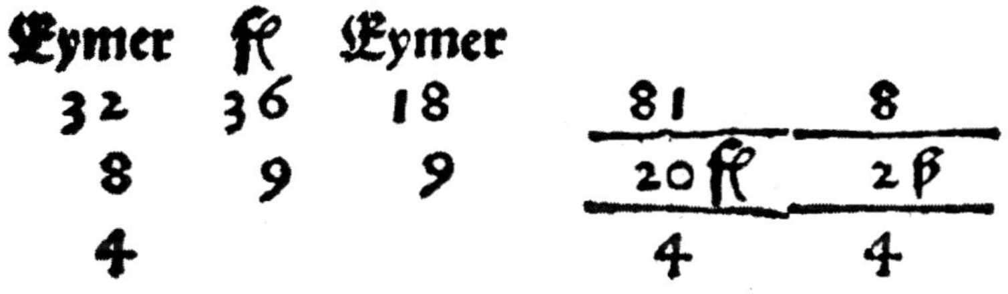

FIGURE 4.19: **Rudolff's reduction of a proportion. No. 81, 1540, f. 46, v.**[29]

This same procedure is used by Clavius, Alsted, Chauvet, and Monier.

[29]Ed. Sister Leontius used the 1532 edition of Rudolff's *Kunstliche rechung* and for this example shows:

32.	36.	18
8.	9.	18
4.	9.	9

The ordinary method of finding the mean proportional between two numbers is given by a large majority of authors, if mean proportion is treated at all. Puig (No. 72, p. 38.) finds the mean proportional between 24 and 6 in a second way, as illustrated below [4.20]:

$$24 \div 6 = 4; \qquad \sqrt{4} = 2; \qquad 24 \div 2 \text{ or } 6 \times 2 = 12.$$

De otra manera se puede hallar vn numero medio proporcional entre dos extremos, con que los dichos dos extremos sean, ò numeros quadrados, ò numeros similes: desta manera, sean los dos extremos 24. y 6. parte el mayor por el menor, y vendràn 4. la rayz quadrada de los quales, que es 2. es la denominacion de la proporcion, que estos numeros han de guardar entre si por lo qual no ay mas que hazer, sino partir 24. por 2. ò multiplicar 6. por el mismo 2. que de vna manera, y otra hallaràs 12. por el medio proporcional que se pide, desta manera, 24. 12. 6.

FIGURE 4.20: Puig's second method for finding a mean proportional. No. 72, 1672, p. 38

4.5 Powers

In the treatment of the last two sections we attempt to avoid infringing too much upon the domain of algebra. Operations with powers, outside of squaring and cubing, were not an integral part of the representative arithmetics of this period. Here we shall discuss powers only in so far as they were mentioned by the writers of arithmetic in connection with the extraction of roots. The formation of the names and symbols for exponents has been treated above.[30]

Because it is a rare exception to find involution included among the fundamental operations, we quote its definition from Rahn (No. 73, p. 2.):

> *Involution* is multiplying any quantity into its self, and that Product into the first quantity, and this last into the first, &c.

```
2 · 4 · 8    &c.
2   2   2
-   -   --
4   8   16
```

The significance of the term *involution* may be best gleaned from the following explanation on page 9:

Involution in Single Whole Quantities.

Its Sign is [⟲]

[30] Ed. See section 3.9 on page 154.

R. If a Quantity be involved or drawn into it self, or afterward into that Product, or again thirdly, into the last Product, &c., as manifold as is the Power, so great must the number be that is used to express it, being set after the Sign of Involution in the broader Column of the Margin.

3yz	abcd		1	
9y^{2}zz	aabbccdd		2	1 Ꝍ 2

The Meaning is, let (3 yz) be a *Line,* this twice involved gives (9 yyzz) and is called a *Quadratick* Power or *Square*: Three times involved, it gives (27 yyyzzz) which is a *Cubick* Power, or *Cube*: Four times involved, it gives (81 y^{4}z^4) which is a *Biquadratick*, &c.

The ordinary definition of a *numero quadrato,* according to the definitions of the seventh and eighth books of Euclid, were given by Aurel (No. 3, f. 40.),

> es un numero que procede de la multiplicaciõ de 2 numeros yguales en quãtidad y genero, como 3 y 3 multiplicados hazen 9, q̃ es numero quadrado: y el 3 es su rayz, o el 3 multiplicado en si (que es lo mesmo) y vernã 9, como cõsta por las definiciones del 8° del mesmo Euclides: diziẽdo: Numero quadrado es n° superficial de yguales lados. &c.

In a similar manner Prestet (No. 70, p. 64.) speaks of powers as 'Multiplications reiterees.'

The derivation of square numbers from odd numbers[31] is found in Bassi (No. 6, p. 526.):

> Li numeri quadrati hanno l'origine dalli numeri dispari, che derivano da un 'unità, sequitando in continua Proportione, che è la seconda spetie della Progressione Arithmetica, come saria 1.3.5.7.9.11.13.15, & altri simili, onde se l'I si congiungera con il 3 sequente, componerassi 4 secondo numero quadrato, e così congiunti insieme li tre numeri dispari 1.3.5, seranno 9, che è il terzo numero quadrato, similmente congiungendo insieme 1.3.5.7, daranno 16, per lo quarto numero quadrato, e seguirassi con tal modo ne gli altri.

He follows this later with the example:

> ...la radice di 36, sarà 6, perciò il numero quadrato 36 è composto di sei numeri dispari, cioè 1.3.5.7.9.11, li quali raccolti insieme, fanno 36.

Aurel (No. 3, f. 40.) and Puig (No. 72, pp. 3, 16, 321.) likewise give the composition or generation of square numbers. The latter calls attention to the fact that a square number does not end in 2, 3, 7, or 8.

We may mention a few of the more important writers who give tables of the squares and cubes of the first 9 digits: Aurel (No. 3, f. 41.), Clavius (No. 25, pp. 513, 517.),

[31] Ed. Sister Leontius is alluding to the relationship between the sum of consecutive odd number (starting with one) and perfect squares. Given by the formula: $1 + 3 + 5 + \cdots + (2n - 1) = n^2$

Alsted (No. 2, p. 74.), and Blondel (No. 9, p. 116.). Lucas (No. 58, pp. 205–214.) extends the table so as to include numbers up to 500. Similar tables of higher powers, as a convenience for extracting higher roots, are given by Moore (1660), as noted previously,[32] and also by Bassi (No, 6 pp. 522, 524.), Puig (No. 72, pp. 327, 330.), and Prestet (No. 70, fold. table opposite p. 100.).

Whereas Tropfke (v. II, 3d. ed., p. 158.) has cited Girard as affording a single instance of the use of fractional exponents, he has neglected to include him as employing zero exponents. Girard (No. 39, f. 9, r.) writes [4.21]:

FIGURE 4.21: Girard's notation for an exponent of zero. No. 39, 1884, f. 9, r.

> **toutefois le (0) est de signification finie, suivant un nombre, comme 18 (0) est 18 precisement, car c'est 18 nullement indefiny, quant à (1) 18, c'est le mesme que 18 (0) car ils s'accordent là.**

According to his notation, 18(0) stands for $18x^0$, and (1)18 for the first root of 18, both of which equal 18.

4.6 Roots

Although the extraction of roots is not always considered a fundamental operation in the early arithmetics, it is generally given special treatment. Of the many texts examined, we find among the works examined, its discussion entirely missing only in the works of Rudolff (No. 81.) (1532), Felner (No. 30.) (1525), Cantone (No. 15.) (1606), and Behme (No. 7.) (1695). It is noteworthy here, too, that Vuelpius (No. 98.) (1544) does not attempt to take the square root of a *numerus non quadratus.*

Rahn's definition of the process of root extraction (No. 73, p. 2.) is worth quoting:

> *Evolution* is the discovery of the Ingredients of the involved quantity.

$$\begin{array}{cccc} 8 & 6 & 4 & 2 \\ \underline{2} & \underline{2} & \underline{2} & \underline{2} \\ 6 & 4 & 2 & 0. \end{array}$$

Aurel (No. 3, f. 40, v.) refers to the sixth book of Euclid when he writes that square root is to find the mean proportional between unity and the proposed number.

Tropfke (v. II, 3d. ed., p. 175.) mentions the method of approximation in the extraction of roots as explained by Clavius (1583). We find that Clavius (No. 25, pp. 274-278.) in the later editions examined, explains three old methods of approximation. Two of these methods he obtains from Ptolemy's *Almagest,* finding one value less and one greater than the true root by the further development of the formula of Archimedes,

$$a \pm \frac{b}{2a} > \sqrt{a^2 \pm b} > a \pm \frac{b}{2a \pm 1}$$

[32] Ed. See section 2.17 on page 95.

A part of the above formula for approximating the root of a number which is an imperfect power, namely,

$$\sqrt{a^2 + b} = a + \frac{b}{2a + 1}$$

was used by Aurel (No. 3, f. 42, r.), Micyllus (No. 60, p. 61.), Steinmetz (No. 86, f. 58.), Bonvicino (No. 12, p. 31.), Puig (No. 72, p. 321.), and Blondel (No. 9, p. 145.). Puig explicitly gives the reason for the form of the denominator of the remainder. In a similar way Monier (No. 62, pp. 426, 463.) approximates the cube root of a number by the formula,

$$\sqrt{a^3 + b} = a + \frac{b}{3a + 3a^2}$$

(Günther, *Geschichte des Mathematik,* Leipzig, 1908, v. I, p. 343, refers to the use of these formulas by Ortega in 1512.).

Both Clavius (No. 25, pp. 274–278) and Prestet (No. 70, pp. 102, 103.) refer to an old method of approximation which corresponds to our modern method of using decimal fractions. By annexing three pairs of zeros to 20, its square root is given by the former as $4\frac{472}{1000}$ whereas Prestet would give the result as $\frac{4472}{1000}$. It is worth adding that Scheubel (No. 83, pp. 64, 65.) devotes considerable space to this decimal method of approximation. Voigtel (No. 97, pp. 5–12) finds the square roots of decimal numbers themselves.

Our modern method of obtaining the square root by use of the principle $(a + b)^2$ is well illustrated in Rahn (No. 73, p. 15.). Authors like Micyllus (No. 60, p. 60.), Chauvet (No. 21, f. 193, r.), and Blondel (No. 9, pp. 119–121.) take pains to illustrate this principle geometrically according to the second book of Euclid. (L. L. Jackson, *Sixteenth Century Arithmetic,* pp. 122, 123, quotes in detail the explanation of this method given by Trenchant in his *L'Arithmetique,* 1578.).

As usual we find an extended treatment of higher roots in Puig (No. 72, pp. 322–332.) and Prestet (No. 70, pp. 83–90.). The former takes roots as high as the fifth by use of the binomial coefficients; Prestet, as also Graffenried (No. 42.), up to the ninth; but they suggest the shorter method for the fourth, sixth, and ninth roots by use of the law

$$\left(a^{\frac{1}{m}}\right)^{\frac{1}{n}} = a^{\frac{1}{mn}}$$

So far as the treatment of operations with radicals is concerned, this properly belongs in the realm of algebra.

Bibliographic Notes

The following table provides some additional bibliographic details regarding the sources of the figures in Chapters 2, 3, and 4. The first column under # lists the index of the title used in the dissertation. In the fourth column under **Year**, S indicates an edition referenced in the dissertation, D indicates an edition used in the preparation of this book, and F indicates a first edition. The **Ed.** information in the sixth column is primarily based upon D.E. Smith's *Rara Arithmetica,* Chelsea Publishing Co., 4th ed., 1970 and on OCLC WorldCat. Entries in the rightmost column **Lib.** note libraries and collections that currently contain the edition as follows:

UM University of Michigan

GAP George Arthur Plimpton Collection, Rare Book and Manuscript Library, Columbia University

DES David Eugene Smith Collection, Rare Book and Manuscript Library, Columbia University

CU Columbia University

NYP New York Public Library

#	Author	Title	Year	City	Ed.	Lang.	Lib.
1	Alexander, Joannes	Synopsis algebraica : opus posthumum Johannis Alexandri, Bernatis-Helvetii. In usum scholae mathematicae apud Hospitium-Christi Londinense, à Carolo II. Fundatae, editum.	1693 S,D,F	London	1st	Latin	UM DES
2	Alsted, Johann Heinrich	Methodus admirandorum mathematicorum Novem libris exhibens universam mathesin	1613 F	Herborn	1st	Latin	GAP
			1641 S,D	Herborn	3rd	Latin	UM GAP
	Barrow, Isaac	Archimedis opera: Apollonii Pergaei conicorum liber IIII...	1675 S,D,F	London	1st	Latin	UM DES CU NYP
5	Bassi, Giulio	Dell' Arithmetica practica di Giulio Bassi... Libri VII	1645 S,D,F	Piacenza	1st	Italian	UM GAP DES
6	Bassi, Giulio	Arithmetica, e geometria practica del dottor Giulio Bassi Piacentino... Libri Otto	1666 S,D,F	Piacenza	1st	Italian	UM GAP
7	Behme, Casper	Arithmetica oder Rechen-kunst	1695 S,D,F	Dantzig	1st	German	UM
9	Blondel, Francois	Cours de mathematique : contenant divers traitez composez et enseignez a Monseigneur le Dauphin (vol. 2)	1683 D,F	Paris	1st	French	DES

TABLE 4.1: **Sources for the Figures in Chapters 2, 3, and 4**

#	Author	Title	Year	City	Ed.	Lan.	Lib.
		Cours de Mathematique, qui contient la mathematique en general, la geometrie speculative, et la geometrie pratiqu... Second Edition augmentee & corrigee	1699 S	Amsterdam	2nd	French	UM
12	Bonvicino, Valeriano	Matematiche discipline per vso della illvstrissima Academia Delia di Padoua dove in sei trattati reuement si ristringono aritmetica, geometria, trigonometria pratiche. Fortificazione, sfera, e geografia	1665 S,D,F	Padua	1st	Italian	UM
14	Calandri, Philippi	Pictoragas arithmetice intruductor	1491 S,D,F	Florence	1st	Italian	UM GAP
16	Cardan, Hieronimo	Hieronimi C. Cardani Medici Mediolanensis, practica arithmetice, & Mensurandi singularis...	1539 S,D,F	Milan	1st	Latin	UM
17	Cataldi, Pietro Antonio	Trattato dell' algebra proportionale	1610 F	Bologna	1st	Italian	UM
		Nuova algebra proportionale...	1619 S,D	Bologna	2nd	Italian	UM
19	Cataneo, Pietro	Le practiche delle due prime Mathematice...	1546 S,D,F	Venice	1st	Italian	UM GAP
			1559 S,D	Venice	2nd	Italian	GAP

TABLE 4.1: Sources for the Figures in Chapters 2, 3, and 4

#	Author	Title	Year	City	Ed.	Lan.	Lib.
20	Ceva, Joannis	Opuscula mathematica de potentiis obliquis...	1682 S,D,F	Milan	1st	Latin	DES
22	Ciacchi, Giuseppe	Regole generali d'abbaco...	1675 S,D,F	Florence	1st	Italian	UM GAP DES CU
24	Clavius, Christopher	Christophori Clavi Bambergensis e Societate Jesu Epitome arithmeticae Practicae	1583 F	Rome	1st	Latin	GAP
		Christophori Clavii Bambergensis e Societate Jesu Epitome arithmeticae Practicae nunc denuo ab ipso auctore recognita	1585 S,D	Rome	3rd	Latin	UM GAP
25	Clavius, Christopher	Arithmetica prattica composta dal molto Rever...	1586 S	Rome	4th	Italian	UM GAP
			1618 S,D	Rome	???	Italian	UM GAP
			1686 S,D	Venice	???	Italian	UM GAP
27	Curtius, Sebastian	Ein Newes Wolgegrundetes Rechunbuch : nach rechter Art der Practic	1604 D	Nuremberg	???	German	
		Arithmetica Practica, das ist: Ein Newes Wolgegrundetes Rechunbuch	1619 S	Leipzig	???	German	DES

TABLE 4.1: **Sources for the Figures in Chapters 2, 3, and 4**

#	Author	Title	Year	City	Ed.	Lan.	Lib.
28	Descartes, Rene	Discours de la methode... la Geometrie...	1637 F	Leiden	1st	French	UM CU NYP
		Geometria a Renato Des Cartes, anno 1637 Gallice edita;... Opera atque studio Francisci a Schooten,...	1649 S,D	Leiden	2nd	Latin	UM CU NYP
29	Feliciano, Francesco	Libro di Arithmetica e Geometria speculatiua e praticale...	1526 D,F	Venice	1st	Italian	GAP
			1550 S,D	Venice	5th	Italian	UM GAP
			1563 S,D	Verona	10th	Italian	UM GAP CU
			1669 S,D	Venice	???	Italian	UM GAP DES
			1692 S,D	Venice	???	Italian	UM
32	Figatelli, Giuseppe	Ristretto arithmetico	1664 F	Modana	1st	Italian	UM DES
		Trattato Arithmetico di Giuseppe Maria Figatelli... In questa seconda impresione aggiontoui l'algebra	1678 S,D	Venice	2nd	Italian	UM DES

TABLE 4.1: **Sources for the Figures in Chapters 2, 3, and 4**

#	Author	Title	Year	City	Ed.	Lan.	Lib.
		Trattato Arithmetico di Giuseppe Maria Figatelli...In questa quinta impresione aggiontoui l'algebra	1699 S,D	Venice	6th	Italian	UM DES
33	Fine, Oronce	Orontii Finaei delphinatis...De arithmetica practica libri quatuor...	1535 F	Paris	1st	Latin	DES NYP
			1555 S,D	Paris	3rd	Latin	UM GAP CU
34	Fine, Oronce	Opere di Orontio Fineo del Delfinato: diuise in cinque parti; arithmetica...	1587 F,S,D	Venice	1st	Italian	UM GAP DES NYP CU
			1670 S	Venice	???	Italian	UM CU
37	Gemma Frisius, Reiner	Arithmeticae practicae Methodus facilis...	1540 F	Antwerp	1st	Latin	UM GAP NYP
			1549 S	Paris	9th	Latin	UM GAP NYP
			1550 S,D	Paris	10th	Latin	UM GAP NYP

TABLE 4.1: Sources for the Figures in Chapters 2, 3, and 4

#	Author	Title	Year	City	Ed.	Lan.	Lib.
38	Ghaligai, Francesco	Summa de arithmetica	1521 F	Florence	1st	Italian	GAP
		Practica d'arithmetica di Francesco Ghaligai…	1552 S,D	Florence	3rd	Italian	UM GAP CU
39	Girard, Albert	Invention nouvelle en l'Agebra	1629 F	Amsterdam	1st	French	UM NYP
		Invention nouvelle en l'Agebra. Reimpression par Dr. D. Bierens de Haan	1884 S,D	Leiden	2nd	French	UM NYP
42	Graffenried, Johann Rudolff von	Arithmeticae Logistica Popularis Libri IIII…	1619 S,D,F	Bern	1st	German	DES
43	Griminelli, Domenico	Novissima prattica d'arithmetica merchantile…	1656 S,D,F	Rome	1st	Italian	UM CU
45	Harriot, Thomas	Artis analyticae praxis…	1631 S,D,F	London	1st	Latin	UM GAP NYP
47	Hemeling, Johann	Arithmetica Historia oder Historische Rechneknunst…	1667 S	Hanover	???	German	UM
48	Hemeling, Johann	Johannis Hemelingii Selbstlehrende Rechneschul: oder: Selbstlehrendes Rechnebuch…	1678 S,D	Frankfurt	???	German	UM

TABLE 4.1: **Sources for the Figures in Chapters 2, 3, and 4**

#	Author	Title	Year	City	Ed.	Lan.	Lib.
49	Herigone, Pierre	Cursus mathematicus. . . (Tomus Primus)	1634 S,D,F	Paris	1st	Latin/French	UM
52	Husswirt, Johannes	Enchiridion novus Algorismi. . .	1501 S,D,F	Cologne	1st	Latin	UM GAP
			1504 D	Cologne	3rd	Latin	CU
54	Jacob, Simon	Ein New und Wolgegrundt Rechenbuch auff den Linien und Ziffern samt der Welschen Practic. . .	1560 F	Frankfurt	1st	German	
			1612 S,D	Frankfurt	???	German	UM
	Kobel, Jakob	Ain new geordent Rechenbiechlin auf den linien mit rechen pfennigen. . .	1514 S,D,F	Augsburg	1st	German	GAP
55	Lambeck, Heino	Heinonis Lambecii vermehrtes und verbessertes Rechen-buch . . .	c.1661 S,D	Stade	???	German	UM
56	Lapazzaia, Georgio	De familiarite arithmeticae et geometriae	1566 F	Naples	1st	Latin	GAP
		Libro d'arithmetica e geometria dell' Abbate Giorgio Lapazzia	1566 F	Naples	1st	Italian	GAP
			1601 S	Naples	5th	Italian	UM GAP

TABLE 4.1: **Sources for the Figures in Chapters 2, 3, and 4**

#	Author	Title	Year	City	Ed.	Lan.	Lib.
			1727 D	Naples	???	Italian	UM GAP DES
58	Lucas a S. Edmundo	Arithmeticus practicus	1697 S,D,F	Tyrnaviae	1st	Latin	UM
59	Manelli, Francesco	Opera d'arithmetica e geometria...	1659 S,D,F	Bologna	1st	Italian	UM GAP
60	Mycillus, Jacobus	Arithmeticae logisticae libri duo...	1555 S,D,F	Basel	1st	Latin	UM GAP
62	Monier de Clairecombe, Jean	Nouvelle pratique d'arithmetique...	1693 D	Paris	???	French	NYP
			1693 S	Amsterdam	???	French	UM
63	Moore, Jonas	Moores Arithmetick: discovering the secrets of that art, in numbers and species. In two bookes. The first teaching (by precept and example) the ordinary operations in numbers whole and broken... The second, the great rule of algebra in species ...	1650 F	London	1st	English	UM
		Moor's Arithmetick. In Two Books. The first treating of the Vulgar Arthmetick... The Second of Arithmetick in Species or Algebra... To which are added two Treatises...	1660 S,D	London	2nd	English	UM NYP

TABLE 4.1: **Sources for the Figures in Chapters 2, 3, and 4**

#	Author	Title	Year	City	Ed.	Lan.	Lib.
	Moore, Jonas	A New Systeme of Mathematicks	1681 S,F	London	1st	English	UM DES NYP
64	Napier, John	Rabdologiae, seu numerationis per virgulas libri duo	1617 F	Edinburgh	1st	Latin	DES NYP
		Raddologia Ouero arimmetica virgolare, in due libri divisa... Tradottore dalla Latina nella Toscana lingua...	1623 S,D	Verona	2nd	Italian	UM
65	Noviomagus, Joannes	De numeris libri duo...	1539 F	Paris	1st	Latin	GAP CU
		De numeris libri II...	1544 S,D	Cologne	2nd	Latin	UM
	Oughtred, William	Arithmeticae in numeris et speciebus institutio: quae tum logisticae, tum analyticae, atoue adeo totius mathematicae, quasi clavis est...	1631 S,D,F	London	1st	Latin	UM DES
67	Peurbach, Georg	Elemanta Arithmetices...	1492 F	s.l.	1st	Latin	
			1539 S,D	Venice	16th	Latin	UM NYP
68	Peverone, Giovanni Francesco	Due brevi e facili trattati il primo d'Arithmetica: l'atro di Geometria:...	1558 S,D,F	Lyons	1st	Italian	UM

TABLE 4.1: **Sources for the Figures in Chapters 2, 3, and 4**

#	Author	Title	Year	City	Ed.	Lan.	Lib.
70	Prestet, Jean	Elements des mathematiques ou princepes generaux…	1675 S,D,F	Paris	1st	French	UM
72	Puig, Andres	Arithmetica especulativa, y practica, y arte de algebra…	1672 S,D,F	Barcelona	1st	Spanish	UM DES
73	Rahn, Johann Heinrich	Teutsche Algebra, oder algebraische Rechenkunst, zusamt ihrem Gebrauch	1659 S,D,F	Zurich	1st	German	UM
		An Introduction to Algebra, Translated out of the High-Dutch into English by Thomas Brancker	1668 S,D	London	2nd	English	UM GAP
74	Ramee, Pierre de La	P. Rami Arithmeticae libri duo: Geometriae septem et viginti	1569 F	Basel	1st	Latin	GAP CU
		P. Rami Arithmeticae libri duo: Geometriae septem et viginti a Lazaro Schonero recogniti & aucti	1599 S,D	Frankfurt	9th	Latin	UM GAP
75	Ramee, Pierre de La	P. Rami Arithmeticae libri duo cum commentariis Wilebrordi Snelli R. F.	1613 S,D	Leiden	???	Latin	UM GAP
76	Regius, Ulricus	Utriusque Arithmetices Epitome…	1536 D,F	Strasburg	1st	Latin	
			1543 S	Freiburg	2nd	Latin	UM DES
	Recorde, Robert	The groūd of artes…	1543 F	London	1st	English	

TABLE 4.1: Sources for the Figures in Chapters 2, 3, and 4

#	Author	Title	Year	City	Ed.	Lan.	Lib.
		The Ground of Artes...	1558 S,D	London	7th	English	GAP
		The whetstone of witte	1557 S,D,F	London	1st	English	UM GAP
77	Reisch, Gregorious	Margarita philosophica	1503 S,D,F	Freiburg	1st	Latin	UM GAP
78	Renaldi, Carlo	Caroli Renaldini Opus Algebraicum...Ars analytica...	1644 F	Ancona	1st	Latin	UM
		Caroli Renaldini...Ars analytica mathematum...pars prima...	1665 S,D	Florence	2nd	Latin	UM
79	Renaldi, Carlo	Caroli Renaldini...Ars analytica mathematum...pars secunda...	1669 S,D	Padova (Patavii)	???	Latin	UM
81	Rudolff, Christoff	Kunstliche rechnung mit der ziffer un mit den zal...	1526 F	Vienna	1st	German	UM GAP
			1532 S	Nuremberg	2nd	German	UM
			1540 D	Nuremberg	5th	German	UM GAP
82	Savonne, Pierre	L'Arithmetique de Pierre Savonne, dict Talon...	1563 F	Paris	1st	French	
			1571 S,D	Lyons	3rd	French	UM GAP

TABLE 4.1: **Sources for the Figures in Chapters 2, 3, and 4**

#	Author	Title	Year	City	Ed.	Lan.	Lib.
			1585 D	Lyons	4th	French	
			1619 S,D	Lyons	???	French	UM GAP
83	Scheubel, Johann	Compendium arithmetique artis...	1549 F	Basel	1st	Latin	GAP
			1560 S,D	Basel	2nd	Latin	UM GAP
85	Sfortunati, Giovanni	Nuovo Lume, Libro di Arithmetica...	1534 F	Venice	1st	Italian	GAP
			1561 S,D	Venice	6th	Italian	UM GAP
86	Steinmetz, Moritz	Arithmeticae praecepta...	1568 F	Leipzig	1st	Latin	NYP
			1575 S,D	Leipzig	3rd	Latin	UM
88	Suevus, Siegmund	Arithmetica historica. Die lobliche Rechenkunst...	1593 S,D,F	Wroclaw (Breslau)	1st	German	UM GAP
89	Tagliente, Girolamo	Libro da abaco...	1515 F	Venice	1st	Italian	
			1520 S,D	Venice	2nd	Italian	UM DES

TABLE 4.1: Sources for the Figures in Chapters 2, 3, and 4

#	Author	Title	Year	City	Ed.	Lan.	Lib.
90	Tartaglia, Niccolo	La prima parte del general trattato di numeri, et misure di Nicolo Tartaglia…	1556 S,D,F	Venice	1st	Italian	UM GAP DES CU
91	Tunstall, Cuthbert	De arte supputandi libri quatot, Cuthberti Tonstalli	1522 F	London	1st	Latin	GAP
			1538 S,D	Paris	4th	Latin	UM GAP
			1544 S,D	Strasburg	6th	Latin	UM GAP
93	Unicorno, Josephus	De l'arithmetica universale del Sig. Joseppo Unicorno…	1598 S,D,F	Venice	1st	Italian	UM GAP
94	Venturoli, Giacomo	Ordini arithmetici…	1663 F	Bologna	1st	Italian	
		Breve compendio di tutte le regole dell' arithmetica practica aggiuntovi…Seconda impressione	1681 S,D	Bologna	2nd	Italian	UM DES
95	Viete, Francois	In artem analyticam isagoge	1591 F	Tours	1st	Latin	
		Zeticorum libri quinque	1591 F	Tours	1st	Latin	

TABLE 4.1: **Sources for the Figures in Chapters 2, 3, and 4**

#	Author	Title	Year	City	Ed.	Lan.	Lib.
		L'algebre nouvelle de Mr. Viete...	1630 S,D	Paris	???	French	UM
96	Viete, Francois	Francisci Vietea Opera mathematica, in unum Volumen congesta, ac recognita, Opera atque studio Francisci a Schooten...	1646 S,D,F	Leiden	1st	Latin	UM GAP CU NYP
97	Voigtel, Nicolaus	Geometria Subterranea, oder Marckscheide-knust...	1686 S,D,F	Eisleben	1st	German	UM DES NYP
98	Vuelpius, Henricus	Libellus decommunibus et usitatis arithmeticae practicae reulis...	1544 S,D,F	Cologne	1st	Latin	UM
99	Waser, Casper	Institutio arithmetica brevis & facilis...	1603 S,D,F	Zurich (Tiguri)	1st	Latin	UM
100	Wolphius, Johannes	Rudimenta Arithmetices...	1527 F	Nuremberg	1st	Latin	
			1534 S,D	Frankfurt	2nd	Latin	UM
101	Zuchetta, Giovanni Battista	Prima parte della arithmetica	1600 S,D,F	Brescia	1st	Italian	UM GAP NYP

TABLE 4.1: **Sources for the Figures in Chapters 2, 3, and 4**

Sister Leontius' Bibliography

[D1] Joannes Alexander of Bern. *Synopsis algebraica, opus posthumum Johannis Alexandri, Bernatis-Helvetii. In usum scholæ mathematicæ apud Hospitum-Christi Londinense, à Carolo II fundatæ, editum.* London: Impensis Hospitii, typis Benj. Motte, 1693

Description: t.-p., (2), 204 pp., (2), diagrs., 17½ cm., London, 1693.

Note: Another edition: Ed. 2, emendatior. Cui adjungitur appendix per Humfredum Ditton. London, 1709. Cajori speaks of an English translation by Sam. Cobb at London in 1709.

[D2] Johann Heinrich Alsted. *Methodus admirandorum mathematicorum Novem libris exhibens universam Mathesin.* Herborn: s.n., 1641

Description: t.-p., 456 pp., diagrs., 12 em., no app., Herborn, 1641 (3d. ed.).

Note: Other editions: 1st. ed., Herborn, 1613; 2nd. ed., ib., 1623; 4th. ed., ib., 1657. Wrote also *Elementa mathematica,* Frankfurt, 1611.

[D3] Marco Aurel, el Aleman. *Libro primero, de arithmetica, Algebratica, enel qual se contiene el arte Mercantiuol, con otras muchas Reglas del arte menor, y la regla del Algebra, vulgarmente llamada Arte mayor, o Regla de la cosa: sin la qual no se podra entender el decimo de Euclides, ni otres muchos primores, assi en Arithmetica como en Geometria:. . . .* Joan de Mey Flandro: Valencia, 1552

Description: t.-p., 4 ff. unnumb. + 140 numb. = 144 ff., 19.3 cm., Valencia, 1552.

Note: There is no other edition.

[D4] Ventura de Ávila. *Elementos de algebra, o sea reglas generales para encontrar lo que vale la incognita en las equaciones de el primero, y segundo grado, en quienes no haya termino irracional, y resolucion de setenta y quatro problemas, distribuido todo en veinte y tres Dialogos de suerte, que en otros tantos dias los puede comprehender el que se halle irnpuesto en los setenta y siete Dialogos dados à luz por el Autor de los presentes. . . .* Burgada; Barcelona: s.n., 1799

Description: t.-p., (4), 107 pp., 3 fold. plates, 14.9 cm., Barcelona, n.d.

Note: The work in hand bears no date; probably of the late seventeenth century.

[D5] Giulio Bassi. *Dell' Arimmetica pratica di Giulio Bassi Libri VII.* Piacenza: G. A. Ardizzoni, 1645

Description: 1 p. l., t.-p., (12), 492 pp., 31 cm., Piacenza, 1645.

Note: Another work, *Abbaco,* appeared in 1791.

[D6] ——. *Arimmetica, e geometria practica del dottor Giulio Bassi Piacentino ... Libri Otto.* Piacenza: Nella stampa ducale di G. Bazachi, 1666

Description: 1 p. l., t.-p., (12), 616 pp., 31 cm., Piacenza, 1666.

Note: A rev. ed. appeared ib., 1765.

[D7] Caspar Behme (?). *Arithmetica oder Rechen-kunst.* Dantzig: s.n., 1695

Description: t.-p., (4), 66, sub. t.-p., 38 pp., 15½ cm. Dantzig, 1695.

[D8] Joannes Martinus Blasius. *Liber arithmetice practice astrologis Phisicis et calculatoribus admodum utilis.* Paris: s.n., 1513

Description: 26 ff. unnumb., 64–66 ll., printed in double columns, 28 cm., Paris, 1513.

Note: Other editions: Paris, 1514; ib., 1519; ib., 1526; Microfilm copy, made in 1964, of the original in Vatican. Biblioteca vaticana. Positive. Negative in Vatican. Biblioteca vaticana. Collation of the original, as determined from the film: 26 unnumb. leaves. illus. Colophon: ... Parisiis ... impressus vero a Thoma Rees ... expensis ... Joannis Parui et Joannis Lambert. Anno domini 1513. Manuscripta, microfilms of rare and out-of-print books.

[D9] Francois Blondel. *Cours de Mathematique, qui contient la mathematique en general, la geometrie speculative, et la geometrie pratique ... Seconde Edition augmentée & corrigée.* Amsterdam: s.n., 1699

Description: t.-p., (22), 198, t.-p., (5), 236 pp., 27 cm., Amsterdam, 1699 (2nd. ed.).

Note: 1st. ed., Paris, 1683 (2 vol.). Blondel also wrote other works on architecture, fortification, and the Roman calendar.

[D10] Johannes Böschensteyn. *Ein New geordnet Rechenbiechlin ....* Augsburg: s.n., 1514

Description: 24 ff. unnumb., 30 ll., 19.2 cm., Augsburg, 1514.

Note: Other editions: Augsburg, 1516; ib., 1518. His son, Abraham, wrote *Ein nützlich Rechenbuchlin,* s. 1., 1536. This work is very rare.

[D11] Rafael Bombelli. *L'algebra opera Di Rafael Bombelli . . . .* Bologna: s.n., 1579

Description: 1 p. l., t.-p., (56), 650 pp., (4), 20½ cm., Bologna, 1579.

Note: 1st. ed., Venice, 1572.

[D12] Valeriano Bonvicino. *Matematiche discipline . . . aritmetica . . . .* Padua: s.n., 1665

Description: t.-p., (12), 201 pp., (2), illus., 22 cm., Padua, 1665.

Note: Another ed., Padua, 1666. Riccardi gives this edition only, and says that the *Bib. militaire* (p. 87) registers an ed. of 1665. We have the 1665 ed. under direct examination.

[D13] Thomas Bradwardin. *Arithmetica thome bravardini. Olivier Senant Venum exponuntur ab Oliuiario senant in vivo diui Jacobi sub si-gno beate Barbare sedente.* Paris: Venum exponuntur ab Oliuiario Senant in vico Diui Iacobi sub signo Beate Barbare sedente, 1510

Description: 6 ff. unnumb., 61 ll., printed in double columns, 27.8 cm., Paris, c. 1510, a unique copy.

Note: Editions: Paris, 1495; ib., 1496; ib., 1498; ib., c. 1550; ib., 1502; Valencia, 1503; Paris, 1504; ib., 1505; ib., 1512; ib., 1530; Wittenberg, 1534; ib., 1536. Probably one of the oldest arithmetics written by an English mathematician after Baeda and Alcuin.

[D14] Philippi Calandri. *Pictagoras arithmetice introductor.* Florence: s.n., 1491

Description: 104 ff. unnumb., 9–26 ll., printed in double columns, 10.5 x 14 cm., Florence, 1491.

Note: Another ed., Florence, 1518. It is the first arithmetic to give division *a danda.* Brunet I:1468 (ed. 1491); Klebs: 236.2; De Morgan, Arithmetical Books from the Invention of Printing to the Present Time, p. 1 (listing Calandri's work as the first printed Arithmetic); Riccardi I 208 (giving the date incorrectly as 1515 and nor mentioning the title, suggesting that he has not seen a copy); Smith, Rara Arithmetica, pp. 47-49 (describing the book as a quarto, suggesting that he has not seen a copy); Sotheran 6673 ("of excessive rarity. . . whereas 6 copies of the edition of 1491 have been sold in the last 10 years, no copy of the present one can be traced at all.")

[D15] Oberto Cantone. *L'uso prattico dell' aritmetica e geometria d.Oberto Cantone ... seconda impressione ....* Naples: Tarqminio Longo, 1606

Description: 1 p. l., t.-p., (8) 292 pp., 19½ cm., Naples, 1606.

Note: 1st. ed. Naples 1599. There are also the 1609 and 1612 editions.

[D16] Hieronimo Cardan. *Hieronimi C. Cardani medici Mediolanensis, practica arithmetice, & Mensurandi singuarlis ....* Milan: s.n., 1539

Description: 1 p. l., t.-p., unpaged, illus., 16½ cm., Milan, 1539.

Note: Besides other works, Cardan published the famous *Ars Magna* in 1545, which was reprinted in Cardan's Opera, 1663.

[D17] Pietro Antonio Cataldi. *Nuova algebra proportionale ....* Bolgona: s.n., 1619

Description: 1 p. l., t.-p., (4) 51 pp., diagrs on 2 pl., 29½ cm., Bolgona, 1619.

Note: Earlier edition, 1610. Cataldi also wrote *Due Lettioni,* 1577, besides several other works on arithmetic, algebra, and geometry. At Bologna were also published *Prima Parte della Practica Arithemtica ...,* 1602, and *Seconda Parte ...,* 1606.

[D18] ——. *Algebra applicata ... di Pietro Antonio Cataldi ....* Bologna: s.n., n.d.

Description: t.-p., (4) 88 pp., diagrs., 31 cm., Bologna, 1622.

[D19] Pietro Cataneo. *Le pratiche delle due prime Mathematiche di Pietro de Catani da Siena.* Venice: G. Griffio, 1546

Description: 64 l. incl. diagrs., 20½ cm., Venice, 1546.

Note: 2nd. ed., 1559. There also are works on architecture, 1554, and 1567.

[D20] Joannis Ceva. *Opuscula mathematica de potentiis obliquis, de pendulis, de vasis, et de fluminibus. Joannis Cevae Mediolanensis.* Milan: Mediolani ex typographia Ludovici Montiae, 1682

Description: t.-p., 2 p. l., 53 pp., (2), 8 fold. tables, 26 cm., Milan, 1682.

Note: Ceva also wrote geometrical works.

[D21] Jacques Chauvet. *Methodiques Institutions de la vraye et parfaite arithmetique de Jacques Chauvet, diuisee en six parties. Reveve, corrigee, et amplifiee d'Exemples Geometriques, ...* Rouen: s.n., 1627

Description: 2 p. l., t.-p., (8), 255 numbered 1. (i.e., 510 p.), (10), 17.2 cm., Rouen, 1627.

Note: This work was a revision, correction, and amplification by P. Taillefer of Chauvet Champenois' *Les Institutions* of 1578. D. E. Smith's *Rara Arithmetica* says there was no other edition, and does not refer to this revised work. A still later ed. appeared at Rouen, 1648. Chauvet also wrote several works on geometry. 1615 edition: Paris: Chez Gilles Hervieu, au mont Sainct Hilaire

[D22] Giuseppe Ciacchi. *Regole generali d'abbaco con le sue dichiarazioni, e prove secondo l'uso practicato da' più periti arimmetici: con un breve trattato di geometria, e modi del misurare le superficie de' terreni, e corpi solidi.* Florence: Nella stamp. del Vang., e Matini, 1675

Description: 1 p. l., t.-p., (16), 398 pp., (1), illus., fold. tab., diagrs., 16 cm., Florence, 1675.

Note: Practical geometry is treated on the last 50 pages.

[D23] Pedro Ciruelo. *Tractatus arithmethice practice qui dicitur algorismus.* Paris: Guy Marchant, 1505

Description: 27 pp., tables, 1 diagr., 20 cm., Paris, 1505.

Note: Other Paris editions: 1495, 1509, 1513. Ciruelo also wrote Cursus quattuor mathematicarum artiū liberaliū, which had nine editions. He also edited Bradwardin's arithmetic, 1495, and 1502.

[D24] Christopher Clavius. *Christophori Clavii Bambergensis e Societate Jesu Epitome arithmeticae Practicae ....* Rome: s.n., 1585

Description: 1 p. l., t.-p., 323 pp., (13), 16½ cm., Rome, 1585.

Note: 1st. ed., ib., 1583; other Latin editions: 1584, 1592.

[D25] ——. *Arithmetica prattica composta dal molto Rever. Padre Christoforo Clauio Bambergense della Compagnia di Jesu. Et tradotta da Latino in Italiano dal Signor Lorenzo Castellano ....* Rome: s.n., 1586

Description: 1 p. l., t.-p., 286 pp., (16), 16.3 cm., Rome, 1586.

Note: Other Italian translations: ib., 1602; ib., 1618; ib., 1626; Venice, 1656; Venice, 1686. In 1611 appeared *Opera omnia;* in 1612, his collected works in 5 volumes at Basel entitled *Opera mathematica,*. Clavius also wrote works on algebra, geometry, the calendar, and the astrolabe.

[D26] Johan Coutereels. *Arithmetica, ... nu nieuwelijcx oversien, alle questien na-gerekent, ... met by-voeginghe van verscheiden andre fraeye questien, en noodige annotatien d. Corn.Fr. Eversdyck, Reken - meester 's Lants ende Graeffelijckheyts van Zeelant.* Middleburg: J. Fierens, 1658

Description: t.-p., 1 p. l., 528 pp., diagrs., 16 cm., Middleburg, 1658.

Note: In the preface Coutereels tells of previous editions: 1st. ed., 1599, 1606, 1610, and 1620. Also *'t konstigh Cyffer-Boek,* Utrecht, 1690, and *Het Konstich Cyffer-Boek,* Middleburg, 1626.

[D27] Sebastian Curtius. *Arithmetica Practica, das ist: Ein Newes Wolgegründetes Rechenbuch.* Leipzig: Kober, 1619

Description: t.-p., unpaged, 20 cm., Leipzig, 1619.

Note: At the end he speaks of examples taken from his 1600 *Schuelrechenbuchleins,* and later from the 1610 ed.

[D28] Rene Descartes. *Geometria, à Renato Des Cartes, anno 1637 Gallicé edita; ... Operâ atque studio Francisci à Schooten,....* Lugduni Batavorum: s.n., 1649

Description: 1 p. l., t.-p., (12), 336 pp., (2), 22½ cm., Lugduni Batavorum, 1649.

Note: His 1st. ed. was in 1637. Another Latin ed., 1659–1661, 1683; French ed., 1886; English ed., 1925.

[D29] Francesco Feliciano. *Libro di Arithmetica e Geometria speculatiua e praticale: ...Nouamente stampato.* Venice: s.n., 1550

Description: t.-p., 159 pp., diagrs., 22 cm., Venice, 1550.

Note: Substantially the same as the 1st. ed., Venice, 1526. The work is a revision of the author's *Libro de abaco,* which was first pub. in Venice, 1517. Other editions: ib., 1527, 1536, 1545, 1550, 1560, 1560, 1561, ib., 1563, Verona, 1563, Venice, 1570, Verona, 1602, and other seventeenth-cent. editions, *Scala grimaldelli,* Padua, 1629, 1669, and 1692.

[D30] Nicolaus Felner von Winnsheim. *Ein new behenndes unnd gantz gründtlichs rechenbuchlin auff den linien unnd federn, mit viel schonen lustparlichen regeln exempeln unnd behendikeitten auff gewin verlust unnd allenn kauffmansschlagk, gantz dinstlich mit vleiss zusamen geordennt unnd gemacht durch Nicolaum Felner von Winsheym, rechenmeister zu Erffürdt.* Erffürt (?): s.n., 1535

Description: t.-p., unpaged, 14½ cm., Erfurt, 1535.

Note: The only copy known. Not mentioned by Cajori, Tropfke, Smith, De Morgan, and Jackson.

[D31] Conrad Feme. *Eyn gut new rechenbuchleyn.* Erfurt: s.n., 1523

Description: Sm. 8, 11 ll., diagrs., Erfurt, 1523.

Note: Of the utmost rarity.

[D32] Giuseppe Figatelli. *Trattato Aritmetico di Giuseppe Maria Figatelli ... In questa seconda impresione aggiontoui l'algebra.* Venice: s.n., 1678

Description: 6 p. l., 251, 8, (6) pp., diagrs., 20½ cm., Venice, 1678.

Note: 1st. ed., 1664. Later editions: 1680, 1686, 1692, 1699, a 6th impression, s.a., 1737, 1738, 1778, 1791, 1797. Riccardi does not mention the above 2nd. impression.

[D33] Oronce Finé. *Orontii Finaei delphinatis ... De arithmetica practica libri quatuor: ....* Paris: s.n., 1555

Description: (8) pp., 72 numb. l. (i.e., 144 p.) incl. tables, 22 cm., Paris, 1555.

Note: 1st. work, *Orontii Finei Delphinatis, Protomathesis,* Paris 1532. A Latin ed. is described above. Other Latin editions: Paris, 1535, 1542, 1544, 1551. Fine also wrote geometrical works.

[D34] ——. *Opere di Orontio Fineo del Delfinato: diuise in cinque parti; arimetica, ... tr. da Cosimo Bartoli ....* Venice: s.n., 1587

Description: 8 p. l., 81, 126, 88, 18 numb. l., 2 l. illus., diagrs., 20½ cm., Venice, 1587.

Note: Another Italian translation, Venice, 1670.

[D35] Lorenzo Forestani. *Pratica d'Arithmetica, e Geometria ....* Venice: Presso F. Franceschi, 1603

Description: t.-p., 10 p. l., 344 ff., (1), illus., diagrs., 20 cm., Venice, 1603.

Note: An ed. at Siena, 1682: Di nuouo Ristampa, e con somma diligenza Ricorretta.

[D36] Pierre Gautruche. *P. Petri Galtruchii ... Mathematicae totius, ...* Cambridge: s.n., 1668

Description: t.-p., (6), 304 pp., (4), 15 cm., Cambridge, 1668.

Note: Gautruche also wrote *Philosophiae ac mathematicae totius institutio* ... Vienna, 1661), and *Cadomi,* 1665, besides works on metaphysics and poetical history.

[D37] Frisius Gemma Reinerus. *Arithmeticae practicae Methodus facilis per Gemmam Frisium, Medicum ac Mathematicum, iam recens ab ipso authore emendata & multis in locis insigniter aucta* .... Paris: s.n., 1549

Description: 96 numb. l. incl. tables, diagrs., 17 cm., Paris, 1549. With this are bound Köbel, Jacob. Astrolabii declaratio. 1550; Cosmographiae introductio. 1550.

Note: 1st. ed., Antwerp, 1540. The editions Witebergae, 1550, and Paris, 1550, are substantially the same. Smith's *Rara* cites 59 or more editions in the 16th cent. besides numerous editions after 1600. We also have these editions: Venice: 1567, and 1576.

[D38] Francesco Ghaligai. *Practica d'arithmetica di Francesco Ghaligai Fiorentino. Nuouamente Riuista, & con somma Dili Diligenza Ristampata.* 1552: s.n., n.d.

Description: t.-p., (2), 114 numb. l., 21 cm., Florence, 1552.

Note: The 1st. ed. was issued in 1521 under the title *Summa de arithmetica.* We also have a 1548 ed.

[D39] Albert Girard. *Invention nouvelle en l'Algebre. Reimpression par Dr. D. Bierens de Haan.* Leyden: s.n., 1884

Description: 4 p. l., t.-p., 64 unnumb. pp., diagrs., 21½ cm., Leyden, 1884.

Note: This is a reprint of the original ed., Amsterdam, 1629. Girard also revised a work of Samuel Warolois in 1628, and wrote on the mathematical works of Simon Stevin, 1634.

[D40] Henricus Glareanus. *De Vl. Arithmeticae practicae speciebus.* Paris: Fribvrgi Brisgoiae. Cum gratia ac priuilegio regio, ad annos sex, 1543

Description: 2 ff. unnumb. + 21 numb., 29 ll., 17 cm., Paris, 1543.

Note: Colophon: Apvd Fribvrgvm Brisgoicvm anno. M.D.XLIII. Excvdebat Stephanvs Melachvs Gravivs. Title vignette (device of Johann Faber (who printed first edition, 1539) with his monogram removed Other editions: Freiburg, 1543; Cracow, 1549; Freiburg, 1550; Paris, 1551; Freiburg, 1555; ib., 1558; Parls, 1558. Glareanus published other works besides this.

[D41] Girjka Gorla Z Gorlssteyna. *Wylifftenow Starém Héfté Praz kém v Girijka Czerneho.* Czerny: s.n., 1577

Description: 9 ff. unnumb. + 89 numb., 21 ll., 15.9 cm., Czerny, 1577.

Note: A very rare arithmetic in the Polish language.

[D42] Johan Rudolff von Graffenried. *Arithmeticae Logistica Popularis Libri IIII.* Bern: bey Abraham Weerli, in verlegung dess authoris, 1618

Description: t.-p., 23 p. l., 704 pp., (15), 20 cm., Bern, 1618.

Note: This volume is from the library of M. Chasles.

[D43] Domenico Griminelli. *Novissima prattica d'arithmetica mercantile Composta dal R. Don Domenico Griminelli ....* Rome: s.n., 1656

Description: t.-p., 4 p. l., 420 pp., (4), 16½ cm., Rome, 1656.

Note: We also have a 2nd. ed., 1670, and Riccardi mentions a probable 3d. ed.

[D44] Cardinaels Sybrand Hansz. *Arithmetica, ofte Reecken-Konst.* Amsterdam: Gedruckt by Ian I. Bouman, 1665

Description: t.-p., unpaged, 15½ cm., Amsterdam, dates on t.-p. of its 4 books, 1665, 1658, 1661, 1660, respectively.

[D45] Thomas Harriot. *Artis analyticae praxis, ...* London: Apud Robertum Barker, typographum regium, 1631

Description: 1 p. l., t.-p., (4), 180 pp., ll., 34½ cm., London, 1631.

Note: Harriot also wrote historical works.

[D46] Andreas Helmreich. *Rechenbuch, erstlich von Vortheil und Behendigkeit, nach der Welschen Practica, mit ihren unterschiedlichen Proportionibus, Grund und vrsach der Regel Detri oder proportionum.* Eisleben: Berwald, 1561

Description: 111 ff. unnumb. + 1 blank, 27–33 ll., 19.3 cm., Eisleben, 1561.

Note: Republished at Leipzig in 1588, 1595, and possibly in 1596. Also wrote *Ein new Visier büchlein,* 1557.

[D47] Johann Hemeling. *Arithmetica Historia oder Historische Rechnekunst ....* Hanover: n.d., 1667

Description: 8 p. l., 148 (i.e., 198) pp., 15½ cm., Hanover, 1667.

Note: Pp. 192–198 erroneously numbered 142–148.

[D48] ——. *Johannis Hemelingii Selbstlehrende Rechneschul oder Selbstlehrendes Rechnebuch ....* Hanover: Grimm Hannover Selbstverl, 1663

Description: 13 p. l., t.-p., 658 + pp., 17½ cm., 20 blank leaves interspersed, Hanover? 1663?.

Note: There is also a revised edition, *Johann Hemelings Kleines rechenbuch* ... by E. A. Talgener, Hanover, 1817.

[D49] Pierre Herigone. *Cursus mathematicus, ...* Paris: s.n., 1634

Description: 1 p. l., t.-p., (80), 983 pp., (21), diagrs., 18 cm., Paris, 1634, v. 1.

Note: 2nd. ed., 1644.

[D50] James Hodder. *Hodder's decimal arithmetick: or, A plain and more methodical way of teaching the said art : than hath hitherto been publish'd. Also, tables of interest upon interest, with the value of all sorts of purchases at any rates, from 5 to 12 per cent. Of rebate, resolving any question by ocular view. Likewise the true use of the said tables. By James Hodder, late writing-master in Lothbury, London, now keeper of a boarding-school in Bromely by Bow.* London: Printed by J.C. for Tho. Rooks, at the Lamb and Ink-bottle next door to the Ship-Tavern in Thredneedle-street, over against the Royal-Echange : Who also makes the best ink for record, 1671

Description: t.-p., 5 p. l., 157 pp., (3), 15 cm., London, 1671.

[D51] ——. *Hodder's Arithmetick ... Eighteenth Edition ... revised, augmented and ... amended ... by Henry Mose.* London: s.n., 1693

Description: Photograph, t.-p., 5 p. l., 216 pp., 14.5 cm., London, 1693.

Note: This is the first arithmetic separate in America. There is also a 1719 ed. On the inside of cover is written: "Not mentioned by DeMorgan; he only says the 1st. was in 1661, the 3d. in 1664, and the 9th. in 1672." There is a 15th. ed., 1685, and a 22nd. ed., 1702.

[D52] Johannes Husswirt. *Enchiridion novus Algorismi ....* Cologne: s.n., n.d.

Description: 2 p. l., t.-p., 39 unnumb. pp., blank ll. bound at end of book, 21½ cm., Cologne, 1501.

Note: Other editions: 1503, 1504, 1507, 1554; and a French ed., (Chasles). Later published with historical notes by Professor Wildermuth, at Tübingen, 1865.

[D53] Sieur C. Irson. *Arithmetique pratique et raisonnée.* Paris: Ghardon, 1692

Description: t.-p., 9 p. l., 557 pp., (2), 25 cm., Paris, 1692.

[D54] Simon Jacob. *Ein Nell und Wolgegründt Rechenbuch auff den Linien und Ziffern samt der Welschen Practic ... zum zweytenmal getruckt.* Frankfurt am Main: Feyerabend & Hüter, 1612

Description: 11 p. l., t.-p., 349 pp., (1), diagrs., 20½ x 17 cm., Frankfurt, 1612. Preface dated 1552; colophon, 1600.

Note: 1st. ed., 1560; other editions: 1565, 1569, 1600, and in the 17. cent. Jacob also wrote *Rechenbüchlein auf den Linian und mit Ziffern,* Frankfurt, 1557, 1574, 1589, 1590, 1599. The two works may be the same.

[D55] Heino Lambeck. *Heinonis Lambecii vermehrtes und verbessertes Rechen-buch.... Neue Auflage. Vermehrt mit der Solution der algebraischen Beschluss-Aufgabe des Johann Hinrich Wohlgemuths.* Stade: Gedruckt bei witwe Pockwitz und Sohn, n.d.

Description: t.-p., 324 pp., 17½ cm., Stade, n.d.

Note: No date on t.-p. According to Poggendorf, he died in 1661. Murhard gives the date 1661 for this work. The present writer, in her research of this book, came across a division problem on page 44 of which the quotient gave 1661 as the year in which Lambeck's *Rechenbuch* was printed.

[D56] Giorgio Lapazzaia. *Libro d'aritmetica e geometria dell' Abbate Giorgio Lapazzia.* Naples: Appresso Mattio Cancer, 1601

Description: 4 p. l., t.-p., 295 (2) pp., 20 cm., Naples, 1601.

Note: Lat. ed., 1566; Ital. editions; 1566, 1569, 1575, 1590, besides several editions after 1600. Riccardi records an ed. in 1784.

[D57] Francois Legendre. *L'arithmetique en sa perfection ... neuvieme edition.* Paris: s.n., 1690

Description: t.-p., 4 p. l., 384 pp., 25 cm., Paris, 1690.

Note: We also have the 11th.ed., 1700.

[D58] Lucas à Sancto Edmundo. *Arithmeticus praticus, utilitati publicae oblatus per patrem Lucam à S. Edmundo è clericis pauperibus matris Dei scholarum piarum, anno computi Christiani 1697.* Tyrnaviae [Nagyszombat]: typis academicis, per J.A. Hörmann, 1697

Description: t.-p., (16), 346 pp., (2), 14½ cm., Tyrnaviae, 1697.

[D59] Faentino Francesco Manelli. *Opera d'arithmetica e geometria ....* Bologna: Gio Battista Ferroni, 1659

Description: t.-p., 3 p. l., 176 pp., port., diagrs., 23 cm., Bologna, 1659.

Note: This work is bound with *Trigonometria* by Cavalieri, 1643, and is followed by a work on the compass by Galilei, 1640.

[D60] Jacobus Micyllus. *Arithmeticæ logisticae libri duo, ...* Basel: s.n., 1555

Description: t.-p., 8 p. l., 294 pp., (9) pp. incl. tables, diagrs., 15½ cm., Basel, 1555.

[D61] Johannes Monhemius. *Methodus arithmetices computatoriae ....* Düsseldorf: J. Oridryus & A. Busius, 1561

Description: 22 ff. unnumb., 25–27 ll., 15cm., Düsseldorf, 1561.

Note: Almost unknown to arithmeticians.

[D62] Jean Monier de Clairecombe. *Nouvelle pratique d'arithmetique ....* Amsterdam: s.n., 1693

Description: t.-p., 8 p. l., 466 pp., (16), 13½ cm., Amsterdam, 1693.

[D63] Jonas Moore. *Moor's Arithmetick. In two books. The first treating of the Vulgar Arithmetick. ... The Second of Arithmetick in Species or Algebra. ... To which are added two Treatises:....* London: Printed by J.G. for Nath. Brook, at the Angel in Cornhil, 1660

Description: 7 p. l., (455) p., front. (port.), diagrs. on 7 pl., (partly fold.) tables, 17 cm., various paging; the 2nd. book and the 2 treatises each have a special t.-p., London, 1660.

Note: 1st. work, *Moor's Arithmetick,* 1650; 2nd. ed., *Moor's Arithmetick in two books,* 1660; 3rd. ed., *Moor's Arithmetick in four books,* 1688. Moore also wrote *Mathematical Compendium,* London, 1674, 1681, and *A New Systeme of the Mathematicks,* London, 1681, in 2 volumes.

[D64] John Napier. *Rabdologia Ouero arimmetica virgolare ... Traddotore dalla Latina nella Toscana lingua ....* Verona: appresso Angelo Tamo, 1623

Description: t.-p., 1 p. l., (14), 269 pp., (1) p. incl. illus., 7 pl. (6 fold.) tables, diagrs., 16 cm., Verona, 1623.

Note: 1st. ed., 1617; other editions; 1626, 1628. Napier also wrote *Descriptio,* 1618, and *Constructio,* 1619, 1620. A ms. of his on algebra was printed in 1839.

[D65] Joannes Noviomagus. *De numeris libri duo, quorum prior logisticen, & ueterum numerandi consuerudinem: prosterior theoremata numerorum complectitur.* Cologne: Joannes Gymnicus, 1544

Description: 2 v. in l., illus., 15½ cm., Cologne, 1544.

Note: Pseud., otherwise, Jan van Bronckhorst. Other editions: 1539, 1551. We also have this same work *'esposti e illustrati dal prof. G. Frizzo, ...* , Verona, ... Drucker, 1901.

[D66] Juan de Ortega. *Suma de Arithmetica: Geometria Practica utilissima....* Rome: S. Guillieri de Lorena, 1515

Description: 2 ff. unnumb. + 114 numb., 32–38 ll., 30.2 cm., Rome, 1515.

Note: 1st. ed. Barcelona, 1512; other editions: Lyons, 1512; ib., 1515; Messina, 1522; 1534; Seville, 1536; ib., 1537; Paris, 1540 (?); Seville, 1542); ib., 1552, s. l. (?), 1552; Granada, 1563.

[D67] Georg Peurbach. *Elementa arithmetices: Algorithmus de numeris integris, fractis, regulis communibus, & de proporcionibus.* n.p.: Impressum Vitebergae per Iosephum Klug, 1539

Description: Special t.-p. (32nd. leaf)., 16 cm., Venice, 1539.

Note: 1st. ed., 1492; other editions: 1500, 1503, 1507, 1510, 1511, 1511 (earlier), 1511, 1512, 1513, 1515, 1520, 1534, 1536, 1538, 1539, 1544. The above work is contained in the 2nd. half of *Elementa Geometriae* by Johann Vogelin. Peurbach wrote many works on geometry and astronomy.

[D68] Giovanni Francesco Peverone. *Due brevi e facili trattati il primo d'Arithmetica: l'altro di Geometria: ne i quali si contengono alcune cose nuoue piaceuoli e utili, si a gentilhuomini coome artegiani.* Lyons: Tournes, 1558

Description: t.-p., 132 pp., (2), illus. (incl. diagrs.), 23½ cm., Lyons, 1558.

Note: There is also a 1581 ed. published at Lyons.

[D69] Monte Regal Piedmontois. *Inuention nouuelle et admirable, pour faire toute sorte de co[m]pte tant de marchandise comme de cha[n]ger monnoyes poids, mesures de diuerse maniere pour vendre & acheter laquelle seruira en tout le monde auec grande facilité, sans geto[n]s ne plume : et a la fin le moyen de mettre vn exercise en bataille de cent saques à quarante mille.* Lyons: il se vend en rue Merciere a la signe de la sphere, 1585

Description: Printed in double columns, 27–28 ll., 9.8 cm., Lyons, 1585.

Note: This book is unknown to most bibliographers.

[D70] Jean Prestet. *Elémens des mathématiques ou principes généraux de toutes les sciences qui ont les grandeurs pour objet, paru anonymement chez A. Pralard à Paris, 1675; 2 nde édition revue sous le titre Nouveaux elemens des mathematiques, ou principes generaux de toutes les sciences qui ont les grandeurs pour objet,.* Paris: Chez Andre Pralard, n.d.

Description: t.-p., 8 p. l., 418 pp., 2 fold. tables, 25½ x 19½ cm., Paris, 1675.

Note: There is a later work of 1689 in 2 volumes.

[D71] Prosdocimo de Beldomandi. *Algorismus de Integris.* Venice: s.n., 1540

Description: 44 ff. unnumb., 30 ll., 14.5 cm., Venice, 1540.

Note: 1st. ed., Padua, 1483. From a note in the Venice ed. of 1540 it appears that it was written in 1410.

[D72] Andres Puig. *Arithmetica especvlativa, y practica, y arte de algebra. En la qval se contiene todo lo que pertenece al arte menor,ò mercantivol, y à las dos algebras, racional,é irracional ; con la explicacion de todas las proposiciones, y problemas de los libros quinto, septimo, octavo, nono, y decimo del principe de la matematica Euclides. Compvesto, ordenado, y hecho imprimir por el maestro Andres Pvig.* Barcelona: A. Lacavalleria,, 1672

Description: t.-p., 8 p. l., 576 pp., (6), 1 l., 19 cm., Barcelona, 1672.

[D73] Johann Heinrich Rahn. *An Introduction to Algebra, Translated out of the High-Dutch into English by Thomas Brancker.* London: Printed by W. G. for Moses Pitt,, 1668

Description: t.-p., 1 p. l., (5), 198 pp., 50 pp., 20 cm., London, 1668.

Note: This is an English translation of Rahn's *Teutsche Algebra,* 1659.

[D74] Pierre de La Ramée. *P. Rami Arithmeticae libri duo: Geometriae septem et viginti a Lazaro Schomero recogniti & aucti.* Frankfurt: apud Andream Wechelum, 1599

Description: t.-p., 8 p. l., 240, 178, (4), 1 l., illus., fold. tab., diagrs., 24½ cm., Frankfurt, 1599.

Note: 1st. ed., Basel, 1569; other editions of the 16th. cent.: 1577; 1580; 1581; 1586; 1591, 1592; 1596; 1599, Schonerus; 1599, Frankfurt, 1627; and many others up to 1636. D. E. Smith in his *Rara Arthmetica* also mentions an English ed. of *The Art of Arithmetike in whole numbers and fractions . . . by P. Ramus . . . translated by William Kempe,* London, 1592. There is also an *Arithmetices et algebrae libri II a L. Schonero emend.,* Frankfurt, 1586.

[D75] ——. *P. Rami Arithmeticae libri duo cum commentariis Wilebror di Snelli R. F.* Leyden: Ex officina Plantiniana Raphelengii, 1613

Description: 2 p. l., 106 pp., 19½ cm., Leyden, 1613.

Note: This comprises Ramus' 2 books of arithmetic with Snell's addition *De re nummaria,* 1635. Also *Arithmeticae libri tres,* 1st. ed., 1555.

[D76] Ulricus Regius. *Utriusque Arithmetices Epitome ....* Freiburg: Gruningerus, 1543

Description: t.-p., 104 numb. l., diagrs., 15½ cm., Freiburg, 1543.

Note: Other editions: 1536, 1550.

[D77] Gregorius Reisch. *Margarita philosophica.* Freiburg: Jo. Schottus, 1503

Description: 2 ff. blank + 289 unnumb., 45 ll., 15.6 cm., Freiburg, 1503.

Note: Further editions: Strasburg, 1504; Freiburg, 1504, ed. s. 1., 1504; Strasburg, 1508; Basel, 1508; Basel, 1512; Strasburg, 1512; ib., 1515; Basel, 1517; Paris, 1523; Basel, 1535; ib., 1583; Venice, 1594; ib., 1599; ib., 1600.

[D78] Carlo Renaldini. *Caroli Renaldinii ... Ars analytica mathematum ... pars prima, ...* Florence: Cocchini, 1665

Description: 1 p. l., port., t.-p., (2), 534 pp., (33), diagrs., 35 cm., Florence, 1665.

Note: There is an earlier ed., Anconnae, 1644, and a later one, Venice, 1684. Riccardi lists 7 works of Renaldini.

[D79] ——. *Caroli Renaldinii ... Ars analytica mathematum, ... pars secunda.* Patavii: ex typographia I. Cocchini, 1669

Description: t.-p., 248 pp., (15), 35 cm., Patavii, 1669.

[D80] Joachim Ringelbergius. *Jochimi Fortii Ringelbergii ... Opera, quae proxima pagina enumerantur.* Leyden: Apud Gryphium, 1531

Description: 687 pp. numb. + 30 = 717 pp., 22–29 ll., 15.7 cm., Leyden, 1531.

Note: Another Leyden ed., 1539; Basel, 1541; Leyden, 1556.

[D81] Christoff Rudolff. *Künstliche rechnung mit der ziffer uñ mit den zal pfennigẽ sampt der Wellischen practica ....* Wien: s.n., 1532

Description: t.-p., (238) pp., 1 diagr., 15½ cm., Wien, 1532. Colophon, Nürnberg.

Note: This is an extension of the 1st. part of the *Coss,* published in 1525. The book is divided into 3 parts, *Das grundbuchlin,* Das *Regelbüchlein, Das exempelbüchlein.* There were many editions: 1st. ed., Vienna, 1526; other editions, 1532, 1534, 1537, 1540, 1546, 1553, 1557, 1561, 1574, 1588. Rudolff also published a collection of problems in several editions.

[D82] Pierre Savonne. *L'arithmetique de Pierre Savonne, dict Talon, ...* Lyons: s.n., 1571

Description: t.-p., 8 p. l., 146 numb. l., (4) pp., 17 cm., Lyons, 1571.

Note: 1st. ed., Paris, 1563; ib., 1565; Lyons, 1585; ib., 1588; ib., 1619, and as late as 1672.

[D83] Johann Scheubel. *Compendium arithmeticae artis, ... Iam denuó ab ipso autore recognitum & emendatum.* Basel: Oporinus, 1560

Description: t.-p., 1 p. l., 193 pp., (11), 1 l., illus., 16 cm., Basel, 1560. Defective copy. All between t.-p. & p. 15 lacking.

Note: 1st. ed., 1549. Scheubel also wrote *De numeris,* 1545, *Algebrae compendiosa,* 1551, and *Euclidis,* 1550.

[D84] Johan van Sesen. *Flores Arithmetici edder Arithmetica (in 2 vs.).* Hamburg: H. Carstens, 1622

Description: t.-p., 1069 pp. in the 2 volumes together, 19 cm., Hamburg, 1622.

[D85] Giovanni Sfortunati. *Nuouo Lumo, Libro di Ari thmetica. Institutio Nuouo lume ....* Venice: s.n., n.d.

Description: t.-p., 129 numb. l., tables, diagrs., 21 cm., Venice, 1561.

Note: 1st. ed., Venice, 1634, very rare, not mentioned by Riccardi. Other editions: 1543, 1544, 1545, 1550, 1568.

[D86] Moritz Steinmetz. *Arithmeticae praecepta ....* Leipsig: J. Rhamba excudebat, 1575

Description: t.-p., (236) pp. incl. tables, diagrs., 15½ x 9 cm., Leipsig, 1575. At the end of the dedication is the date 1568

Note: 1st. ed., 1568. Steinmetz also published *Euclides,* 1577.

[D87] Bernaert Stockmans. *...Arithmetica ....* Dordrecht: s.n., 1609

Description: 4 ff. blank + 211 unnumb., 31–33 ll., 14 cm., Dordrecht, 1609.

Note: 1st. ed., 1589. Of the number of editions after 1600, the library of G. A. Plimpton has the following: 1637, 1644, 1653, 1676. There is also the ed., Utrecht, 1633.

[D88] Siegmund Suevus. *Arithmetica historica. Die löbliche Rechenkunst. Durch alle Species und fürnembste Regeln mit schönen gedenckwirdigen Historien und Exempeln; auch mit Hebraischer, Grichischer, und Römischer Münke, Gewicht und Mass, deren in heiliger Schrifft und gutten Geschichtbüchern gedache wird, der lieben Jugend zu gutte erkleret. Auch denen die nicht Rechnen können, wegen vieler schönen Historien und derselbigen bedeutungen lustig und lieblich zu lesen. Aus viel gutten Büchern und Schrifften mit fleis zusammen getragen. Durch Sigismundum Sueuum.* Breslaw: G. Bawman, n.d.

Description: t.-p., 8 p. l., 455 pp., (50), ll., 19 x 15½ cm., Breslau, 1593.

[D89] Girolamo Tagliente. *Libro da abaco ch insegnia a fare ogni raxone mar cadantile & apertegare le terre con larte di la giometria & altre nobilissime raxone straordi narie con la tarifa come raspondeno li pexi & monete de molte terre del mondo con la incli ta citta de Venetia. El qual libro se chiama tesauro vniuersale. Concesso per lo serenissi mo dominio venetiano per anni diexe. Cũ g̃rã Restampata nouamente del 1520.* Venice: s.n., 1520

Description: t.-p., (157) pp., 1 l. ill. (incl. tables, diagrs.), 15 cm., Venice, 1520, (2nd. ed.).

Note: Smith *Rara Arithmetica* (1908), v. I, pp. 114, 115 lists 36 editions; 1st. ed., Venice, 1515. Giovanni Antonio Tagliente is joint author of above work.

[D90] Niccoló Tartaglia. *La prima parte del general trattato di nvmeri, et misvre di Nicolo Tartaglia ....* Venice: Navo, Curzio Troianoi, 1556

Description: t.-p., (8), 277 numb. l., illus., ports., tables, diagrs., 30 cm., Venice, 1556.

Note: There were subsequent editions of this work in 1560, 1592-1593, 1578 (Paris), 1578 (Antwerp). He also published *Tutte l'opera arithmetica,* 1592, and *L'arithmetique,* Paris 1613, besides other works on Euclid's *Elements,* etc.

[D91] Cuthbert Tunstall. *De arte supputandi libri quatuor, Cutheberti Tonstulli ....* Paris: s.n., 1538

Description: t.-p., 259 pp., 20½ cm., Paris, 1538.

Note: 1st. ed., London, 1522; other editions: 1529; 1535, 1543, 1544, (Strasburg), 1548, 1551.

[D92] Theodor Tzwivel. *Arithmetice: opuscula duo Theodorici Tzwiuel : de numero[rum] praxi (que algorithmi dicunt) vnu[m] de integris per figuraru[m] (more alemano[rum]) deletione[m] : alteru[m] de proportionibus cuius vsus frequens in musica harmonica Seuerini Boetij.* Cologne: Quentell, 1507

Description: 10 ff., 46 ll., 20 cm., Cologne, 1507.

Note: There is also a Cologne ed. in 1505.

[D93] Josephus Unicorno. *De l'arithmetica universale del Sig. Joseppo Unicorno ....* Venice: Appresso Francesco de' Franceschi, 1598

Description: t.-p., 2 v. in 1, 8 p. l., 395 numb. l. (1) p., 22 cm., Venice, 1598.

Note: He also wrote *De utilitate mathematicarum artium,* Venice.

[D94] Giacomo Venturoli. *Breve compendio di tutte le regole dell' aritmetica pratica aggiuntovi nuove osservazioni con tutte le regole della geometria pratica. Seconds Impressione.* Bologna: Nella stamperia del Longhi, 1681

Description: t.-p., 6 p. l., 208 pp., 140 diagrs. on 9 pl., 20 cm., Bologna, 1681.

Note: This says 2nd. ed. The 1st. ed. was published in 1663 with title: *Ordini aritmetici ...* There are also 3d. and 4th. ed., 1705 and 1754. The *Dialogo aritmetico* was first published separately in 1666 with title: *Scorta di economia, ó sia dialogo aritmetico.* (cf. Riccardi. Bibl. mat. ital.). Riccardi records 1666 and 1672 editions, but gives no date for the Bologna ed. just mentioned. We also have the 4th. ed., Bologna, 1745.

[D95] Francois Viéte. *L'algebre nouvelle de Mr. Viete ... Traduicte en Francois Par A. Vasset.* Paris: s.n., 1630

Description: t.-p., 15 p. l., 186 pp., (2), diagrs., 21½ cm., Paris, 1630.

Note: This is a French translation of *In artem analyticam isagoge,* Turoris, 1591, and of *Zeticorum libri quinque,* 1593. Vieta also wrote on works of Apollonius. There is also Hume's *L'algébre de Viéte,* Paris, 1636.

[D96] ——. *Francisci Vietae Opera mathematica, in unum Volumen congesta, ac recognita, Operâ atque studio. Francisci à Schooten ....* Lugduni Batavorum: ex Officina Bonaventurae & Abrahami Elzeviriorum, 1646

Description: t.-p., 6 p. l., 554 pp., illus., diagrs., 31 cm., Lugduni Batavorum, 1646.

[D97] Niolaus Voigtel. *Geometria Subterranea, oder Marckscheide-Kunst : darinnen gelehret wird Wie auff Bergwercken alle Klüffte und Gänge in Grund und am Tag gebracht auch solche von einander unterschieden werden sollen ; so wohl Was bey Durchschlägen in Ersparung Kosten Bringung Wetters und Benehmung Wassers denen Zechen oder Gebäuden mit zubeoachten ; Item Wie Streitigkeiten so sich unter miteinander schnürenden Gewercken offters zuereignen pflegen dem Maaße nach aus einander zusetzen ; Sambt noch andern in nechstfolgendem Indice enthaltenen und zu dieser Kunst dienlichen Sachen ; Allen Bergwercks-Liebenden zum Unterricht und versicherlichen Nutzen ....* Eisleben: Selbstverlag, 1686

Description: t.-p., and half-title, 6 p. l., 152 pp., incl. tables, illus., 9 fold. pl., diagrs., 32 cm., Eisleben (?), 1686.

Note: Murhard lists editions corresponding to this as follows: Eisleben, Gedruckt Johann Dietzeln 1686; 1688; Leipsig, 1692. Date in ms. on t.-p.: 1691. We also have a later ed., Eisleben, 1713.

[D98] Henricus Vuelpius. *Libellus de communibus et usitatis arithmeticae practicae regulis, ...* Cologne: s.n., 1544

Description: t.-p., (115) pp., illus., diagrs., 19½ cm., Cologne, 1544.

[D99] Caspar Waser. *Institutio arithmetica brevis & facilis ....* Tiguri: s.n., 1603

Description: t.-p., 92 numb. l., illus., II fold. pl., 16 cm., Tiguri, 1603.

Note: Waser also wrote geometrical works.

[D100] Johannes Wolphius. *Rudimenta arithmetices ....* Frankfurt: Apud H⊠r. Chr. Egen, 1534

Description: t.-p., (111) pp., diagrs., 14 cm., Colophon: Frankfurt, 1534.

Note: 1st. ed., Nürnberg, 1527: other Frankfurt editions: 1537, 1548, 1561; Strasburg editions: 1539, 1540. This arithmetic is followed by geometry by Johannes Vogelin.

[D101] Giovanni Battista Zuchetta. *Prima parte della arimmetica di Gio. Battista Zuchetta cittadino Genouese: Per la quale con mirabile ordine, & nuoue regole si risolue con marauigliosa facilità ogni dubbio mercantesco. Con vn trattato che risolue qualunque quesito bisongnoso à zecchieri, orefici, & argentari. Copiosa di postille, & tauole.* Brescia: Vincenzo Sabbio, 1600

Description: t.-p., 1 p. l., (24), 412 pp., (4), (10) pp. incl. tables, diagrs., 36 cm., Brescia, 1600. 10 pp. of ms. tables at end of book.

Additional Reading on Mathematical Notation

[O1] Hossam M.K. Aboulfotouh, "The geometric grids of the hieratic numeral signs," *Mediterranean Arhaeology and Archaeometry*, Vol. 12, No. 1, pp. 129–138, 2012

[O2] Virgil C. Adrich, "Symbolization and similarity," *The Monist*, Vol. 42, pp. 564–576, 1932

[O3] Désiré André. *Des notations mathématiques. Enumération, choix esage.* Paris: Gauthier-Villars, 1909

[O4] W. W. Rouse Ball, "On the use of a single symbol to denote the incommensurable 3.14159...," *Bibliotheca Mathematica*, Vol. 13, pp. 105–106, 1894

[O5] Petr Beckmann. *A History of Pi.* New York: St. Martin's Press, 1971

[O6] Suzan R. Benedict, "The development of algebraic symbolism from Paciuolo to Newton," *School Science & Mathematics*, Vol. 9, pp. 375–384, 1909

[O7] H. Brocard, "2549. (**1903**, 70) (G. Peano).— Origine du signe C." *L'Intermédiaire des mathématiciens*, Vol. 13, p. 18, 1906

[O8] Florian Cajori, "Plus and minus again," *Science*, Vol. 38, No. 967, pp. 51–52, July 11 1913

[O9] ——, "On the history of a notation in trigonometry," *Nature*, Vol. 94, No. 2363, p. 642, 1915

[O10] ——, "The spread of Newtonian and Leibnizian notations of the calculus," *Bulletin of the American Mathematical Society*, Vol. 27, No. 9, pp. 453–458, 1921

[O11] ——, "History of Symbols for $\underline{n}$ = Factorial," *Isis*, Vol. 3, No. 3, pp. 414–418, 1921

[O12] ——, "The St. Andrew's Cross (x) as a mathematical symbol," *Mathematical Gazette*, Vol. 11, No. 160, pp. 136–143, October 1922

[O13] ——, "The origin of the symbols for "degrees, minutes and seconds"," *American Mathematical Monthly*, Vol. 30, No. 2, pp. 65–66, February 1923

[O14] ——, "The St. Andrew's Cross $\times$ as a mathematical symbol," *Bulletin of the American Mathematical Society*, Vol. 29, No. 1, p. 12, 1923

[O15] ——, "Mathematical signs of equality," *Isis*, Vol. 5, No. 1, pp. 116–125, 1923

[O16] ——, "Origin of the names arithmetical and geometrical progressing and proportion," *Bulletin of the American Mathematical Society*, Vol. 29, No. 1, p. 12, 1923

[O17] ——, "The evolution of the modern exponential notation," *Bulletin of the American Mathematical Society*, Vol. 29, No. 5, p. 195, 1923

[O18] ——, "Note on our sign of equality," *Isis*, Vol. 6, No. 4, pp. 507–508, 1924

[O19] ——, "Rahn's algebraic symbols," *American Mathematical Monthly*, Vol. 31, No. 2, pp. 65–71, February 1924

[O20] ——, "The history of notations of the calculus," *Annals of Mathematics*, Vol. 25, pp. 1–46, 1924

[O21] ——, "Empirical generalizations on the growth of mathematical notations," *Isis*, Vol. 6, pp. 391–394, 1924

[O22] ——, "American contributions to mathematical symbolism," *American Mathematical Monthly*, Vol. 32, No. 8, pp. 414–416, October 1925

[O23] ——, "Leibniz, the master-builder of mathematical notations," *Isis*, Vol. 7, pp. 412–429, 1925

[O24] ——, "Leibniz, the master-builder of mathematical notations," *Isis*, Vol. 7, pp. 412–429, 1925

[O25] ——, "The unification of mathematicsl notations in the light of history," *Mathematics Teacher*, Vol. 17, No. 2, pp. 87–93, February 1925

[O26] ——, "A Notable Case of Finger=Reckoning in America," *Isis*, Vol. 8, No. 2, pp. 325–327, 1926

[O27] ——. *A History of Mathematical Notations*. Chicago: The Open Court, 1928

[O28] Karine Chemla, Eberhard Knobloch, and Antoni Malet, "History and philosophy of mathematical notations and symbolis," *Oberwolfach Reports*, Vol. 6, No. 4, p. Oberwolfach Reports, 2009

[O29] Stephen Chrisomalis. *Numerical notation : a comparative history*. New York: Cambridge University Press, 2010

[O30] R.B. Clifton, "Note on Professor De Morgan's paper entitled "on the early history of the signs $+$ and $-$"," *Transactions of the Cambridge Philosophical Society*, Vol. 11, pp. 213–218, 1864

[O31] Joseph V. Collins, "Defects remaining in the notation and nomenclature of elementary mathematics," *Mathematics Teacher*, Vol. 16, No. 3, pp. 157–161, March 1923

[O32] Agustus De Morgan, "On the invention of the signs + and −; and on the sense in which the former was used by Leonardo da Vinci," *London, Edinburgh, & Dublin Philosophical Magazine*, Vol. 20, pp. 135–136, 1842

[O33] Augustus De Morgan, "On the early history of the signs + and −," *Transactions of the Cambridge Philosophical Society*, Vol. 11, pp. 203–212, 1864

[O34] Salih Zéky Efendi, "Notation algébrique chez les orientaux," *Journal Asiatique ou Recueil de Mémoires, d'Extraits et de Notices relatifs à l'Histoire, à la Philosophie, aux Langues et à la Littérature des Peuples Orientaux*, Jan 1898

[O35] Eneström G, "Sur le premier emploi du symbole π pour 3,14159...," *Bibliotheca Mathematica*, Vol. 12, p. 28, 1889

[O36] G. Eneström, "Réponse à la question de M. Cantor: Quand et par qui les *mots plus* et *minus* ont-ils été introduits pour *prononcer* les signes + et −?" *L'Intermédiaire des mathématiciens*, Vol. 1, pp. 119–121, 1894

[O37] Franz M. Feldhause. *Geschichte des technischen Zeichnens*. Wilhelmshaven: Franz Kuhlmann, 1953

[O38] Robert Feys and Frederic B. Fitch. *Dictionary of symbols of mathematical logic*. Amsterdam: North-Holland, 1969

[O39] Steven R. Finch. *Mathematical Constants*. Cambridge: Cambridge University Press, 2003

[O40] Graham Flegg. *Numbers: Their History and Meaning*. New York: Schocken Books, 1983

[O41] G.B.M., "A history of arithmetical notation," *Nature*, Vol. 77, No. 1998, pp. 347–348, February 13 1908

[O42] Midhat Gazale. *Number: From Ahmes to Cantor*. Princeton: Princeton University Pres, 2000

[O43] J. W. L. Glaisher, "On the early history of the signs + and − and on the early German arithmeticians," *Messenger of Mathematics*, Vol. 51, pp. 1–148, 1921

[O44] Mary Geraldine Green, "A Trip to Notation Land," *Arithmetic Teacher*, Vol. 6, No. 4, pp. 215–216, October 1959

[O45] George Bruce Halsted, "Plus and minus," *Science*, Vol. 37, No. 961, pp. 836–837, May 30 1913

[O46] Julian Havil. *Gamma: Exploring Euler's Constant*. Princeton: Princeton University Press, 2009

[O47] Albrecht Heeffer, "From the second unknown to the symbolic equation," in *Philosophical Aspects of Symbolic Reasoning in Early Modern Science and Mathematics, Ghent, August 27–29, 2009*, Albrecht Heeffer and Maarten Van Dyck, Eds. London: College Publications, 2009, pp. 57–102. [Online]. Available: http://www.pasr.ugent.be/proceedings.php

[O48] Jens Høyrup, "Hesitating progress — The slow development toward algebraic symbolization in Abacus and related manuscripts, ca. 1300 to ca. 1550," in *Philosophical Aspects of Symbolic Reasoning in Early Modern Science and Mathematics, Ghent, August 27–29, 2009*, Albrecht Heeffer and Maarten Van Dyck, Eds. London: College Publications, 2009, pp. 3–56. [Online]. Available: http://www.pasr.ugent.be/proceedings.php

[O49] Robert William Brierley Jackson. *Inconsistencies in mathematical notation : the use of the full stop as a decimal point*. Toronto: University of Toronto, 1961

[O50] Eberhard Knobloch, "Leibniz between ars characteristica and ars inveniendi: Unknown news about Cajori's 'master-builder of mathematical notations'," in *Philosophical Aspects of Symbolic Reasoning in Early Modern Science and Mathematics, Ghent, August 27–29, 2009*, Albrecht Heeffer and Maarten Van Dyck, Eds. London: College Publications, 2009, pp. 289–302. [Online]. Available: http://www.pasr.ugent.be/proceedings.php

[O51] Lattin H P, "The origin of our present system of notation according to the theories of nicholas bubnov," *Isis*, Vol. 19, pp. 181–194, 1933

[O52] Mario Livio. *The Golden Ratio: The Story of Φ, the World's Most Astonishing Number*. New York: Broadway Books, 2008

[O53] Madeleine Seymour Loomis. *The writing of Braille mathematical notation; a guide to the Taylor system of mathematics*. New York: Columbia Univ, 1941

[O54] Dietrich Mahnke, "Die indexbezeichnung bei liebniz als beispiel seiner kombinatorischen charakteristik," *Bibliotheca Mathematica*, Vol. 13, pp. 250–260, 1913

[O55] ——, "Leibniz als Begrunder der symbolischen Mathematik," *Isis*, Vol. 9, No. 2, pp. 279–293, 1927

[O56] Eli Maor. e*: The Story of a Number*. Princeton: Princeton University Press, 2009

[O57] Joseph Mazur. *Enlightening Symbols: A Short History of Mathematical Notation and Its Hidden Powers*. Princeton: Princeton University Press, 2014

[O58] Paul J. Nahin. *An Imaginary Tale: The Story of* $\sqrt{-1}$. Princeton: Princeton University Press, 2010

[O59] Abraham Nemeth. *The Nemeth Braille code for mathematics and science notation*. Louisville: American Printing House for the Blind, 1973

[O60] Eric Harold Neville, "Mathematical notation," *The Mathematical Gazette*, Vol. 48, pp. 145–163, 1964

[O61] Antoine-Paulin Pihan. *Exposé des Signes de Numération Usités Chez les Peuples Orientaux Anciens et Modernes*. Paris: L'Imprimerie Imperiale, 1860

[O62] Henry A. Pogorzelski, "A note on mathematical notation," *Mathematics Magazine*, Vol. 33, No. 1, p. 184, October 1959

[O63] Alfred S. Posamentier. *Pi: A Biography of the World's Most Mysterious Numbe.* Amherst: Prometheus Books, 2004

[O64] C.B. Read, "Bibliography for the origins of various mathematical terms and symbols," *American Mathematical Monthly*, Vol. 46, No. 4, p. 233, April 1939

[O65] György Révéz. *On translating ordinary mathematical notation.* Yorktown Heights: IBM, 1989

[O66] J.D. Roberts. *Notation.* London: Methuen, 1968

[O67] George Sarton, "Extreme slowness of the introduction of elementary algebraic symbols," *Isis*, Vol. 27, p. 328, 1937

[O68] Edward R. Scheinerman. *Mathematical Notation.* San Fransisco: CreateSpace, 2011

[O69] Sister Mary Leontius Schulte Schulte. *Additions in arithmetic, 1483-1700, to the sources of Cajori's "History of mathematical notations" and Tropfke's "Geschichte der Elementar-Mathematik".* Ann Arbor: Lithoprinted by Edwards Brothers, 1935

[O70] Michel Serfati, "Symbolic revolution, scientific revolution: mathematical and philosophical aspects," in *Philosophical Aspects of Symbolic Reasoning in Early Modern Science and Mathematics, Ghent, August 27–29, 2009*, Albrecht Heeffer and Maarten Van Dyck, Eds. London: College Publications, 2009, pp. 103–122. [Online]. Available: http://www.pasr.ugent.be/proceedings.php

[O71] Lao G. Simons, "Bibliography for the origins of various mathematical terms and symbols," *American Mathematical Monthly*, Vol. 47, No. 2, p. 107, February 1940

[O72] David E. Smith, "The NABLA," *Scripta Mathematica*, Vol. 1, No. 4, p. 304, 1932

[O73] Dorothy G. Stewart. *A guide to the typing of mathematical notation.* Santa Monica: Rand Corp, 1964

[O74] Paul Tannery, "Sur le symbole de soustraction chez les Grecs," *Bibliotheca Mathematica*, Vol. 5, pp. 5–8, 1904

[O75] H. M. Taylor. *Braille Mathematical Notation.* Louisville: American Printing House for the Blind, 1942

[O76] Johannes Tropfke. *Geschichte der Elementar-Mathematik in Systematischer Darstellung.* Leipzig: Veit, 1902

[O77] Roy Wagner, "Symbolic revolution, scientific revolution: mathematical and philosophical aspects," in *Philosophical Aspects of Symbolic Reasoning in Early Modern Science and Mathematics, Ghent, August 27–29, 2009*, Albrecht Heeffer and Maarten Van Dyck, Eds. London: College Publications, 2009, pp. 229–269. [Online]. Available: http://www.pasr.ugent.be/proceedings.php

Index

Made in United States
North Haven, CT
05 November 2026

21768323R10143